D1349027

Genetically Modified Crops:
Assessing Safety

Genetically Modified Crops: Assessing Safety

Edited by
Keith T. Atherton

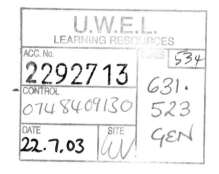

London and New York

First published 2002 by Taylor & Francis
11 New Fetter Lane, London EC4P 4EE

Simultaneously published in the USA and Canada
by Taylor & Francis Inc,
29 West 35th Street, New York, NY 10001

Taylor & Francis is an imprint of the Taylor & Francis Group

Typeset in 10/12pt Sabon by Graphicraft Limited, Hong Kong
Printed and bound in Great Britain by
TJ International Ltd., Padstow, Cornwall.

British Library Cataloguing in Publication Data
A catalogue record for this book is available from the British Library

Library of Congress Cataloging in Publication Data
Atherton, Keith T., 1949–
   Genetically modified crops : assessing safety / Keith T. Atherton.
      p. cm.
   Includes bibliographical references (p.).
   ISBN 0-7484-0913-0 (hb : alk. paper)
   1. Transgenic plants. 2. Crops—Genetic engineering. I. Title.

SB123.57 .A84 2002
631.5′233—dc21   2002071958

ISBN 0-748-40913-0

Contents

Contributors   vii
Preface   ix

1 The regulatory and science-based safety evaluation of genetically
  modified food crops – a USA perspective   1
  EARLE NESTMANN, TODD COPELAND AND JASON HLYWKA

2 The regulatory requirements for novel foods – a European perspective   45
  NICK TOMLINSON

3 The concept of substantial equivalence: an overview   63
  PAUL R. MAYERS, PETER KEARNS, KAREN E. MCINTYRE AND
  JENNIFER A. EASTWOOD

4 Strategies for analysing unintended effects in transgenic food crops   74
  H.P.J.M. NOTEBORN, A.A.C.M. PEIJNENBURG AND R. ZELENY

5 Allergenicity of foods produced by genetic modification   94
  DEAN D. METCALFE

6 Biosafety of marker genes – the possibility of DNA transfer from
  genetically modified organisms to the human gut microflora   110
  KIERAN TUOHY, IAN R. ROWLAND AND PAUL C. RUMSBY

7 Case study: canola tolerant to Roundup® herbicide: an assessment
  of its substantial equivalence compared to non-modified canola   138
  THOMAS E. NICKSON AND BRUCE G. HAMMOND

8 Case study: Bt crops – a novel mode of insect control   164
  BRIAN A. FEDERICI

9 Case study: recombinant baculoviruses as microbial pesticidal agents   201
  IVAN E. GARD, MICHAEL F. TREACY AND JOHN J. WRUBEL

10 Case study: virus-resistant crops   219
    HECTOR QUEMADA

    Index   241

Contributors

Keith Atherton
Central Toxicology Laboratory, Alderley
Park, Macclesfield, Cheshire SK10 4TJ, UK

Todd Copeland
Regulatory Sciences Manager, Cantox
Health Sciences International, 2233 Argentia
Road, Suite 308, Mississauga, ON L5N 2X7
Canada

Jennifer A. Eastwood
Bureau of Food Policy Integration, Food
Directorate, Health Protection Branch,
Health Canada, Ottawa, Ontario K1A 0L2,
Canada

Professor Brian Federici
Department of Entomology &
Graduate Programs in Genetics &
Microbiology, University of California,
Riverside, Riverside, CA, USA

Ivan Gard
Entomological Associates of Northampton,
7 Ponderosa Drive, Holland, PA 19866,
USA

Bruce Hammond
Monsanto Company, 700 Chesterfield
Parkway North, St Louis, Missouri 63198,
USA

Jason Hlywka
Cantox Health Sciences International, 2233
Argentia Road, Suite 308, Mississauga, ON
L5N 2X7 Canada

Peter Kearns
Environmental Health and Safety Division,
Environmental Directorate, Organisation for
Economic Cooperation and Development, 2
rue André Pascal, 75775 Paris, France

Paul Mayers
Office of Food Biotechnology, Health
Canada, Postal Locator 2204A1, Tunney's
Pasture, Ottawa, Ontario, K1A 0L2 Canada

Karen E. McIntyre
Bureau of Food Policy Integration, Food
Directorate, Health Protection Branch,
Health Canada, Ottawa, Ontario K1A 0L2
Canada

Dean Metcalfe
Chief, Laboratory of Allergic Diseases,
National Institute of Allergy and
Infectious Diseases, National Institute
of Health, Building 10, Room 11C205,
10 CENTER DR MSC 1881, Bethesda,
MD 20892-1881, USA

Earle Nestmann, Cantox Health Sciences
International
Suite 308, 2233 Argentia Road,
Mississauga, Ontario,
L5N 2X7 Canada

Thomas E. Nickson
Monsanto Company, 700 Chesterfield
Parkway North, St Louis, MO 63198,
USA

Hubert P.J.M. Noteborn
RIKILT-DLO, Department of Food
Safety & Health, Bornsesteeg 45, PO
Box 230, NL-6700 AE Wageningen,
The Netherlands

A.A.C.M. Peijnenburg
RIKILT-DLO, Department of Food
Safety & Health, Bornsesteeg 45, PO
Box 230, NL-6700 AE Wageningen,
The Netherlands

Hector Quemada
Crop Technology Consulting Inc., 2624
East G. Avenue, Kalamazoo, NI 49004,
USA

Ian R. Rowland
Northern Ireland Centre for Food and
Health, School of Biomedical Sciences,
University of Ulster, Coleraine BT52 1SA,
UK

Paul C. Rumsby
MRC Institute for Environment and Health,
University of Leicester, 94 Regent Road,
Leicester, LE1 7DD, UK

Nick Tomlinson
Food Standards Agency, 73 Grove Road,
Harpenden, Herts, AL5 1EN, UK

Michael F. Treacy
Agricultural Products Research Division,
American Cyanamid Company, Princeton,
NJ 08543, USA

Kieran M. Tuohy
Food Microbial Science Unit, School of
Food Biosciences, University of Reading,
Whiteknights PO Box 226, Reading RG6
6AP, UK

John J. Wrubel
Agricultural Products Research Division,
American Cyanamid Company, Princeton,
NJ 08543, USA

R. Zeleny
Institute for Agrobiotechnology, Center for
Analytical Chemistry, A-3430 Tulin, Austria

Preface

Man has cultivated plants for thousands of years, during which time crop plants have been continually selected for improved yield, growth, disease resistance or other useful characteristics. Plant breeding is an exceptionally successful enterprise that has fashioned the raw material of unimproved germplasm into the modern high-yielding crop and pasture varieties on which we now depend. Until recently plant breeders had to depend on empirical methods to reach their goals. However, the discovery of plant transformation is changing the way that breeders approach the challenge of creating new varieties to fulfil specific needs. Directed genetic changes provide an important new tool and allow the use of genetic information from almost any life form to be introduced into crop plants and produce desirable new characteristics.

Improvements in agronomic traits such as yield and disease resistance continue to be driving forces behind today's seed industry, but increasingly, attention is also focussed on speciality traits, including high oilseed grains, low saturated fat oilseeds and delayed ripening fruits and vegetables. Such traits command premium values in the market place. Whilst plant breeders have used conventional breeding methods coupled with techniques such as tissue culture and mutagenesis to produce new commercial lines, genetic engineering now provides a real alternative. The first generation of genetically modified (GM) crop varieties, which have been altered for agronomic traits are generally those encoded by a single gene, such as virus-, insect-, or herbicide resistance. A comprehensive list of these can be found on the Biotechnology Industry Organisation website (www.Bio.org).

A long history of producing new varieties of crop plants by conventional breeding has rarely resulted in forms that have had to be withdrawn from the market because of health concerns. Plant breeders have introduced thousands of new crop varieties that have had little, if any, effect on food safety. The concern over GM crops stems mainly from the fact that plant breeders now have access to a much wider range of genetic information from any living organism or synthetic DNA sequences. The developers of new plant varieties using genetic engineering have the responsibility of establishing that the newly introduced varieties, and the food products derived from them are as safe and nutritious as their traditional counterparts.

This book attempts to lead the reader through the main issues associated with the safety evaluation of GM crops. Chapters 1 and 2 deal with the regulatory requirements for the registration of GM crops both in the USA and in the European Community. Since there are significant limitations in the use of conventional animal toxicology studies for the safety assessment of whole foods, a new approach has had to be established. Chapters 3 and 4 discuss the concept of Substantial Equivalence, which is at the heart of current safety assessment strategies, and some of the emerging methodology that might be applied to this area in the future. Chapters 5 and 6 then deal with two of the most commonly voiced concerns with the safety of GM crops – food allergenicity and the biosafety of marker genes. The final four chapters look in detail at case studies describing the safety evaluation strategies for a number of GM crops developed to date.

It is left to the reader to decide if the developers of this technology have adequately established that these new GM varieties and the food products developed from them are as safe and nutritious as their traditional counterparts.

Keith T. Atherton
January 2002

Chapter 1

The regulatory and science-based safety evaluation of genetically modified food crops

A USA perspective

Earle Nestmann, Todd Copeland and Jason Hlywka

Introduction

Genetic modification, otherwise referred to as recombinant DNA (rDNA) technology or gene-splicing, has proven to be a more precise, predictable and better understood method for the manipulation of genetic material than previously attained through conventional plant breeding. To date, agricultural applications of the technology have involved the insertion of genes for desirable agronomic traits (e.g. herbicide tolerance, insect resistance) into a variety of crop plants, and from a variety of biological sources. Examples include soybeans modified with gene sequences from a *Streptomyces* species encoding enzymes that confer herbicide tolerance, and corn plants modified to express the insecticidal protein of an indigenous soil microorganism, *Bacillus thuringiensis* (Bt). A growing body of evidence suggests that the technology may be used to make enhancements to not only the agronomic properties, but the food, nutritional, industrial and medicinal attributes of genetically modified (GM) crops.

Regulatory supervision of rDNA technology and its products has been in place for a longer period of time in the United States than in most other parts of the world. The methods and approaches established to evaluate the safety of products developed using rDNA technology continue to evolve in response to the increasing availability of new scientific information. As our understanding of the potential applications of the technology is broadened, the safety of products developed using rDNA technology and the potential effects of introduced gene sequences on human health or the environment will be more closely scrutinized. In fact, much of the knowledge acquired during the commercialization of the products of rDNA technology in agriculture is now finding application in evaluating the safety of products developed through more conventional means.

The objective of this chapter is to provide the reader with an overview of the significant events leading up to the present, science-based, regulatory framework that exists for the safety evaluation of GM food crops within the United States. An attempt has been made to discuss concerns over the sufficiency of existing regulations, as well as to highlight recent initiatives taken by federal regulatory agencies to address them. Through better communication of how the regulatory process functions within the United States, it is anticipated that current and future applications of rDNA technology in agriculture will be met with a greater level of understanding and acceptance.

Historical perspective

rDNA technology was first developed in the 1970s. The initial response of the scientific community, including members of the National Academy of Science (NAS), to the prospects of rDNA technology, was to postpone any further research involving the technology until the potential risks to human health and the environment could be evaluated. Researchers attending the International Conference on Recombinant DNA Molecules in 1975, otherwise known as the Asilomar Conference, tried to establish a scientific consensus on how best to self-regulate emerging applications of the technology. The conditions and restrictions that were proposed at this conference have formed the basis by which federal guidelines and policies for rDNA technology research were drafted within the United States.

National Institutes of Health (NIH)

The National Institutes of Health (NIH) was the first federal regulatory agency to publish their interests in evaluating the safety of rDNA technology in 1976, in the form of guidelines for the conduct of research (NIH, 1976). Because of the uncertainties that existed at the time, all research into the potential applications of rDNA technology was limited to the confines of federally funded laboratories under NIH control. After continued research, and a more careful assessment and monitoring of the risks, a set of less restrictive guidelines was published in 1978 (NIH, 1978). However, the environmental release of organisms developed using rDNA technology outside the confines of controlled laboratory conditions was prohibited unless otherwise approved by the NIH director (NIH, 1978). In the early 1980s, the NIH established an rDNA Advisory Committee (RAC) to review all data and experience gained with applications of the technology under its control. Based on recommendations of the RAC, a more relaxed set of research guidelines was published by the NIH in 1983 (NIH, 1983).

The NIH approved the first environmental release of an organism developed using rDNA technology (ice-minus strain of *Pseudomonas*) in 1983. In response, they were criticized for failing to prepare a statement or assessment of the environmental impact of their regulatory decision as required under the National Environmental Policy Act (NEPA) (US Congress, 1969; NIH, 1983). Once the legal controversy had subsided, all responsibility that the NIH had for regulating the environmental introduction of GM organisms was relinquished. Nevertheless, NIH guidelines continue to be referenced in assessing the safety of rDNA research performed within industry, federal and other state laboratories. However, it was unclear which federal regulatory agencies would be responsible for ensuring the safety of the products developed using rDNA technology.

Office of Science and Technology Policy (OSTP)

In response to a need for clarification, the Office of Science and Technology Policy (OSTP) began work on the development of a policy to establish a federal regulatory framework for evaluating the safety of products developed using rDNA technology (OSTP, 1984). Following an opportunity for public comment, OSTP published a

Table 1.1 Overview of responsible agencies under the coordinated framework

Responsible agency	Products regulated	Reviews for safety
FDA	Food, feed, food additives, veterinary drugs	Safe to eat
USDA	Plant pests, plants, veterinary biologic	Safe to grow
EPA	Microbial/plant pesticides, new uses of existing pesticides, novel microorganisms	Safe for the environment Safety of a new use of a companion herbicide

final version of the 'Coordinated Framework for Regulation of Biotechnology' (Coordinated Framework) in 1986 (OSTP, 1986). The policy provided the basis by which federal regulatory agencies got involved in evaluating the safety of products at later stages of commercial development at that time.

According to the Coordinated Framework, the products of rDNA technology should be regulated on the basis of the unique characteristics and features that they exhibit, not their method of production. The products of rDNA technology were considered to pose risks to human health and the environment similar to those posed by conventional products already regulated within the United States. As a result, no new federal regulatory agencies or regulations were required. The Coordinated Framework did not, however, rule out the possibility of the development of new guidelines, procedures, criteria or even regulations to supplement or alter the scope of existing statutes for the products of rDNA technology.

The Coordinated Framework identified three federal regulatory agencies within the United States: the US Food and Drug Administration (US FDA), the US Department of Agriculture (USDA) and the US Environmental Protection Agency (US EPA), as having primary responsibilities for evaluating the products of rDNA technology under development at that time. An overview of federal regulatory agency responsibilities as outlined within the Coordinated Framework is provided within Table 1.1.

In 1992, the OSTP released another document entitled, 'Exercise of Federal Oversight within the Scope of Statutory Authority: Planned Introductions of Biotechnology Products Into the Environment', outlining the proper basis by which federal regulatory agencies were expected to exercise their regulatory authority (OSTP, 1992). As with conventional products, dependent upon the intended use and function, more than one federal regulatory agency may share an interest in evaluating the safety of a product developed using rDNA technology. If more than one federal regulatory agency has an interest, lead agencies are identified as being responsible for coordinating activities to limit any potential duplication of efforts. Although federal regulatory agencies worked independent of one another, it was realized that close working relationships would need to be established in order to evaluate effectively the safety of products developed using rDNA technology.

Recently, the OSTP teamed up with the White House Council on Environmental Quality (CEQ) to perform a six-month inter-agency evaluation of the federal regulatory agency responsibilities in evaluating the environmental safety of products developed using rDNA technology (CEQ/OSTP, 2001). A case-study approach for a variety of different classes of products developed using rDNA technology was used to evaluate the level of federal regulatory agency involvement, to identify strengths, weaknesses and areas of potential improvement. The review concluded that none of the previously approved products of rDNA technology has had any significant negative impact on the environment (CEQ/OSTP, 2001). Although all the case studies were published, OSTP/CEQ failed to reach a consensus on issues relating to the relevant strengths and weaknesses of the existing regulatory structure within the time allotted for the completion of its review. A review of the case studies published provides a comprehensive interpretation of the responsibilities of each federal regulatory agency in ensuring the safety of the products developed using rDNA technology.

National Academy of Sciences (NAS)

The National Academy of Sciences (NAS), and its operating arm, the National Research Council (NRC), have served as a primary source of scientific, technological, human health and environmental policy advice during the development of regulatory approaches for the safety evaluation of products developed using rDNA technology within the United States.

In 1987, the NRC published a report concerning the potential human health and environmental hazards associated with the commercial introduction of GM organisms, entitled *Introduction of Recombinant DNA-Engineered Organisms into the Environment: Key Issues* (NAS, 1987). The risks associated with the introduction of GM organisms were considered to be essentially the same in kind as those associated with unmodified organisms (NAS, 1987). In other words, rDNA technology did not appear to introduce any unique risks as compared to the products that had been developed using more conventional methods of genetic modification. In reaching these conclusions, the NRC performed an evaluation of the similarities and differences in the properties exhibited by products developed using a variety of different techniques. To this day, the conclusions of this report continue to be referenced by the regulators and developers of GM crop varieties worldwide.

A subsequent NRC report, entitled *Field Testing Genetically Modified Organisms: Framework for Decisions*, also reached similar conclusions, but provided additional guidance as to how regulatory decisions concerning the introduction of GM organisms should be made (NRC, 1989). The NRC recommended that regulatory decisions concerning the introduction of GM organisms should be made on a case-by-case basis. Consistent with the Coordinated Framework (OSTP, 1986), the NRC did not consider the nature of the process used for the genetic modification of an organism to be a useful criterion for determining whether a product requires less or more regulatory oversight (NRC, 1989). As a result, no valid reason existed to regulate organisms genetically modified via modern techniques (e.g. rDNA technology) any differently from organisms genetically modified via more conventional means (NRC, 1989). Similar conclusions have been published in the reports of international standards-setting organizations (OECD, 1993; FAO/WHO, 1996).

In retrospect, within both reports, the NRC acknowledged that modern and conventional methods of genetic modification are not without risks to human health or the environment, and as a result, neither could be considered inherently more risky. As a result, regulatory decisions concerning the safety of the products of rDNA technology need only to take into consideration the specific characteristics exhibited by a particular GM organism and the environment in which it is to be introduced, and not the method by which it has been produced. The NRC has further articulated their conclusions into what is now commonly referred to as the 'Concept of Familiarity'.

Although familiarity with the characteristics of a particular organism or the environment to which it will be introduced would not necessarily mean it was safe, it can be expected to provide a sufficient amount of information to allow for a judgement to be made of the risks. For example, familiarity with a new GM plant variety could be established based on comparisons between characteristics of the parent line or other crop species exhibiting similar traits, as well as through the results of actual field tests involving the GM plant. These principles were further elaborated upon by the Organisation for Economic Co-operation and Development (OECD) (OECD, 1993), and as a result, have been referenced in the development of regulatory policies for evaluating the safety of GM crops on a global basis.

More recently, a committee established by the NRC published a report entitled *Genetically Modified Pest-Protected Plants: Science and Regulation*, based on a review of all scientific and regulatory data collected during the regulatory approval process for GM crops within the United States (NAS, 2000). The primary objective was to assess independently the effectiveness of existing and proposed regulations for the safety evaluation of GM crops expressing plant pesticides (i.e. plant-incorporated protectants). No new evidence was identified to suggest plants expressing plant pesticides posed any greater risk to human health or the environment as a result of their genetic modification (NAS, 2000). In fact, the NRC concluded, 'with careful planning and appropriate regulatory oversight, commercial cultivation of GM plants is not expected to pose higher risks and may pose less risk than other commonly used chemical and biological pest-management techniques'. However, the NRC report included requests for federal regulatory agencies to further strengthen the current regulatory approval process through better coordination and communication between agencies, on-going investment in the research and monitoring of potential human health (e.g. allergenicity) and environmental impacts (e.g. insect resistance), and by providing greater access to information evaluated in support of regulatory decisions (e.g. websites).

US federal regulations for agricultural biotechnology – introduction

Regulatory systems for the products of agricultural biotechnology have been in existence since the mid- to late 1980s within the United States. The regulatory approach to the safety evaluation of plants developed using rDNA technology has evolved in the best interests of research scientists, industry and the general public. The agricultural products of rDNA technology, such as GM foods and crops, may require approvals from up to three regulatory agencies; the US FDA, the USDA and the US EPA, depending upon the characteristics exhibited by the GM plant, its proposed use and introduced traits. The same standards of safety are applied to all products regardless of the technology used in their development.

The US FDA is responsible for ensuring the human safety of all new foods and food components, including products developed using rDNA technology, under the Federal Food, Drug, and Cosmetic Act (FFDCA). The USDA evaluates the potential of a GM plant to become a plant pest following its environmental introduction under the Federal Plant Pest Act (FPPA). The US EPA evaluates pesticides, including plant systems modified to express pesticides (e.g. insect-protected or virus resistance), under the Federal Insecticide, Fungicide, and Rodenticide Act (FIFRA). As a result, the expression of an insecticidal protein in a food crop would undergo review by the USDA, US EPA and US FDA; a GM food crop exhibiting a modified oil content would be evaluated by the USDA and US FDA; and a non-food horticultural plant developed using rDNA technology for any other purpose (e.g. flower color) would be subject to review by the USDA alone.

In most instances, obtaining all necessary approvals for the commercialization of an agricultural crop developed using rDNA technology takes a decade or more. However, the exact amount of time required will depend on the need to confirm performance, to evaluate characteristics of the food, environmental effects, and to produce the required amount of seed before the product can be distributed and commercially grown by farmers. Up to five years of field trials (5–10 generations of plants) are required for the developer of a new plant variety to collect sufficient data to meet the reporting requirements of the USDA. An additional five months to two years may be required for the US FDA, USDA and/or US EPA to complete all necessary product consultations, reviews and approvals.

Approval for the first commercial planting of a GM food crop (i.e. Flavr Savr™ tomato) was not issued until 1995. Since 1995, more than 40 new agricultural crops developed using rDNA technology have received approval for commercial planting within the United States. In 1999, approximately 72% of the total 39.9 million hectares (more than 98 million acres) of GM crops grown worldwide were planted in the United States (James, 1999). Herbicide-tolerant soybeans (54%), Bt corn (19%) and herbicide tolerant canola (9%) accounted for approximately 82% of the GM plants cultivated. With an increasing number of agricultural biotechnology products reaching later stages of commercial development, it is anticipated that the overall area planted with GM crops will continue to rise.

The regulatory approach to evaluating the human health and environmental safety of GM crops within the United States is best described as a science-based, case-by-case assessment of hazards and risks. This approach has provided the flexibility required to reduce the regulatory burden placed on products that have been determined to be of low risk or concern. All agencies involved in the regulation of plants developed using rDNA continue to implement and develop policies based on recommendations made within the Coordinated Framework (OSTP, 1986). In the future, the US FDA, USDA and the US EPA will be dedicating additional resources towards communicating how GM food and food components are regulated within the United States, and how these regulations function to be protective of both human health and the environment.

US Food and Drug Administration (US FDA)

Authority

The US FDA is responsible for ensuring the safety and wholesomeness of all food and food components, including the products of rDNA technology, under the FFDCA.

The US FDA has the authority for the immediate removal of any product from the market that poses potential risk to public health or that is being sold without all necessary regulatory approvals. As a result, a legal burden is placed on developers and food manufacturers to ensure the commodities utilized and foods available to consumers are safe and in compliance with all legal requirements of the FFDCA.

In order to understand the regulatory approach followed by the US FDA in the safety evaluation of GM crops, it is useful to consider food and food safety from a historical context.

People had been consuming foods derived from agricultural crops for many years prior to the existence of any food laws or regulations within the United States. Based on this experience, agricultural crops have been accepted as being safe for consumption as food, without additional testing to demonstrate their safety. As long as the new crop variety has exhibited similar agronomic properties, and an appropriate taste and appearance, it has been considered safe to consume (Kessler *et al.*, 1992). As a result, most foods consumed today, in particular whole foods (i.e. fruits, grains and vegetables) and conventional foods, have not been subject to any kind of premarket review or approval by the US FDA.

Nevertheless, food scientists have a good understanding that many of the commonly consumed agricultural crops contain natural toxicants (e.g. tomatine in tomatoes, solanine in potatoes, cucurbiticin in cucumber, psoralens in celery, etc.). As a result, new plant varieties may be subject to routine chemical analyses to ensure that none of these substances is present at potentially harmful levels. This type of general approach has been used in assessing the safety of thousands of new plant varieties that have been developed over a number of decades of crop breeding without compromising the safety of whole foods (Munro, 2000).

Role

Consistent with recommendations within the Coordinated Framework (OSTP, 1986), the US FDA considered existing provisions of the FFDCA to be sufficient for the regulation of foods and food components developed using rDNA technology. It was concluded that the scientific and regulatory issues posed by the products of rDNA technology were not significantly different from those posed by conventional products. As a result, GM foods and food components have been subject to the same standards of safety as already exist for the regulation of other foods and food components under the FFDCA. In order to better communicate interpretations of existing provisions of the FFDCA as they relate to the safety evaluation of foods derived from new plant varieties, including the products of rDNA technology, the US FDA released a policy statement in 1992 entitled 'Statement of policy: foods derived from new plant varieties' (1992 US FDA Statement of Policy) (US FDA, 1992).

The US FDA has considered the use of genetic modification (i.e. rDNA technology) in the development of new plant varieties to represent a continuum of conventional plant breeding practices (e.g. mutagenesis, hybridization, protoplast fusion, etc.), and as a result, the safety evaluation of all new plant varieties, not just those developed using rDNA technology, have been evaluated based on an objective analysis of the characteristics of a food or its components, and not on its method of production (Kessler *et al.*, 1992).

Table 1.2 Chronology of FDA regulation of new plant varieties modified by rDNA technology

Date	Event
1984–1994	Public meetings and invitation to comment
1986	Publication of Coordinated Framework for Regulation of Biotechnology (OSTP, 1986)
1992	FDA Policy statement on 'Foods Derived From New Plant Varieties' (57 FR 22984)
1994	Food Advisory Committee discussion of voluntary consultation
1994	Decision on FLAVR SAVR™ tomato
1996	Guidance to industry for consultation
2000	Public meetings
2001	Proposed rule on 'Premarket Notice Concerning Bioengineered Foods'

More recently, the US FDA has been requesting public comment on a newly proposed rule for the mandatory pre-market notification of bioengineered foods based on existing provisions under the FFDCA (US FDA, 2001a). Table 1.2 outlines the chronology of events leading up to the publication of this proposed rule by the US FDA.

Mandate

The US FDA relies on two primary sections of the FFDCA, Section 402 (21 USC §342) and Section 409 (21 USC §321), to ensure the safety of all foods and food components (e.g. ingredients, additives, contaminants), including the products of rDNA technology (Kessler *et al.*, 1992).

- *Section 402 – Adulteration Provisions*   Section 402 protects consumers from the intentional and unintentional adulteration of foods and food components with substances considered poisonous or hazardous to human health. A food or food component is considered adulterated if a reasonable certainty exists that its consumption may have deleterious effects on human health.
- *Section 409 – Food Additive Provisions*   Section 409 pertains to substances intentionally added to foods that are not generally recognized as safe (GRAS) based on prior scientific testing or historical use, or that are not otherwise exempt (e.g. pesticides, etc.). Therefore, the concept of GRAS was established indirectly in the definition of food additives (Vetter, 1996). A substance may be considered a food additive if determined to be significantly different in structure, function or amount from a substance already consumed as a part of the diet or if it lacks a sufficient history of safe use. Pre-market review and approval are required for substances considered to be food additives under the FFDCA. The GRAS status is reviewed when a question arises concerning the regulatory status of a substance intentionally added to foods.

Within this statutory context, the 1992 US FDA Statement of Policy included interpretations of these provisions as they relate to GM foods and food components (US FDA, 1992). At the time of publication, the US FDA anticipated that most substances introduced into or produced by GM plants would be comparable to substances already consumed as a component of existing foods, and as such, would be considered

GRAS. If the levels of an introduced substance were determined to be significantly different, or if the nutritional composition was significantly altered, introduced substances would be subject to requirements for pre-market review and approval by the US FDA.

The 1992 US FDA Statement of Policy established a voluntary consultation process with the US FDA for the developers of new plant varieties (US FDA, 1992). Although considered voluntary, all those involved with the development of GM food and food components took advantage of the opportunity provided. Following extensive consultation and a series of public meetings, the US FDA has recently proposed that a mandatory pre-market consultation process be followed for bioengineered foods (US FDA, 2001a). As proposed, 120 days prior to a planned introduction, the developers of bioengineered foods would be expected to submit a notification to the US FDA (US FDA, 2001a).

Definition and scope of bioengineered foods

The recently proposed rule of the US FDA (2001a) concerns 'bioengineered foods' which have been defined as 'foods derived from plant varieties that are developed using *in vitro* manipulations of DNA (generally referred to as rDNA technology)'. As a result, the proposed rule has a much narrower focus than the 1992 US FDA Statement of Policy. The US FDA has explained the need for a change in emphasis based on their expectations that many of the new plant varieties exhibit a greater potential to 'contain substances that are significantly different from, or that are present in food at a significantly different level than before' (US FDA, 2001a). As a result, the substances present in foods and food components derived from new plants developed using rDNA technology are less likely to be considered GRAS, and as a result, will require pre-market approval from the US FDA.

Requirements for commercialization

A proposed rule that outlines the reporting requirements to be addressed in a pre-market biotechnology notice (PBN) for the commercialization of bioengineered foods has recently been published by the US FDA (US FDA, 2001a). All required parts have been identified in Table 1.3, and have been established based on; (a) the 1992 US FDA Statement of Policy (US FDA, 1992); (b) the 1997 procedures document (US FDA, 1997); and, (c) the extensive experience of the US FDA in voluntary consultation with developers of GM foods and food components.

Safety and nutritional evaluation

The primary objective of the safety and nutritional evaluation is to demonstrate that the food derived from a new plant variety is as safe or nutritious as foods already consumed as a part of the diet. For new plant varieties, including those developed using rDNA technology, a science-based approach is used to focus the evaluation on the demonstrated characteristics of the food or food component. The evaluation of a GM food or food component typically involves reviewing information or data on any newly introduced substances, the known levels of toxicants, as well as the nutritional composition of the plant following modification. Substances that raise safety concerns

Table 1.3 Pre-market biotechnology notice (PBN): required parts

Part	Part title	Contents
I	Letter	1 Statements regarding the notifier's responsibility and the balanced nature of the notice 2 Statements regarding the availability of data and information for FDA's review 3 Statement regarding public disclosure
II	Synopsis	
III	Status at other federal agencies and foreign governments	
IV	Method of development	
V	Antibiotic resistance	
VI	Substances in the food	1 Covered substances 2 Identity, function, level, and dietary exposure 3 Allergenicity 4 Other safety issues
VII	Data and information about the food	1 Comparable food 2 Historic uses of the comparable food 3 Comparing the composition and characteristics of the bioengineered foods to that of comparable food 4 Other relevant information 5 Narrative

Source: FDA (2001a).

(e.g. toxicants, allergens) would be subject to more extensive evaluation, since both intended and unintended changes may affect the levels of toxicants and nutrients in a food following the modification.

Guidance for performing a safety and nutritional evaluation was provided in the 1992 US FDA Statement of Policy (US FDA, 1992), in a series of flow charts and text that cover: (i) the crop that has been modified; (ii) source(s) of the introduced genetic material; and (iii) new substances intentionally added to the food as a result of the genetic modification (e.g. proteins, but also fatty acids, and carbohydrates). A schematic of the safety and nutritional evaluation process performed by the US FDA, as summarized by Kessler *et al.* (1992), has been reproduced in Figure 1.1. Documentation required to support the evaluation typically includes: the purpose or intended technical effect of the modification on the plant, together with a description of the various applications or uses, a molecular characterization of the modification including the identities, sources and functions of introduced genetic material; information on the expressed protein products encoded by introduced genes; information relating to the known or suspected allergenicity and toxicity of any expressed gene products; for foods known to cause allergy, information on whether the endogenous allergens have been altered by the genetic modification; information on the compositional and nutritional characteristics of the foods, including anti-nutrients; and in some instances, comparative results of feeding studies involving the foods derived from plants modified using rDNA technology and the non-modified counterpart.

In performing its evaluation, the US FDA is particularly interested in the identification of inherent toxicants, known or potential allergens, assessing the concentration

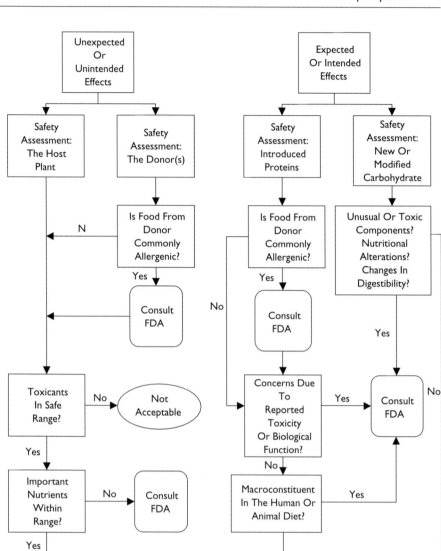

Figure 1.1 Decision analysis for safety and nutritional assessment of new plant varieties. (As depicted in Kessler et al., 1992.)

and bioavailability of essential nutrients, the safety and nutritional value of any newly introduced proteins, and the identity, composition, and nutritional value of modified carbohydrates, fats and oils. If additional questions of safety remain following this evaluation, further toxicological studies may need to be performed. It is recognized that absolute assurance of the safety of any food does not exist. As a result, the goal of the safety evaluation is to establish a reasonable certainty of no harm under anticipated conditions of consumption. With this in mind, experience with the existing food supply has provided the basis for evaluating the safety of new food or food components. Both the Food Advisory Committee and Committee for Veterinary Medicine have been extensively involved in the development of approaches for the safety and nutritional evaluation of foods and food components derived from new plant varieties, including those developed using rDNA technology.

PRODUCT CHARACTERIZATION

Product characterization takes into consideration information relating to the modified food crop, the introduced genetic material and its expression product, and acceptable levels of inherent plant toxicants and nutrients. All characteristics of the gene insert must be known, including the source(s), size, number of insertion sites, promoter regions, and marker sequences. It must be established that the transferred genetic material does not come from a pathogenic source, a known source of allergens, or a known toxicant-producing source. The introduced genetic material should be well characterized to ensure that the introduced gene sequences do not encode harmful substances and are stably inserted within the plant genome to minimize any potential opportunity for undesired genetic rearrangement.

Analytical data are required to evaluate the nutritional composition, the levels of any known toxicants, anti-nutritional and allergenic substances, and the safety-in-use of antibiotic resistance marker genes. Any new substances introduced into crops through rDNA technology (e.g. proteins, fatty acids, carbohydrates) will be subject to pre-market review as food additives by the US FDA, unless substantially similar to substances already safely consumed as a part of foods, or that are considered GRAS. To date, substances that have been added to foods through rDNA technology have been previously consumed or have been determined to be substantially similar to substances already consumed as a part of the diet. As such, introduced substances have been considered exempt from the requirement for pre-market approval as food additives with the US FDA.

A more rigorous safety evaluation of a GM crop is warranted if the introduced gene sequence(s) has not been fully characterized, the nutritional composition has been significantly altered, antibiotic resistance marker genes have been used during its development, or if an allergenic protein or toxicant has been detected at levels higher than what is typically observed in edible varieties of the same crop species. In any event, determinations as to the safety of substances that have been introduced into new plant varieties through rDNA technology are made on a case-by-case basis.

Although an evaluation of the introduced gene sequences and expression product(s) provides assurance as to their safety, further studies may be required to predict whether unexpected effects may result following their interaction with other genes within the plant. In addition, the product characterization of a GM plant involves assessing sequence homology to known toxicants and allergens, thermal and digestive stability, and if required, the results of both *in vitro* and *in vivo* assays to demonstrate lack of toxicity.

COMPOSITIONAL ANALYSIS

The results of field trials performed over several years serve to characterize the phenotypic and agronomic characteristics exhibited by the plant (e.g. height, color, leaf orientation, susceptibility to disease, root strength, vigor, fruit or grain size, yield, etc.), as well as to provide the materials required for the compositional analysis. Any anomalies in the phenotypic or agronomic characteristics exhibited by a plant may result in a requirement for additional information. Protein, fat, fibre, starch, amino acid, fatty acid, ash and sugar levels are determined, as well as the levels of anti-nutrients, natural toxicants or known allergens. Studies of the nutritional composition are performed to determine whether the levels of any key nutrients, vitamins or minerals have been altered as a result of the genetic modification.

Based on the results of these studies, a determination is made as to whether the phenotypic and agronomic characteristics of a GM crop or the concentrations of inherent constituents fall within ranges typical of its conventional counterpart. If inserting a new gene causes no change in any of the assessed parameters, the US FDA can conclude with reasonable assurance that the GM crop is as safe as the conventional crop (FAO/WHO, 1991; OECD, 1993; US FDA, 1997). If the levels of essential nutrients or inherent toxicants are found to be significantly different in the GM crop, the US FDA may recommend additional action prior to commercialization, such as obtaining food additive status, or the use of specific labels to alert consumers of an altered nutritional content, etc.

ALLERGENICITY

In consultation with scientific experts in the areas of food safety, food allergy, immunology, biotechnology and diagnostics, the US FDA published guidelines for assessing the allergenicity of GM foods or food components in 1994 (US FDA, 1994). These guidelines follow a decision-based approach that provides for an assessment of both the source of the genetic material incorporated into the GM food, and the expressed protein product of the introduced gene sequence as summarized in Figure 1.2. The strategy outlined in this figure reflects that developed by the Allergy and Immunology Institute of the International Life Sciences Institute and the International Food Biotechnology Council (Metcalfe *et al.*, 1996), and further recommended by the Joint FAO/WHO Expert Consultation on Foods Derived from Biotechnology (FAO/WHO, 2000).

The approach to assessment is multi-faceted, incorporating data regarding the origin of the genetic material, and the biochemical, immunological and physicochemical properties of the expressed protein. The overall assessment is reliant upon the fact that all known food allergens are proteins and, notwithstanding the number of shared properties between allergenic and non-allergenic food proteins, food allergens tend to exhibit a number of similar characteristics. In general, food allergens share a number of common properties: they have a molecular weight of over 10 000 Da; they represent more than 1% of the total protein content of the food; they demonstrate resistance to heat, acid treatment, proteolysis and digestion; and they are recognized by IgE (Lehrer *et al.*, 1996; Metcalfe *et al.*, 1996; Taylor and Lehrer, 1996).

For gene sequences derived from known allergenic sources (e.g. peanuts), the developers of GM plants are expected to demonstrate that allergenic proteins have not been introduced into the food. For assessment purposes, it is assumed that any genetic material derived from a known allergenic source will encode for an allergen. To demonstrate otherwise, the amino acid sequence of an expressed protein must be compared with that of known allergens using protein sequence databases. Furthermore, *in vitro* and/or *in vivo* immunologic analyses using the sera of allergic patients

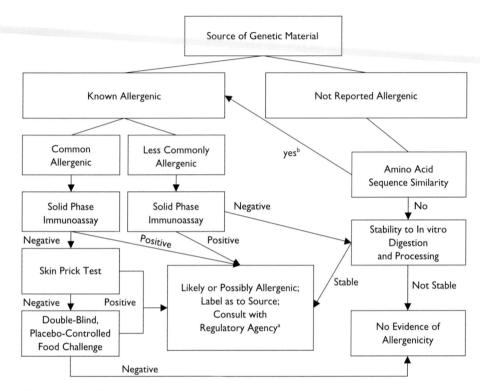

a Proposed to be Mandatory
b Identity of 8 contiguous amino acids in expressed protein and known allergen in protein database.

Figure 1.2 Decision-based assessment of the potential allergenicity of bioengineered food. (Modified from Metcalfe *et al.*, 1996, and FAO/WHO, 2000.)

sensitive to the source of the genetic material may need to be performed to determine whether or not a potentially allergenic protein is being expressed in the GM food.

Some GM foods may be modified to express genes from a source that is not known to be allergenic when consumed. Under these circumstances, the US FDA follows a similar decision tree-based approach to determining the allergenic potential of the expression product. In assessing these proteins, should any amino acid sequence exhibit homology with a known allergen, the expressed protein would then be evaluated in immunologic tests using the sera of patients known to be allergic to the identified homologous protein. Regardless of the origin of the genetic material, physicochemical studies are performed *in vitro* to provide information concerning the expected stability of the expressed protein (Lehrer *et al.*, 1996; Metcalfe *et al.*, 1996; Taylor and Lehrer, 1996). All known food allergens tend to be resistant to digestive degradation, as demonstrated in simulated gastric fluid models, or to decomposition under conditions of food processing (Metcalfe *et al.*, 1996).

In making a determination regarding a GM food, it is the totality of the biochemical, immunological and physicochemical properties of the introduced protein that provides guidance as to the allergenic potential of such a protein being expressed in food. The level of protein expressed, produced and consumed as a part of the diet is also a primary indicator of the allergenic potential, since nearly all food allergens are known to be major proteins in their respective foods. If the results of any of these studies suggest an allergenic potential, the US FDA may recommend further scientific evaluation, require special labelling to alert sensitive consumers, or alternatively caution the developer about proceeding with the development of a particular GM food.

Most recently, a joint FAO/WHO Expert Consultation on Allergenicity of Foods Derived from Biotechnology (FAO/WHO, 2001), recommended a revised decision tree, as outlined in Figure 1.3. This decision-tree strategy was modified from the previous version (Figure 1.2) by Metcalfe *et al.* (1996) and FAO/WHO (2000) to include a revised definition of sequence homology for gene product comparison; a greater emphasis on serum testing, even with gene products without homology to known allergens and not derived from an allergenic source; and animal models to assess potential allergenicity, despite acknowledgement by the Expert Consultation that these models are currently under development and at present, not predictive of food allergies in humans. More recently, the *ad-hoc* Open-Ended Working Group on

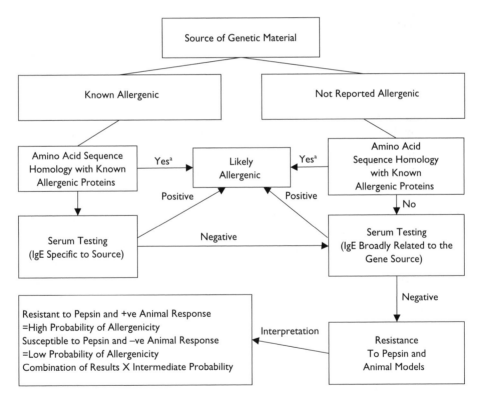

[a] >35% identity in the amino acid sequence of expressed protein and known allergen using a window of 80 amino acids and suitable alignment program, or identity of 6 contiguous amino acids in expressed protein and known allergen.

Figure 1.3 Revised decision-based assessment of the allergenic potential of foods derived from biotechnology. (As recommended by FAO/WHO Expert Consultation, 2001.)

Allergenicity (Codex, 2001) established by the *ad-hoc* Intergovernmental Codex Task Force on Foods Derived from Biotechnology, considered the FAO/WHO (2001) strategy in drafting an approach to assessing the potential allergenicity of foods derived from rDNA plants. The Codex Working Group recognized the absence of a definitive predicitve test for allergenicity in humans to a newly expressed protein and recommended an integrated, stepwise assessment strategy. The strategy recommended by the Working Group, and accepted by the Codex Task Force (Codex, 2002) for inclusion in the draft forwarded for final adoption by the Codex Alimenarius Commission, is consistent with that of FAO/WHO (2000). However, the Working Group suggested that some of the modifications included in FAO/WHO (2001) could contribute to the overall weight of evidence of any conclusion of potential allergenicity (e.g., allergen specific serum depositories, animal models), pending development and validation.

ANTIBIOTIC RESISTANCE

The use of antibiotic-resistance genes as selectable markers has been common practice in the development of new plant varieties using rDNA technology. Concerns relate to the potential transfer of antibiotic-resistance genes from GM plants to pathogens in the environment or to the gut of humans consuming foods or food components. Issues relating to the use of antibiotic resistance genes were identified in the 1992 US FDA Statement of Policy Statement (US FDA, 1992) as well as discussed in additional guidance entitled *Guidance for Industry: Use of Antibiotic Resistance Marker Genes in Transgenic Plants* (US FDA, 1998). The guidance provided within these documents was established in consultation with experts in the fields of microbiology, medicine, food safety, bacterial and mycotic diseases, and includes suggestions with respect to the continued safe use of antibiotic-resistance marker genes by the developers of new plant varieties.

The use of marker genes that encode resistance to clinically important antibiotics has raised questions as to whether their presence in food could reduce the effectiveness of oral doses of the antibiotic or whether the gene present in the DNA could be transferred to pathogenic microbes, rendering them resistant to treatment with the antibiotic. The risk of transfer of antibiotic-resistance genes from plants to microorganisms considered to be pathogenic to humans, however, is considered to be minimal if not insignificant. Furthermore, the potential risks are becoming less of a concern as more developers are beginning to research the use of alternative technologies (e.g. non-resistance-based markers) in plant breeding. The conclusions with respect to the safe use of antibiotic-resistance marker genes are consistent with the findings of other national and international food safety organizations (WHO, 1993; US EPA, 1994; FAO/WHO, 1996; Karenlampi, 1996).

Consultation and filing process

The submission of a Pre-market Biotechnology Notification (PBN) has recently been proposed as a mandatory requirement for the commercialization of bioengineered foods and food ingredients within the United States (US FDA, 2001a). Minimal differences exist between the information to be submitted as a part of a PBN and that previously presented in voluntary consultations with the US FDA. As proposed, developers are still being encouraged to consult with the US FDA as early and as often as necessary in the development of a bioengineered food, such that any potential scientific or regulatory concerns can be identified and addressed prior to the submission of the PBN.

Guidelines for performing consultations with the US FDA were released in a publication entitled *Guidance On Consultation Procedures: Foods Derived from New Plant Varieties* in 1997 (US FDA, 1997). The publication recommended an approach for developers to proceed by submitting a request for consultation, outlined the internal process by which all requests would be handled, and included additional guidance as to the type of safety and nutritional information to be presented during consultation with the US FDA.

Once sufficient safety and nutritional information has accumulated to demonstrate that a product is safe and in compliance with the FFDCA, developers typically schedule a consultation to present their scientific findings and conclusions to the US FDA. Consultations prior to notification not only serve to keep the US FDA informed of advances made in the application of rDNA technology in food production, but also keep the developers of bioengineered foods aware of emerging safety, nutritional or regulatory concerns of the US FDA.

Consultations are considered complete when all safety and regulatory concerns between the US FDA and the developer have been resolved. A brief chronological summary of the consultations considered complete up to and including the year 2001 has been provided in Table 1.4.

Under the newly proposed rules, the developer of bioengineered food would be required to submit a PBN addressing all required data parts as previously outlined in Table 1.3 at the end of consultations. The US FDA proposes to perform an initial evaluation of a PBN within 15 days of receipt to determine completeness, at which point, if considered complete, the PBN will be filed, and a response can be expected within 120 days. The US FDA does not issue a product approval *per se*, but informs the developer by letter that: (a) the evaluation period has been extended; (b) the notice is not complete and why; or (c) it has no further questions 'at this time' based on the information that has been presented (US FDA, 2001a).

Labelling

The FFDCA defines what information must be disclosed to consumers on a food label, such as the common or usual name, and other limitations concerning the representations or claims that can be made or suggested about a food product. All foods must be labelled truthfully and not be misleading to consumers. Taking this into consideration, the FFDCA does not stipulate the disclosure of information on the basis of consumer desire to know. Labelling may be considered misleading if it fails to reveal material facts in light of representations that are made with respect to a product.

The labelling of foods derived from new plant varieties, including plants developed using rDNA technology, was originally addressed in the 1992 US FDA Statement of Policy (US FDA, 1992), and most recently discussed in Draft Guidance for Industry for the voluntary labelling of bioengineered foods (US FDA, 2001b). To date, the US FDA is not aware of any information that would distinguish foods developed using rDNA technology (e.g. bioengineered foods) as a class from foods developed through other methods of conventional plant breeding, and as such, have not considered the

Table 1.4 Summary of completed consultations for new plant varieties with the FDA

Firm	New variety
2000	
Aventis Crop Science	Male sterile and glufosinate-tolerant corn
Monsanto Co.	Glufosinate-tolerant rice
	Glyphosate-tolerant corn
1999	
Agritope Inc.	Modified fruit-ripening cantaloupe
BASF AG	Phytaseed canola
Rhone-Poulenc Ag Company	Bromoxynil-tolerant canola
1998	
AgrEvo, Inc.	Glufosinate-tolerant soybean
	Glufosinate-tolerant sugar beet
	Insect-protected and glufosinate-tolerant corn
	Male sterile or fertility restorer and glufosinate-tolerant canola
Calgene Co.	Bromoxynil-tolerant/insect-protected cotton
	Insect-protected tomato
Monsanto Co.	Glyphosate-tolerant corn
	Insect and virus-protected potato
	Insect and virus-protected potato
Monsanto Co. / Novartis	Glyphosate-tolerant sugar beet
Pioneer Hi-Bred	Male sterile and glufosinate-tolerant corn
University of Saskatchewan	Sulfonylurea-tolerant flax
1997	
AgrEvo, Inc.	Glufosinate-tolerant canola
Bejo Zaden BV	Male sterile and glufosinate-tolerant radicchio
Dekalb Genetics Corp.	Insect-protected corn
DuPont	High oleic acid soybean
Seminis Vegetable Seeds	Virus-resistant squash
University of Hawaii / Cornell University	Virus-resistant papaya
1996	
Agritope Inc.	Modified fruit-ripening tomato
Dekalb Genetics Corp.	Glufosinate-tolerant corn
DuPont	Sufonylurea-tolerant cotton
Monsanto Co.	Insect-protected potato
	Insect-protected corn
	Insect-protected corn
	Glyphosate-tolerant/insect-protected corn
Northrup King Co.	Insect-protected corn
Plant Genetic Systems NV	Male sterile/fertility-restorer and glufosinate-tolerant oilseed rape
	Male sterile and glufosinate-tolerant corn
1995	
AgrEvo Inc.	Glufosinate-tolerant canola
	Glufosinate-tolerant corn
	Bromoxynil tolerant cotton
Calgene Inc.	Laurate canola
Ciba-Geigy Corp.	Insect-protected corn
Monsanto Co.	Glyphosate-tolerant cotton
	Glyphosate-tolerant canola
	Insect-protected cotton
	Glyphosate-tolerant soybean
	Improved ripening tomato
	Insect-protected potato
Asgrow Seed Co.	Virus-resistant squash
DNA Plant Technology Corp.	Improved ripening tomato
Zeneca Plant Science	Delayed softening tomato
1994	
Calgene Inc.	FlavorSavr™ tomato
2001	
Monsanto Co.	Insect and antibiotic resistant corn
Dow Agro Sciences LLC	Insect resistant and glufosinate-tolerant corn

Source: http://www/cfsan.fda.gov/~lrd/biocon.html (as of March, 2002)

method of development a material fact requiring disclosure on product labels. Nevertheless, after extensive consultation, including thousands of written comments and a series of public meetings, the US FDA has observed 'a general agreement that providing more information to consumers about bioengineered foods would be useful' (US FDA, 2001b).

Requirements for labelling

Special labelling is required if the composition of the bioengineered food differs significantly from its conventional counterpart. For example, for a food that has been genetically modified to contain a new major sweetener, a new common or usual name or other labelling may be required. Similarly, if a GM food contains an allergen that consumers would not expect to be present in that food, special labelling may be necessary to alert sensitive consumers. If a protein commonly associated with an allergic reaction (e.g. peanut protein) is transferred to another food through genetic modification, the US FDA would evaluate whether labelling would provide sufficient consumer protection. If labelling would not be considered to provide a sufficient level of protection, the US FDA would take appropriate steps to ensure the GM food would not be marketed. Therefore, current policy requires a GM food to be labelled when the resulting product poses a safety issue or is substantially different from its conventional counterpart, and as a result, could be considered to pose a misrepresentation to consumers.

The 1992 US FDA Statement of Policy (US FDA, 1992) and the recent Draft Guidelines (US FDA, 2001b) do not consider the use of rDNA technology in the development of food products to be a material fact requiring specific disclosure on the label. Rather it is a method of development, similar to other methods of plant breeding, which have not required disclosure on the label (US FDA, 1992). Bioengineered foods cannot be distinguished compositionally from foods modified through more conventional methods and thus do not require specific disclosure through labelling. The Draft Guidelines (US FDA, 2001b) reaffirm the US FDA position that bioengineered foods do not require special labelling.

Voluntary labelling

To provide guiding principles for voluntary labelling, in recognition of the desire of certain manufacturers to label foods as produced either with or without bioengineering, the US FDA published a 'Draft Guidance for Industry: voluntary labelling indicating whether the foods have or have not been developed using bioengineering' (US FDA, 2001b). Emphasizing that the use of rDNA technology 'is not a material fact', the US FDA recognizes that some consumers want disclosure of bioengineered content and that some manufacturers wish to provide it. In response, Draft Guidance was issued with suggestions concerning the use of labelling statements that are not considered misleading.

PRESENCE OR USE OF rDNA

The US FDA provided several examples of how disclosure of bioengineering can be accomplished, be informative and not be misleading (US FDA, 2001b).

- Example 1   'Genetically engineered' or 'This product contains corn meal that was produced using biotechnology'. These disclosures reveal the minimum amount of optional information about bioengineering.
- Example 2   'This product contains high oleic acid soybean oil from soybeans developed using biotechnology to decrease the amount of saturated fat.' This statement explains the proper and required name of this type of soybean oil, since it differs from what would be considered as soybean oil. The optional comments about biotechnology and decreasing saturated fat provide information that could be seen as benefits of the product.
- Example 3   'These tomatoes were genetically engineered to improve texture.' This example was one included to illustrate how it could be misleading to consumers if they cannot discern a difference in texture, but would not be misleading if they can tell a difference. If the former, and the new texture is to facilitate processing, then this intention should be made clear, i.e. 'to improve texture for processing'.
- Example 4   'Some of our growers plant tomato seeds that were developed through biotechnology to increase crop yield.' This is another example of optional information that would explain an indirect, agricultural benefit.

ABSENCE OR NOT BIOENGINEERED

The US FDA provides an important commentary in the Draft Guidance (US FDA, 2001b) about 'genetically modified organisms (GMO)' versus 'bioengineered'. It would be technically inaccurate to use the phrase 'not genetically modified' or 'GMO-free' to mean that bioengineering was not used. This is because most conventional foods have been genetically modified over the years by traditional crop breeding practices. Examples of acceptable voluntary statements would include:

- 'We do not use ingredients that were produced by biotechnology.'
- 'This oil is made from soybeans that were not genetically engineered.'
- 'Our tomato growers do not plant seeds developed using biotechnology.'

Another important point is made about a term such as 'GMO-free' that is misleading for two reasons: (1) many foods do not contain organisms anyway and therefore should not be labelled as 'organism-free'; and (2) 'free' implies complete absence or 'zero' amount of bioengineered material. In a practical sense, it is impossible to demonstrate analytically the complete absence of anything. Therefore, in reality, a threshold for possible adventitious presence of a low level of bioengineered material may be necessary and has been the subject of much debate.

The Agency also provides additional guidance about misleading statements that: (1) could be interpreted to suggest that the absence of bioengineering would make the food superior to the bioengineered alternative; (2) claim the absence of one bioengineered ingredient when the food contains another ingredient that is bioengineered; and (3) claim that a food is not bioengineered when in fact this type of food (e.g. green beans) has never been modified through rDNA technology.

Table 1.5 Contents of a synopsis prior to a pre-submission consultation meeting

Notifier's name and address
Name of bioengineered food and originating plant species
Type(s) of transformation event(s)
Origin of introduced genetic material
Purpose or intended effect in the bioengineered food
Intended uses
Potentially unsuitable uses
Other

Future considerations

Presubmission consultation

As mentioned previously, the US FDA recommends early consultation prior to the submission of a Pre-market Biotechnology Notice (PBN). This consultation process allows the US FDA and the manufacturer to discuss expectations and requirements, and to identify potential issues, before initiation of the PBN process. As a result, the US FDA expects the 120-day period, between the date of filing the PBN and the intended date of introduction to the market, sufficient for a favourable response, since no unanswered questions should remain at this point.

A written request for consultation should be submitted by the manufacturer, along with a synopsis of information sufficient to allow for 'a meaningful dialogue' (US FDA, 2001a). The elements to be addressed in a synopsis submitted to the US FDA along with a request for consultation are listed in Table 1.5.

The consultation meeting is expected to proceed by allowing (1) the prospective notifier to first explain their intentions to satisfy potential concerns of the US FDA, based on prior experience with similar foods; and (2) the US FDA responding with questions regarding issues that may be considered new.

Public disclosure

It is recognized that the majority of information and data presented or submitted in support of consultations with the US FDA would not be considered exempt from requests under the Freedom of Information Act (FOIA). As a result, unless a developer has submitted a specific request for the US FDA to withhold specific materials considered confidential or trade secrets, the public would be provided with full access to all information or data contained within the file. In an attempt to increase the transparency of the existing regulatory process, the US FDA has recently revised their website to provide the public with access to the results of prior consultations with the developers of plants developed using rDNA technology. In the future, the public may be provided with full access to the content of a PBN that is submitted for a bioengineered food. Of course, requests for exemption of specific components of a PBN under the FOIA would be reviewed by the US FDA and considered on their own merits.

US Department of Agriculture (USDA)

Authority

The Animal and Plant Health Inspection Service (APHIS) of the USDA is responsible for the protection of domestic agricultural resources and the environment from the threats posed by pests and diseases. The Federal Plant Pest Act (FPPA) has been the primary statute adapted by APHIS in regulating the products of agricultural biotechnology. Under the FPPA, APHIS has the authority to restrict the import of plant pests, to introduce measures (e.g. quarantines) to prevent their movement, or to destroy plant pests cultivated in violation of the FPPA. The requirements under the FPPA are equally applicable to scientists within company, academic, research, and private organizations.

Role

Following publication of the Coordinated Framework (OSTP, 1986), APHIS was the first federal regulatory agency to communicate their interests in evaluating products developed using rDNA technology (USDA, 1987). Under the FPPA, it reviews GM plants containing, or that are produced using, potential plant pests.

The regulatory authority of APHIS under the FPPA was considered to extend to the introduction of organisms meeting the definition of a plant pest or suspected to be plant pests. It was recognized that both conventional and GM plant species shared the potential to act as plant pests following their introduction. As such, organisms developed using rDNA technology would be evaluated in a manner similar to that used for organisms developed through conventional plant breeding.

Mandate

APHIS exercises its regulatory authority under the FPPA through a permit and notification system. In 1987, the USDA published a list of organisms considered to be plant pests (USDA, 1987).

A regulated article was considered any organism modified using rDNA technology, from a donor organism, recipient organism, vector or vector agent, that is a plant pest or contains plant pest components (as listed in 7 CFR 340.2) (USDA, 1991). For a GM plant to be considered a regulated article, a plant pest component must be involved during the modification process, an organism used in its development must be unclassified, or a reason must exist to believe that the GM plant should be considered a plant pest.

Definition of plant pests

As previously mentioned, a plant pest is any agent, substance or organism, or parts thereof, that may cause injury, disease or damage, either to plants or to the environment in which it is introduced. APHIS maintains a list of organisms considered to be plant pests, and therefore, subject to regulation under the FPPA. Table 1.6 serves as a summary of the types of organisms that have been identified as plant pests. If an organism is not on the list, it may still be subject to regulation as a plant pest if there is reason to

Table 1.6 Types of organisms considered plant pests[a] or that contain plant pests (7 CFR 340.2)

Types of organism(s)	Representative genus or family
Viruses	All members of groups containing plant viruses, and all other plant and insect viruses
Bacteria	*Pseudomonas, Xanthomonas, Rhizobium, Agrobacterium, Bradyrhizobium, Phyllobacterium, Erwinia, Streptomyces, Actinomyces, Clavibacter, Arthrobacter, Curtobacterium, Corynebacteria,* Rickettsiaceae
Algae/Fungi	*Cephaleuros, Rhodochytrium, Phyllosiphon, Eumycota,* Lagenidiaceae, Olpidiopsidaceae, Albuginaceae, Peronosporaceae, Pythiaceae, Saprolegniaceae, Leptolegniellaceae, Choanephoraceae, Mucoraceae, Entomophthoraceae, Protomycetaceae, Taphrinaceae, Elsinoeaceae, Myriangiaceae, Parmulariaceae, Phillipsiellaceae, Ophiostomataceae, Hysteriaceae, Ascocorticiceae, Hemiphacidiaceae, Dermataceae, Sclerotiniaceae, Sarcosomataceae, Auriculariaceae, Ceratobasidiaceae, Corticiaceae, Hymenochaetaceae, Echinodontiaceae, Fistulinaceae, Clavariaceae, Polyporaceae, Tricholomataceae
Plants[b]	Balanophoraceae, Cuscutaceae, Hydnoraceae, Krameriaceae, Lauraceae (*Cassytha*), Lennoaceae, Loranthaceae, Myzodendraceae, Olacaceae, Orobanchaceae, Rafflesiaceae, Santalaceae, Scrophulariaceae (*Alectra, Bartsia, Buchnera, Buttonia, Castilleja, Centranthera, Cordylanthus, Dasistoma, Euphrasia, Gerardia, Harveya, Hyobanche, Lathraea, Melampyrum, Melasma, Orthantha, Orthocarpus, Pedicularis, Rhamphicarpa, Rhinanthus, Schwalbea, Seymeria, Siphonostegia, Sopubia, Striga, Tozzia*), Viscaceae
Protozoa	*Phytomonas,* Anguinidae, Belonolaimidae, Caloosiidae, Criconematidae, Dolichodoridae, Fergusobiidae, Hemicycliophoridae, Heteroderidae, Hoplolaimidae, Meloidogynidae, Nacobbidae, Neotylenchidae, Nothotylenchidae, Paratylenchidae, Pratylenchidae, Tylenchidae, Tylenchulidae, Aphelenchoididae, Longidoridae, Trichodoridae
Molluscs	Planorbacea, Strophocheilacea, Succineidae, Achatinacae, Arionacae, Limacacea, Helicacea, Veronicellacea
Insects/Mites	Ascoidea, Dermanyssoidea, Superfamily Eriophyoidea, Tetranychoidea, Eupodoidea, Tydeoidea, Erythraenoidea, Trombidioidea, Hydryphantoidea, Tarsonemoidea, Pyemotoidea, Hemisarcoptoidea, Acaroidea, Sminthoridae, Acrididae, Gryllidae, Gryllacrididae, Gryllotalpidae, Phasmatidae, Ronaleidae, Tettigoniidae, Tetrigidae, Thaumastocoridae, Aradidae, Piesmatoidea, Lygaeoidea, Idiostoloidea, Coreoidea, Pentatomoidea, Pyrrhocoroidea, Tingoidea, Miroidea, Anobiidae, Apionidae, Anthribidae, Bostrichidae, Brentidae, Bruchidae, Buprestidae, Byturidae, Cantharidae, Carabidae| Cerambycidae, Chrysomelidae, Coccinellidae, Epilachninae, Curculionidae, Dermestidae, Elateridae, Hydrophilidae (*Helophorus*), Lyctidae, Meloidae, Mordellidae, Platypodidae, Scarabaeidae, Melolonthinae, Rutelinae, Cetoniinae, Dynastinae, Scolytidae, Selbytidae, Tenebrionidae, Diptera, Agromyzidae, Anthomyiidae, Cecidomyiidae, Chloropidae, Ephydridae, Lonchaeidae, Muscidae (*Atherigona*), Otitidae (*Euxeta*), Syrphidae, Tephritidae, Tipulidae, Apidae, Caphidae, Chalcidae, Cynipidae, Eurytomidae, Formicidae, Psilidae, Siricidae, Tenthredinidae, Torymidae, Xylocopidae

Source: http://www.orcbs.msu.edu/biological/7cfr340.html#340.2

[a] Any genetically engineered organism composed of DNA or RNA sequences, organelles, plasmids, parts, copies, and/or analogs, of or from any of the organisms listed is deemed a regulated article if it also meets the definition of plant pest in 7 CFR 340.1.
[b] Organisms listed in the Code of Federal Regulations as noxious weeds are regulated under the Federal Noxious Weed Act.

believe it is or will act as a plant pest. Organisms considered plant pests, or that exhibit the potential to be plant pests, are treated as 'regulated articles' under the FPPA.

The regulations extend to the introduction of GM organisms meeting the definition of a plant pest or for which there is reason to believe they are plant pests (USDA, 1987). APHIS considers the environmental release of a GM crop to be equivalent to the introduction of new organisms. Therefore, until proven otherwise, a GM crop is considered a 'regulated article' under the FPPA. The use of the term 'plant pest' in reference to GM plants means the 'non-pest' nature of the crop has yet to be determined. To date, all field tests conducted have demonstrated that GM crops exhibit no more 'pest-like' traits than their conventional counterparts.

All GM plants developed to date have involved the use of at least one designated 'plant pest' as either promoters or vectors, and as a result, have been subject to regulatory review under the FPPA. APHIS takes the position that an entire plant can be designated as a regulated article, even if it was not developed using plant pests, if the plant is the product of genetic engineering, and the agency determines or has reason to believe it is a plant pest. Given this position, APHIS would likely challenge any attempt to introduce a GM plant into commerce unless the developer has first gone through the APHIS regulatory review process.

Requirements for commercialization

A number of years of field testing are required to evaluate the agronomic and product quality characteristics exhibited by new plant varieties developed within a laboratory or greenhouse. APHIS regulations outline procedures for obtaining a permit or providing notification prior to the importation, interstate movement or release of a regulated article, which of course, includes field testing. Prior to an introduction, a proponent must provide notification to APHIS of its intentions or submit an application to obtain a permit. The notification and permit requirements have evolved as an extension of the long-standing program for the regulation of plant pests within the United States. Since 1987, APHIS has issued permits or acknowledged the receipt of notifications for the field release of more than 6500 new crop varieties.

Permits

For APHIS approval, developers must demonstrate that a new plant variety poses no significant risk to other plants in the environment or is at least as safe as similar crop varieties. APHIS review and approval are required for the shipment of seed and for the conduct of field trials involving new crop varieties. APHIS issues site-specific permits for field tests or releases into the environment.

Prior to field testing, APHIS may seek further clarification of study objectives, specify how, when or where research may be conducted, establish the data to be generated, collected and reported, stipulate additional requirements for the monitoring or securing of test sites, and specify methods for the disposal of crop residues that remain at the end of a study. The results of field trials are often used in making a determination as to whether a GM plant exhibits any effects on non-target species, or poses any unique plant pest problems. For field trials conducted under either notification or permit, developers of GM plant varieties are required to submit reports to

APHIS that include: details regarding the methods of observation; the data collected; and interpretations concerning the effects on plants, non-target organisms, or the environment. In the event that a GM crop is likely to be the subject of a petition for a determination of non-regulated status, developers must provide a description of known and potential differences, and substantiate that the regulated article is not likely to pose a greater plant risk than the organism from which it was derived.

An environmental assessment is a requirement of the NEPA, Council on Environmental Quality regulations, and USDA procedures for issuing permits under the FPPA. Under the permit process, developers of GM plant varieties must disclose information concerning the plant, test facilities, and control measures in place for its transport and field testing. Based on this information, the USDA performs an assessment of the potential environmental impact following a release. If the agency reaches a 'Finding of No Significant Impact' (FONSI), a permit may then be issued.

Notifications

Depending on the plant species and intended introduction, applying for a permit from APHIS can be a complicated and time-consuming process. In the spring of 1993, APHIS introduced a simplified notification process as an alternative to applying for a permit, which in most instances applies to the introduction of GM plants (USDA, 1993). To qualify for notification, a GM plant must be introduced in accordance with the eligibility criteria as outlined in Table 1.7 (USDA, 1993).

Originally, six GM crops (corn, cotton, potatoes, soybean, tobacco and tomatoes) qualified for notification (USDA, 1993), however, the process was more recently expanded to include most other crop species not capable of becoming a noxious

Table 1.7 Eligibility criteria for USDA field permit notification

1 Plant species are corn, cotton, soybean, tobacco, tomato, or other additional approved crops.
2 The introduced genetic material is stably integrated.
3 The function of the introduced genetic material is known and its expression in the regulated article does not result in plant disease. If the nucleotide sequence encodes a protein, then its function should be known.
4 The introduced genetic material does not cause the production of an infectious entity; encode substances that are known or likely to be toxic to non-target organisms known or likely to feed on or live on the plant species; or encode products intended for pharmaceutical use.
5 The introduced genetic sequences do not pose a significant risk of creating any new plant virus. Plant virus sequences eligible include: non-coding regulatory sequences of known function; sense or antisense constructs derived from viral coat protein genes; or antisense constructs derived from noncapsid viral genes.
   Sense and antisense constructs must be derived from plant viruses that are prevalent and endemic in the area where the introduction will occur and that infect plants of the same host species. For a non-coding regulatory sequence to be eligible, the DNA sequence must be a promoter, enhancer, intron with enhancer activity, upstream activating sequence, polyadenylation signal, transcription terminator, or other known regulatory sequence. This criterion excludes the release of plants expressing sense constructs for viral genes other than the coat protein.
6 The plant has not been modified to contain certain genetic material derived from an animal or human pathogen.

weed (USDA, 1997). The developers of new GM plant varieties are encouraged to contact APHIS when making a determination as to whether a GM plant would qualify for notification. If a particular GM plant does not qualify for notification, a more involved regular permit process must be followed.

For a GM plant to qualify for notification, it and its predecessor must not be a noxious weed, the introduced gene sequence(s) must be stably integrated within the host, not originate from a plant or animal pathogen, exhibit a known function that will contribute to plant disease, or encode for substances that could be potentially toxic or infective to non-target organisms, such as pharmaceuticals, or viral components other than those coat proteins already established as safe (USDA, 1997).

Performance standards also exist for the conduct of field trials to ensure that an adequate level of containment is provided for all introductions. It is the responsibility of the developer to ensure that an adequate level of containment is provided for field trials involving GM plants under notification. The performance standards utilized will vary according to the biology of the plant and the nature of the introduction. In general, this includes specifying methods for the proper handling, shipment, field monitoring and disposal of plant material at the end of the study.

As a part of the notification process, an applicant must provide information about the plant, identify the source of any genes used; the method of genetic modification; and the size, date and location of any proposed field test or introduction. Notification is required at least 10 days prior to the interstate movement or 30 days in advance of the field testing or importation of a regulated article that qualifies for notification with APHIS.

Determination of non-regulated status

Once sufficient data have accumulated through laboratory and field trials performed under permit or notification, a developer may petition APHIS for a determination of non-regulated status (DONRS) (USDA, 1993; 7 CFR 340.6). A DONRS allows a GM plant to be grown, tested or used for crop breeding without further regulatory oversight by APHIS (subject to limitations imposed by either US FDA or US EPA). APHIS will grant a DONRS if sufficient evidence can be provided to demonstrate that no plant pest potential exists (i.e. that the GM plant is not expected to become a pest, poses no significant risk to the environment, or is as safe as conventional plant varieties). The developer is required to demonstrate a lack of change in disease or pest resistance status, an absence of any potential for contributing to development of a new plant pathogen or pest, as well as to address any questions relating to potential environmental consequences of the release.

APHIS examines several parameters when making a determination as to whether a GM crop should be considered a plant pest. Reporting requirements include submitting a detailed rationale for development; overview of the biology of the crop (competitiveness, survivability, dormancy); molecular characterization; protein expression; morphological/phenotypic characteristics; outcrossing/geneflow; weediness; insect and disease susceptibility; and, recombination potential (for viral genes).

Other considerations include assessing whether a GM crop could itself become a weed. In making this determination, factors that are considered include: how the seeds are dispersed in the environment; whether the seeds are capable of surviving over the winter; the potential for the development of volunteer plants; and whether these plants could reproduce into viable offspring. Another important consideration is whether the seeds or plants are capable of surviving outside of a managed agricultural environment (e.g. watering, fertilizer, etc.), and how the introduced traits could influence the viability of the crop under marginal conditions.

APHIS also takes into consideration potential adverse effects on wildlife, including birds, beneficial insects and mammals, following the introduction of a GM crop. Some of this information comes from field trials that are conducted in multiple locations for several years. By observing crops growing under actual conditions of use in the field, scientists can compare insect populations existing within a field planted with GM crops to those coexisting in fields planted with a non-modified variety of the same plant. Field trials are also useful in identifying any changes in the plant physiology, such as height, color, leaf placement, time of flowering, etc., that result from modifications in the genome of the plant. Monitoring of these changes during field trials allows APHIS to evaluate how these changes might benefit or adversely impact the behavior of wildlife in contact with these crops. Knowing that wildlife, such as deer, often feed on agricultural crops, APHIS also takes into consideration the nutritional content of the GM crop. Comparisons of the essential nutrient content are often made between the GM plant and its conventional counterpart. Other interests may include determining the impact of the modification on the levels of adverse factors (e.g. natural toxicants or anti-nutrients) present in the plant.

APHIS maintains a list of all crops that have received a DONRS. At any time, APHIS retains the authority to prohibit the commercial planting of a GM crop if it is determined that it is becoming a plant pest.

PETITION PROCESS

Before a GM plant can be freely transported and commercialized, the developer must first petition the APHIS for a DONRS. A petition for a DONRS requires the submission of extensive data on the introduced gene construct, effects on plant biology and effects on the ecosystem, including the spread of the gene to other crops or wild relatives. Developing all data necessary to support a petition for a DONRS may take months or years depending upon the scientific issues that need to be addressed. Based on the information provided within the petition, APHIS evaluates the potential impact (e.g. toxicological) that the introduced modifications will have on non-target organisms in the environment, including threatened or endangered species. A petition for a DONRS must provide a rationale for the development and introduction of the GM crop, starting with genotypic and phenotypic information on both the host and donor organism(s) (USDA, 1996). This includes a detailed description of the methods of transformation, identifying all gene sequences (e.g. promoters, leader sequences, introns, selectable markers, etc.), their function, and potential for plant risk following introduction. A combination of different analytical methods (e.g. Southern, Northern, Western, PCR, ELISA) may be used to demonstrate the stable integration of an introduced gene sequence within the genome, its inheritance and expression in host plants. To the extent possible, the gene(s) of interest (e.g. resistance, marker genes), levels of expression, and impacts on levels of inherent plant toxicants in plant tissues, under experimental and actual growth conditions in the field, must be characterized.

Field performance studies (e.g. leaf morphology, pollen viability, seed germination, viability, insect susceptibilities, disease resistance, yield, etc.) are used to identify any differences or similarities in phenotypic characteristics exhibited by the GM and non-modified crop. The extent of field performance data required in support of a petition for non-regulated status depends not only on the nature of introduced gene sequences, but also on the physiology of the host plant. Less extensive field performance studies are required for plants that are: highly domesticated (e.g. corn); self-pollinating (e.g. soybean); male sterile; those with high seed germination rates (>90%); or unlikely to influence their potential weediness or fitness (e.g. delayed ripening, oil seed content). Whereas, GM crops exhibiting a tolerance or resistance to cold, salt, biotic (e.g. insects, pathogenic agents), or other abiotic (e.g. herbicide) stresses, are subject to more extensive field performance testing.

APHIS considers several parameters in reaching a DONRS, including whether the GM plant can cross-pollinate with other plants in the wild, and if it can, the ecological consequences (e.g. insect resistance, herbicide tolerance) that might result. Extensive databases are maintained on species capable of cross-pollinating with GM crops (e.g. corn, soybean, cotton, canola, etc.), based on the results of years of breeding experiments, biological surveys and research conducted by agricultural and weed specialists within APHIS. In instances where out-crossing might be expected to occur, APHIS may stipulate the planting of refuge areas to limit the potential for adverse ecological effects following commercial planting.

AGENCY RESPONSE

APHIS will grant a determination of non-regulated status if the developer can sufficiently demonstrate a lack of plant risk. In making a determination that a GM plant will no longer be regulated, APHIS prepares two documents, an environmental assessment and a DONRS to satisfy regulatory requirements under NEPA and FPPA, respectively. Both documents are developed based on the review of data submitted as a part of the petition for a DONRS. Once the DONRS has been granted by APHIS, a GM crop no longer requires notification or permit prior to its movement or release within the United States.

A review for DONRS is generally completed within 10 months of submitting a formal application to the APHIS, allowing sufficient time for publication in the Federal Register, and opportunity for public comment. Notices of all petitions for a DONRS are published in the Federal Register, and the public is given 60 days to comment for or against the petition, after which APHIS has up to an additional 180 days to either approve or deny it. APHIS provides access to a considerable amount of information relating to approved field trials and petitions received for GM crops on their website. The petition for a DONRS must include scientific details regarding the genetics of the plant, the nature and origin of the genetic material used, the potential for indirect effects on other plants, and any other information that could be considered unfavorable to the petition.

Extensions of non-regulated status

In certain instances, the non-regulated status of a GM crop may be extended to include other varieties of the same crop species, provided that changes in introduced gene sequences are insignificant, and no new plant pest issues would be expected

(USDA, 1997). For an extension, developers must substantiate that the new regulated article poses no serious issues meriting review under separate petition, by providing a precise description of the genetic modifications to the regulated article and a detailed comparison to modifications made in the GM plant previously granted non-regulated status by APHIS. Field trials must be performed under notification or permit to demonstrate the similarities in the phenotypic properties. In support of multiple extensions, product identity standards may be submitted describing the genotypic and phenotypic properties exhibited by new plant varieties considered extensions of an existing DONRS for a GM crop.

Future considerations

The USDA recently appointed an Advisory Committee for Agricultural Biotechnology that will work in conjunction with the NAS (Standing Committee on Biotechnology Food and Fiber Production, and the Environment) on the completion of a critical review of the regulations and policies used by APHIS in the commercial approval of GM crops. The committee comprises scientists, farmers and representatives of consumer groups and seed companies, and is anticipated to take two years to complete its assignment.

In addition to APHIS, other functional areas within the USDA are beginning to get more actively involved with the regulation of agricultural biotechnology. For example, the Grain Inspection Packers and Stockyards Administration (GIPSA) has made a commitment to work with farmers and industry groups in the development and validation of reliable test methods and quality assurance programs to differentiate GM from non-GM commodities. As a part of this commitment, GIPSA intends to publish a proposed rule on the standardization of test methods and approaches for the detection and identity preservation of grains. GIPSA is also intending to offer accreditation services for laboratories testing grains for the presence of GM content, and to evaluate commercial test kits to ensure they can be considered accurate and reliable. It is expected that an administrative fee will apply to the accreditation services being offered by GIPSA.

The USDA also has made continuous commitments to funding research that will assist federal regulatory agencies in making science-based decisions concerning the safe introduction of GM organisms into the environment. Active areas of research include assessing the potential risks associated with the environmental introduction of GM plants (e.g. gene flow), the cumulative effects of large-scale commercial plantings, the potential interactions between GM and non-GM crop species, programmed resistance, and the development of statistical methodology and quantitative approaches to measuring the risks associated with the field testing of GM plants, etc.

US Environmental Protection Agency (US EPA)

Authority

The US EPA has authority over the registration of chemical and biological pesticides under the Federal Insecticide, Fungicide and Rodenticide Act (FIFRA). A pesticide is any product 'intended for preventing, destroying, repelling, or mitigating any pest'.

Table 1.8 Chronology of EPA regulation of biopesticides and the products of rDNA technology

Date	Event
1965–75	Increasing interest in biological pesticides
1974	EPA Workshop and Symposia – Unique Safety Concerns
1978	OPP Policy on Biorational Pesticides (distinct from chemicals)
1983	Testing guidelines (Subdivision M) published by NTIS:
	Biorational pesticides
	Genetically modified microbial pesticides (case by case)
1989	Testing guidelines (Subdivision M) reviewed by SAP and published
1994	Proposed rule for plant-pesticides
2001	Final rule for plant-incorporated protectants (PIPS)

All pesticides require registration by the US EPA prior to their distribution, sale or use within the United States. As a part of the registration process, an applicant is required to submit information and data in support of the safety of a pesticide in its intended use.

In addition, US EPA is responsible for establishing acceptable tolerance levels for registered pesticides on or in raw agricultural food commodities under the FFDCA. At any point, US EPA has authority to amend or revoke a registration or residue tolerance, if an adverse effect is observed, or if the risks associated with the use of the pesticide are determined to be unacceptable.

Role

Under the Coordinated Framework (OSTP, 1986), the US EPA was identified as being responsible for the review, assessment and registration of the products of rDNA technology that act as pesticides. Leading up to this publication, the US EPA had acquired considerable expertise in the registration of biological pesticides (e.g. microbial). The US EPA was first involved in discussions regarding the potential risks of pesticides developed using rDNA technology in the early 1980s, during a series of public meetings involving the FIFRA Scientific Advisory Panel (SAP) and the Biotechnology Scientific Advisory Committee (BSAC). A chronology of significant events in US EPA's evolving history of regulating biopesticides and the products of rDNA technology has been provided in Table 1.8.

Mandate

It was recognized that the potential risks posed by plants modified to express pesticidal components through rDNA technology were unique and warranted additional consideration relative to conventional pesticides. For example, the potential for introduced genetic material to out-cross from a modified plant to a sexually compatible wild or weedy relative through pollen spread was one of the unique, theoretical risks posed by plant-pesticides.

As a result, the US EPA proposed a plant-pesticide rule in 1994, outlining their interests in the regulation of products (US EPA, 1994). A package of three proposed final rules was released early in 2001, redefining regulations for plant-incorporated protectants (PIPS), previously referred to as plant-pesticides. Essential elements of the proposed rules for the regulation of GM plants expressing pesticidal traits have been followed since 1994. The final rules provide further clarification of how PIPS will be regulated under FIFRA and FFDCA, including specific exemptions (US EPA, 2001a,b,c) and became effective in September 2001.

Definition of plant-incorporated protectants (PIPS) or plant-pesticides

US EPA's final rule (US EPA, 2001a) clarifies the regulation of whole plants that contain PIPS. When plants are used intentionally for controlling pests in some way, such uses meet the definition of a pesticide under FIFRA. Nevertheless, the agency exempts the plants themselves from regulation, focussing instead on the substances they contain, previously referred to as plant-pesticides and now known as PIPS (US EPA, 2001a). A plant-incorporated protectant (PIP) includes the substance (e.g. *Bacillus thuringiensis* or Bt toxin, etc.) produced by the living plant, and all genetic material necessary for its production (including promoters, enhancers, etc.). Thus, genes and substances (e.g. enzymes) responsible for the conversion of natural plant constituents into pesticidal substances would be considered PIPS. To date, PIPS registered by the US EPA have been proteins and the genes required to make these proteins within the plant. Crops involved include potatoes, cotton, field corn, sweet corn and popcorn. A listing of PIPS that have been registered by the US EPA appears in Table 1.9.

Clearly, Bt toxins represent the most predominant class of PIPS that have been registered by the US EPA. Major reasons for this include the success of microbial products containing Bt and their long history of use for nearly 40 years. It has been estimated that combined, Bt corn, potato and cotton were cultivated on approx-

Table 1.9 Registered plant-incorporated protectants (PIPS)

Active ingredient	Crop	Trade name	Company	Date registered
Bt Cry 3A	Potato	NewLeaf	Monsanto	1995
Bt Cry 1Ab	Field corn	NatureGuard Corn Borer Control Protein	Mycogen	1995
Bt Cry 1Ab	Field corn	NK Brand Bt Corn with KnockOut	Novartis	1995
Bt Cry 1Ab	Popcorn	Hybrid Popcorn with KnockOut		1998
Bt Cry 1Ac	Cotton	BollGuard	Monsanto	1995
Bt Cry 1Ab	Field corn	YieldGuard	Monsanto	1996
Bt Cry 1Ab	Field corn	NK Brand Bt Corn with YieldGuard	Novartis	1996
Bt Cry 1Ab	Sweet corn	Attribute	Novartis	1998
Bt Cry 1Ac	Field corn	DEKALBt	Dekalb	1997
Bt Cry 9c	Field corn	Starlink	AgrEvo	1998
PLRV virus resistance gene	Potato	NewLeaf Plus	Monsanto	1998

Source: www.epa.gov/oppbppd1/biopesticides

Table 1.10 Cotton bollworm/budworm insecticide use reductions after the introduction of Bt-protected cotton[a]

Insecticide	Usage in 1998, relative to 1995
Amatraz (Ovasyn)	−42
Cyfluthrin (Baythroid)	−35
Cypermethrin (Ammo)	−81
Deltamethrin (Decis)	+11
Esfenvalerate (Asana)	−19
Lambdacyhalothrin (Karate)	−58
Methomyl (Lannate)	−156
Profenofos (Curacron)	−1014
Spinosad (Tracer)	+19
Thiodicarb (Larvin)	−665
Tralomethrin (Scout)	−4
Cypermethrin (Fury)	+1
Total	**−2044**

[a] Estimated uses in five agricultural states in the US (Arizona, Arkansas, Louisiana, Mississippi and Texas), as modified from Betz *et al.* (2000).

imately 10 million acres in 1997, 20 million in 1998, and 29 million in 1999 (Betz *et al.*, 2000). As a result, less chemical insecticide has been applied, and there have been significantly higher crop yields. In addition, due to reduced opportunity for opportunistic fungal infections of insect-damaged corn, lower levels of mycotoxins have been produced, reducing the toxicological risk to both humans and animals. The most significant reductions in chemical applications of insecticides have been reported in cotton-producing regions within the United States, following the registration of Bt cotton in 1995, as summarized in Table 1.10.

This trend can be further visualized in Figure 1.4, which depicts annual reductions in the number of insecticide applications over a four-year period within the same growing regions. A review by Betz *et al.* (2000) also discusses a number of other additional advantages of PIPS.

Exemptions

US EPA has concluded that living plants with pesticidal activity do not require a high level of regulatory scrutiny by the agency and thus, are exempt from regulation under FIFRA (US EPA, 2001a). The agency commented that focussing on the substance produced within the plant, i.e. the PIP eliminates any need for the US EPA to register plants, thus conserving agency resources and reducing potential overlap with other agencies, notably the USDA. For example, although whole plants that are used as biological control agents themselves, such as chrysanthemums, are exempt, substances that are extracted from plants, such as the insecticidal material pyrethrum, are not excluded from regulation under FIFRA.

Further exemptions have been proposed (US EPA, 1994), based on familiarity and presence of the pesticidal substances in the food supply. Only one of the additional proposed exemptions has been included in the final rule (US EPA, 2001a), and all others have been proposed for further public comment (US EPA, 2001d). Within the

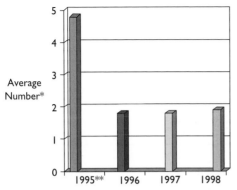

*   Average number of insecticide treatments in 6 states (Alabama, Arizona, Florida, Georgia,
    Louisiana, Mississippi): 4.8 (1995), 1.8 (1996,1997), 1.9 (1998).
** Before introduction of Bt-protected cotton.

Figure 1.4 Number of insecticidal treatments per year for pest control in cotton. Average
             number of insecticide treatments in six states (Alabama, Arizona, Florida, Georgia,
             Louisiana, Mississippi): 4.8 (1995), 1.8 (1996, 1997), 1.9 (1998). Bt-protected cotton
             was introduced between 1995 and 1996. (Modified from Betz *et al.*, 2000.)

final rule, US EPA (2001a) has exempted PIPS and encoding genetic material originating from plants sexually compatible through conventional breeding but not through
rDNA technology. The rationale for this exemption is that such substances could
be bred, either naturally or through conventional plant breeding, from close plant
relatives that share genetic material from a common gene pool. On the other hand,
genetic modifications that were never before possible can be made using rDNA technology, involving the genes from sexually incompatible plant species. Therefore, the
exemption of plant-pesticides developed using rDNA technology, involving genes
from sexually compatible plants has not been finalized as a part of the supplementary
proposal to the final rules for PIPS (US EPA, 2001d).

The other exemptions for further comment have also been identified within the supplemental proposal (US EPA, 2001d). The first group are PIPS that 'act primarily
by affecting the plant'. The group have in common some sort of defense mechanism
such that a pest would be less able to attach to, penetrate, or successfully invade a
plant. Examples include: (1) some sort of structural barrier such as wax or plant
hairs; (2) inactivation of or resistance to a pest or substance (e.g. toxin) produced by
a pest; (3) creation of a deficiency in a nutrient or growth factor required by a pest;
(4) hypersensitivity response that would contain or limit spread of the pest (e.g.
necrosis of surrounding plant tissue, creating a functional barrier); and (5) plant
hormones that are naturally occurring but with change can affect a plant in a variety
of ways with potentially dramatic results.

The other class of PIPS proposed for exemption include viral proteins and their
encoding genes (US EPA, 1994, 2001d). The expression of genes for viral proteins
within a host plant confers resistance to infection by that virus as well as certain
related viruses. These PIPS are plant virus specific, and as such, would not be expected
to pose any risk to the public. Table 1.11 identifies viral proteins that act as plant
pesticides, and that are considered exempt from regulatory review by the US EPA.

Table 1.11 Viral proteins

Protein	Virus
Viral coat protein	Cucumber mosaic
	Papaya ringspot
	Potato Y
	Watermelon mosaic
	Zucchini yellow mosaic
Replicase	Potato leaf roll

Modified from NRC (2000).

Herbicide-tolerant crops

In the case of herbicide-tolerant crops, US EPA determines whether the property of increased resistance may encourage higher rates or more extensive application of an agricultural chemical to food crops. The higher exposures to herbicides would be subject to evaluation through risk assessment, requiring the submission and review of detailed data. If the risk is determined to be acceptable, a herbicide product would require label extensions for the addition of new tolerant crop varieties.

Experimental Use Permit (EUP)

Small-scale research involving PIPS is covered under the permit and notification process that exists for new plant varieties administered by APHIS. The US EPA requires that an Experimental Use Permit (EUP) be obtained for all research and field testing involving greater than 10 acres of a plant modified to express a pesticidal substance. The EUP process allows a developer to gather the product use, performance, residue and other types of data, required in support of the registration of PIPS. All US EPA decisions concerning applications for EUPs are published within the Federal Register for comment. As noted in Table 1.12, insect and virus resistance are the most prevalent types of pest-resistant crops grown under EUP within the United States.

Registration requirements

For the registration of a plant-incorporated protectant, the US EPA evaluates the potential impact, on human health and the environment, of the pesticide substance

Table 1.12 Field trials of pest-resistant crops

Pest type	Number of trials
Insect	1505
Virus	1013
Fungal	378
Bacterium	78
Nematode	7

Source: www.nbiap.vt.edu/cfdocs/fieldtests1.cfm, as compiled in NRC (2000).

and of the genetic material necessary for its expression within the modified plant. With this in mind, the US EPA developed their guidance for evaluating the safety of PIPS based on data requirements that were established for microbial pesticides. A number of modifications to the reporting requirements were required to focus the assessment on non-target toxicity rather than pathogenicity, as well as to evaluate the expression of the pesticide substance within the plant itself, rather than by spraying, etc.

Although specific reporting requirements for the registration of PIPS are established on a case-by-case basis by the US EPA, the general categories of information and data required in support of the registration of a plant-pesticide (Kough and Vaituzis, 1999), include:

1   product characterization
2   toxicology
3   environmental fate and exposure
4   non-target organism effects
5   if necessary, plans for pest resistance management.

The US EPA continues to hold meetings of their SAP to re-evaluate the reporting requirements that have been established for PIPS and to ensure that all potential human health and environmental concerns are being addressed as new scientific information becomes available. In addition, the developers of PIPS are encouraged to consult with US EPA during all phases of the research and development process, particularly when it concerns an environmental release.

As a part of their review, scientific evaluators within the US EPA weigh the potential hazards inherent to the generic consideration of PIPS and the potential for exposure to the protectant substance. The characterization of possible hazards, in the context of exposure, allows for estimates of potential environmental risk to be established, as outlined in Table 1.13.

The commercial approval of a PIP may take between 15 and 18 months following receipt of a full data package by the US EPA, and it involves a public notification and consultation process. However, this time frame would be considered an ideal, since in reality, deciding what constitutes a 'full' data package may take several rounds of discussions with the US EPA.

Table 1.13 Environmental risk assessment of plant-incorporated protectant

Potential hazard	(X)	Exposure potential
Inherent nature of gene(s)		Feeding on plant or pollen
Effect on host plant		Gene transfer to other plant relatives
Effect on non-target susceptible species		Concentration in plant part of gene product
(e.g. plants; invertebrate and vertebrate animals; microbes)		Degradation/persistence of gene product
Disruption of ecosystem checks and balances		Release/movement of gene product
Weediness		Production of pollen
		Distribution of seed
	(=) Risk Assessment	

Product characterization

Product characterization allows for consideration of any potential hazard(s) that might be associated with anticipated exposure to a plant-incorporated protectant. This process includes a review of the source of introduced genetic material, including all regulatory sequences, the modifications that have been made, methods of transformation, inheritance, stability and expression in the host plant. For the introduced pesticidal substance, the anticipated modes of action, specificity and toxicity to both target and non-target organisms are evaluated. Target organisms may include weeds, insects and diseases (viral, bacterial, fungal) affecting a crop. The crop itself, as a volunteer weed, may also be of concern following crop rotations. Once a PIP and any potential for exposure have been characterized, the data required in support of an environmental and human health risk assessment can be defined.

Ecological effects

Environmental fate and exposure are important considerations when evaluating the potential impact of PIPS on the environment or on non-target organisms (e.g. wildlife, beneficial insects). Depending on the crop and introduced pesticide substance, data on the level of expression and rate of degradation may be required, first for the tissues of the modified plant, and secondly, the fate in soil, water, or any other environmental compartment. Such data allow the potential for exposure to the pesticide substance to be evaluated for non-target organisms via feeding on, or through decomposition of, the modified plants.

   The data reporting requirements for assessing the potential effects of a PIP on non-target organisms are outlined in Table 1.14. Studies are arranged in tiers, starting with Tier I, consisting of acute toxicology and basic characterization studies that help predict environmental fate. Depending on the results obtained in Tier I, and the level of concern, additional studies may be requested from higher tiers (Andersen, 1999; Kough and Vaituzis, 1999). It is expected that most PIPS will only undergo Tier I testing, since negative results are expected at relatively high exposure levels. The 'maximum hazard approach' when used in Tier I provides a high level of assurance that adverse effects would not be expected to occur under actual conditions of use within the field. If the results of Tier I reveal a potential hazard at doses within the range of expected concentrations in the environment, additional studies would be requested in higher tiers to better characterize the potential for any kind of effect.

   Data requirements may be satisfied by generating test data or by requests for data waivers with credible justification. The type and extent of dietary studies required will depend on properties of the crop and on the introduced pesticide substance. Dietary studies typically involve feeding plant tissues (e.g. grain, pollen) that express the PIP or that have been spiked with the pesticidal substance. However, when determined necessary, the pure pesticidal substance may be administered to non-target species at doses of up to 10 to 100 times the expected level of exposure in the field. Dependent upon the potential for exposure, additional studies may also be requested involving any number of other non-target organisms (e.g. honeybees, green lacewing, ladybird beetles, parasitic wasps, earthworms, etc.). When toxicity is observed, the

Table 1.14 Testing requirements for non-target organisms

Tier	Test[a]	Required[b]
I	Avian oral toxicity	R
	upland game bird	
	waterfowl species	
	Avian injection	CR
	Avian inhalation	CR
	Wild mammal oral toxicity (rodent species)	R
	Freshwater fish oral toxicity	R
	cold water species	
	warm water species	
	Freshwater invertebrates (*Daphnia* or aquatic insect)	R
	Estuarine and marine animal	R
	grass shrimp	
	fathead minnow	
	Non-target plant	
	terrestrial	
	aquatic	
	out-crossing issues	
	Non-target insect	R
	predator	
	parasite	
	Honey bee (larval and adult)	R
	Earthworm and springtail	R
	Fate and degradation in soil	R
	Plant tissue degradation	R
II[c]	Fate in freshwater/marine environment	
III	Chronic	
	Reproduction	
	Life cycle	
IV	Microcosm	
	Field study	

[a] Satisfied by data or requests for waiver.
[b] R = required; CR = conditionally required.
[c] Adverse effects in Tier I at expected field application doses might result in requirements for other Tier I tests or higher tier levels.

potential for exposure becomes an important consideration when ascertaining whether an adverse effect might be expected under actual conditions of use in the field.

The requirement for provision of a resistance management plan is not routine practise for PIPS, but depends on an evaluation of the potential for resistance to develop and the threat to existing or functionally equivalent pest management practises. To date, resistance management plans have only been placed on Bt PIPS. If considered appropriate, the US EPA may request that a resistance management plan be submitted as a part of a condition for registration. A resistance management plan may require a commitment towards the generation of additional research data, to implement structured refuges, to perform annual resistance monitoring, to initiate remedial action plans, and to educate growers. The US EPA has worked closely with academia, other federal agencies, public interest groups, industry, and growers, to establish resistance management plans based on the most current science.

Human health effects

For agricultural food crops, special consideration is given to evaluating the safety of the pesticidal substance introduced as a component of consumed foods. Rodents (often mice, to conserve test material) are orally administered the pesticide substance at doses of up to 100 000 times the levels of expected human consumption as a part of the diet. Although the modified plant tissues would be considered the most appropriate vehicle for determining the acute toxicity of a PIP, it is not typically feasible, since only small amounts of the pesticidal substance are present in plant materials. Therefore, an alternative source (e.g. microbial) is often used to generate a sufficient amount of identical pesticide substance for testing purposes. In this case, the equivalency of the test substance obtained from an alternative source to the PIP (e.g. composition, identity, etc.) being expressed within the plant is an important consideration that must be well documented.

For pesticide substances that are proteins, the principal concern relates to the potential for the introduction of a new allergen or toxin into the food supply. Since most allergenic food proteins are stable to digestion, *in vitro* digestibility studies are used to predict how long a pesticidal substance may persist in digestive fluids (e.g. gastric, intestinal) following ingestion. In addition, heat stability studies are performed to provide an indication of the fate of the PIP during food processing. Commonly, assays of amino acid sequence homology are performed to determine similarity to substances that are known to pose unique safety concerns (e.g. toxins, allergens). The US EPA uses a 'decision tree' approach similar to that followed by the US FDA in assessing the potential allergenicity of a new food protein (see Figure 1.2). In many instances, US EPA and US FDA collaborate with each other when evaluating issues of safety for new pesticidal proteins being expressed in GM food crops.

Residue tolerances for plant-incorporated protectants

Analogous to the consideration of tolerances for residues of conventional pesticides, the US EPA reviews data on the human, animal and environmental safety of PIPS to determine whether residue tolerances should be established for the amounts expected to be present in foods derived from the GM plant. Under Section 408 of the FFDCA, if the GM plant expressing a PIP is a food crop, the US EPA is obligated to establish the 'safe level' of pesticide residue allowed, or tolerance level.

The Food Quality Protection Act (FQPA) served to amend the FFDCA and FIFRA by outlining the process by which the safety of pesticide residues on raw or processed foods can be established. Under this process, in order for a pesticide residue to be considered safe, a 'reasonable certainty of no harm' must exist assuming aggregate exposure of the public, taking into consideration sensitive children within the population. To date, all pesticidal proteins that have been registered as PIPS have been determined to be non-toxic, and as a result, have been considered exempt from the requirement to establish a tolerance under the FFDCA.

Consistent with categorical exemptions as originally proposed for plant-pesticides under FIFRA (US EPA, 1994), the US EPA also proposed that tolerances might not be required for pesticidal substances derived from sexually compatible plants, regardless of the process used to introduce the PIP, provided that the genetic material encoding the pesticidal substance was from a plant commonly used as food, and that the presence of the pesticidal substance in a host plant did not result in any new or significantly different dietary exposures. Final rules include exemptions from tolerances for the residues of PIPS, and their encoding nucleic acids (US EPA, 2001b,c), but only for PIPS derived through the conventional breeding of sexually compatible plants. US EPA is requesting further comment on the possible exemption from the requirement of a tolerance for: (1) PIPS derived through rDNA technology; (2) PIPS that act by affecting plant defenses; and (3) PIPS developed based on viral coat proteins (US EPA, 2001d).

Future considerations

The US EPA anticipates that the requirements for the registration of PIPS will continue to evolve as the science and policies relating to biotechnology continue to mature. In light of on-going criticisms that the plant-pesticide rule has received since its publication in 1994, a number of stakeholder workshops have been hosted by the US EPA. In March 1999, the House Agricultural Subcommittee hosted a congressional hearing to address a number of these concerns. One of the primary questions addressed was whether the definition of a pesticide under FIFRA provides the US EPA with the authority to regulate pest-protected plants that have been developed using rDNA technology. Another concern was that the proposed rule was too broad in scope because all plants have natural defense mechanisms, and as such, would be classified as 'plant-pesticides'. In addition, an exemption from the plant-pesticide rule for those plants not developed using rDNA technology would serve to regulate the process used in the development of the GM plant, rather than on the pesticidal substance being produced within the plant. Further criticism related to the use of the term 'plant-pesticide' in relation to the products of the genes that had been introduced into GM crops.

As a part of its final rule (US EPA, 2001a), the US EPA recommended that plant-pesticides be renamed PIPS, an alternative name which distinguishes them from the other types of pesticide products requiring registration with the US EPA. A new FIFRA section (Part 174) has been added to the Code of Federal Regulations (Part 174) specific to the regulation of PIPS. In doing so, the US EPA has acknowledged that the new final rule would be making a distinction between the products of conventional breeding and rDNA technology, i.e. regulating on the basis of process not product. One reason given for this divergence was that the use of this criterion would be expected to provide the public with increased confidence that an appropriate level of regulatory oversight was in place for assessing the risks of rDNA technology (US EPA, 2001a), which is a societal not scientific issue.

SAPs are charged with providing the Office of Pesticide Programs (OPP) of the US EPA with advice and guidance on human health and environmental issues relating to proposed regulatory decisions for pesticides, including PIPS, based on the latest scientific data. For example, the SAP for Bt Plant Pesticides met three times in the year 2000 to consider issues related to the safety of plants that had been genetically modified to express insect-specific Bt toxins.

The US EPA released a decision to temporarily extend registrations for all currently registered pest protected plants throughout the 2001 growing season (US EPA,

2000). Additional provisions, such as the recently strengthened resistance management requirements for plant-pesticides, along with the original registration conditions, were considered to provide adequate protection during the extended time period. In addition, the reassessment process for these products has been designed to assure maximum transparency and opportunity for public comment in the regulatory decision-making process.

Concluding comments

The US FDA has been praised for its science-based regulatory approach to evaluating the safety of products developed using rDNA technology based on the characteristics of the food product, and not the means by which it has been created. On the other hand, the regulatory approach of both the USDA and US EPA has been criticized for discriminating against the products of rDNA technology.

A consortium of 11 major international scientific societies (IFT, 1996) and the Council on Agricultural Science and Technology (CAST), an organization representing 36 scientific and professional groups (CAST, 1998), have strongly criticized the US EPA for their proposed approach to regulating plant-pesticides on the basis of the process used in its development rather than the characteristics exhibited by the plant. More recently, the US Congressional Committee on Biotechnology released a report entitled 'Seeds of Opportunity' recommending revisions to the USDA and US EPA approaches to the regulation of GM plants (US Congressional Committee, 2000). It was felt that the proposals by both agencies failed to take into consideration the current scientific consensus on the potential risks associated with agricultural biotechnology. Overall, the report was supportive of the science-based policy of regulation, based on the characteristics of a food product and not on the means by which it was created.

As a part of their role, each federal regulatory agency involved in assessing the human health and environmental safety of GM crops attempts to ensure that only the most current scientific data, information, and methods are referenced. The US FDA consults a Food Advisory Committee, the US EPA refers questions to Scientific Advisory Panels, and the USDA has recently established its own advisory panel. Aside from periodic reviews by credible independent research organizations such as NAS, the individual policies of each agency have been subject to regular review for harmonization with approaches recommended by international standard-setting bodies, such as OECD, WHO and FAO.

It is expected that many of the future food products of rDNA technology in agriculture may not be equivalent in composition to their conventional counterparts. However, the existing decision-tree approach to regulation anticipates these types of improved food products, and provides for the case-by-case assessment of their safety. Field testing and pre-market review for food safety provide the required level of assurance that the foods derived from the application of rDNA technology in agriculture are at least as safe as existing foods, and are consistent with all existing standards of food safety. All new varieties of vegetables, fruit, corn and soybeans developed using rDNA technology have been subject to considerable scientific and regulatory review by the US FDA, the USDA and/or the US EPA. As such, the developers of new crop varieties have had to take into consideration not only the interests of farmers, food processors and consumers, but also the expectations of regulatory authorities.

Both regulators and consumers have had a number of opportunities to influence the acceptance of a new crop variety, and continue to have significant influence beyond final regulatory approval and commercial introduction. It is expected that the issues of human health and environmental safety will continue to be of major concern in the future as a greater number of GM food products reach later stages of commercial development. A recent trend has been observed towards the commercialization of products that offer more transparent benefits and value to consumers. The new products of rDNA technology will be expected to further challenge the flexibility that has been exhibited by the US regulatory process. As a result, both industry and regulatory authorities will need to be responsive to uphold the human health and safety record achieved for GM product introductions. The vast majority of the new products of rDNA technology are not expected to be commercially available until after 2005.

The crops and foods improved through rDNA technology have been developed with much more precision, and their human health and environmental safety has been assessed in more depth and detail than any other crops and foods developed using more conventional means. They are routinely evaluated for any potential impact on both the environment and food safety. Despite years of experience and research, no verifiable data implicating rDNA technology as a food safety or environmental hazard have presented themselves. The research and development of GM crops have been performed under extremely controlled conditions and tight regulatory oversight from research and development within the laboratory through field trials and commercial planting for food-use. Based on evaluations conducted to date, no evidence exists to suggest that GM foods currently sold on the market pose any significant human health concerns or that they are in any way less safe than foods derived from crops developed through more conventional means.

Although the foods developed using rDNA technology have been subject to rigorous scientific review for safety, the US regulatory system has allowed for considerable advancement in the application of the technology, and significant product innovation. To continue to be successful, it is recognized that the US FDA, USDA and US EPA will need to cooperate: the approval process must be clear, timely and transparent; emphasis must be on science-based/risk-based regulatory assessments and decisions; the regulatory process must be flexible to adapt to new information; and, the agencies must be staffed by full-time, highly qualified scientists in many diverse fields.

Over the past 25 years, the approach to regulating the products of agricultural biotechnology has evolved in response to the experience gained during the research and development of applications for rDNA technology. In the future, as the number of products developed using rDNA technology continues to increase, additional research will be needed to better evaluate any potential effects on human health and the environment, and to further refine the scientific basis of making regulatory decisions. Within the United States, the regulatory approach to evaluating the safety of products developed using rDNA technology continues to evolve as new applications continue to emerge and our understanding of the potential risks improves. To date, existing regulations have proven adequate to assure the safety of all new products developed using rDNA technology, and additional innovation has not been discouraged. Accordingly, the National Research Council has recommended that any new rules for the regulation of GM plants be sufficiently flexible enough to reflect improvements in scientific understanding (NAS, 2000).

In the future, as rDNA technology finds broader application, and the new prospects for this technology become commercial realities, the regulatory approaches to evaluating the safety of GM food crops are expected to evolve accordingly.

References

Andersen, J.L. (1999) Testimony of Janet L. Andersen, PhD, Director, Biopesticides and Pollution Prevention Division, Office of Pesticide Programs, Office of Prevention, Pesticides, and Toxic Substances, US Environmental Protection Agency Before the Committee on Agriculture, Nutrition, and Forestry, United States Senate, October 7, 1999. US Environmental Protection Agency (US EPA), Office of Pesticide Programs, http://www.epa.gov/oppbppd1/biopesticides/otherdocs/testimony-wsenate.htm

Betz, F.S., Hammond, B.G. and Fuchs, R.L. (2000) 'Safety and advantages of Bacillus thuringiensis-protected plants to control insect pests', *Regulatory Toxicology and Pharmacology* 32(2): 156–73.

CAST (1998) 'The proposed EPA Plant Pesticide Rule', Council for Agricultural Science and Technology, CAST Issue Paper no. 10, http://www.cast-science.org/epar_ip.htm

CEQ/OSTP (2001) *Case Studies of Environmental Regulation for Biotechnology*. Washington, DC: Council on Environmental Quality/Office of Science and Technology Policy, http://ostp.gov/html/012201.html

Codex (2001) *Report of the Working Group. Ad Hoc Open-Ended Working Group on Allergenicity. Codex Ad Hoc Intergovernmental Task Force on Foods Derived from Biotechnology*. Vancouver, British Columbia. September 10–12.

Codex (2002) *Ad Hoc Intergovernmental Codex Task Force on Foods Derived from Biotechnology. 3rd Session*. Yokohama, Japan.

FAO/WHO (1991) *Strategies For Assessing The Safety of Foods Produced by Biotechnology. Report of a Joint FAO/WHO Consultation*. Geneva: World Health Organization.

FAO/WHO (1996) *Biotechnology and Food Safety: Report of a Joint FAO/WHO Expert Consultation*. Geneva: World Health Organization.

FAO/WHO (2000) *Safety Aspects of Genetically Modified Foods of Plant Origin. Report of a Joint FAO/WHO Expert Consultation on Foods Derived from Biotechnology. May 19 to June 2*. Geneva: World Health Organization.

FAO/WHO (2001) *Evaluation of Allergenicity of Genetically Modified Foods. Report of a Joint FAO/WHO Expert Consultation on Allergenicity of Foods Derived From Biotechnology. 22–25 January*. Geneva: World Health Organization, http://www.who.int/fsf/GMfood/Consultation_Jan2001/report20.pdf

IFT (1996) *Appropriate Oversight for Plants with Inherited Traits for Resistance to Pests, A Report from 11 Professional Scientific Societies*. Coordinating Society, Institute of Food Technologies, July 1996.

James, C. (1999) *Global Review of Commercialized Transgenic Crops: 1999*. Ithaca, NY: International Service for the Acquisition of Agri-Biotech Applications. ISAAA Briefs no. 12, http://www.isaaa.org/publications/briefs/Brief_12.htm

Karenlampi, S. (1996) *Health Effects of Marker Genes in Genetically Engineered Food Plants*. Copenhagen, Denmark: Nordic Council of Ministers.

Kessler, D.A., Taylor, M.R., Maryanski, J.H. *et al.* (1992) 'The safety of foods derived by biotechnology', *Science* 256: 1747–832.

Kough, J. and Vaituzis, Z. (1999) 'Characterization and non-target organism data requirements for protein plant-pesticides', paper presented at the FIFRA Scientific Advisory Panel, 8 December 1999.

Lehrer, S.B., Horner, W.E. and Reese, G. (1996) 'Why are some proteins allergenic? Implications for biotechnology', *Critical Reviews in Food Science and Nutrition* 36(6): 553–64.

Metcalfe, D.D., Astwood, J.D., Towsend, R. *et al.* (1996) 'Assessment of the allergenic potential of foods derived from genetically engineered crop plants', *Critical Reviews in Food Science and Nutrition* 36(Suppl.): S165–S186.

Munro, I.C. (2000) 'Society of Toxicology/EUROTOX Debate Presentation', paper presented at the Proceedings of the Society of Toxicology Annual Meeting, Philadelphia, Pennsylvania, 21 March.

NAS (1987) *Introduction of Recombinant DNA-Engineered Organisms into the Environment: Key Issues.* Washington, DC: National Academy of Sciences.

NAS (2000) *Genetically Modified Pest-Protected Plants: Science and Regulation.* Committee on Genetically Modified Pest-Protected Plants, National Research Council (NRC), National Academy of Sciences. Washington, DC: National Academy Press (NAP).

NIH (1976) 'Recombinant DNA research: guidelines', *US Federal Register* 41: 27901.

NIH (1978) 'Guidelines for research involving recombinant DNA molecules', *US Federal Register* 43: 60108.

NIH (1983) 'Recombinant DNA research; action under guidelines', *US Federal Register* 48: 16459.

NRC (1989) *Field Testing Genetically Modified Organisms: Framework for Decisions.* Committee on Scientific Evaluation of the Introduction of Genetically Modified Microorganisms and Plants into the Environment, National Research Council. Washington, DC: National Academy Press (NAP).

OECD (1993) *Safety Evaluation of Foods Derived by Modern Biotechnology. Concepts and Principals.* Paris: Organisation for Economic Co-operation and Development.

OSTP (1984) 'Proposal for a coordinated framework for regulation of biotechnology', *US Federal Register* 49(50): 856.

OSTP (1986) 'Coordinated framework for the regulation of biotechnology: announcement of policy and notice for public comment', *US Federal Register* 51: 23302–93.

OSTP (1992) 'Exercise of Federal oversight within scope of statuatory authority: planned introductions of biotechnology products into the environment', *US Federal Register* 57: 6753–62.

Taylor, S.L. and Lehrer, S.B. (1996) 'Principles and characteristics of food allergens', *Critical Reviews in Food Science and Nutrition* 36(Suppl.): S91–S118.

US Congress (1969) *National Environmental Policy Act*, 42 U.S.C. §4321, as amended.

US Congressional Committee (2000) *Seeds of Opportunity: An Assessment of the Benefits, Safety, and Oversight of Plant Genomics and Agricultural Biotechnology.* Washington, DC: Subcommittee on Basic Research, US Congressional Committee on Science. 106th Congress, 2nd Session, Committee Print 106-B.

US EPA (1994) 'Proposed policy; plant-pesticides subject to the Federal Insecticide, Fungicide, and Rodenticide Act and the Federal Food, Drug and Cosmetic Act', *US Federal Register* 59(225): 60496–518.

US EPA (2000) 'Time extension for B.T. corn and B.T. cotton plant-pesticides expiring registrations; registration process and public participation opportunity', *US Federal Register* 65(154): 48701–705.

US EPA (2001a) 'Regulations under the Federal Insecticide, Fungicide, and Rodenticide Act for plant-incorporated protectants (Formerly Plant-Pesticides). Final rule. 40 CFR Parts 152 and 174 [OPP-300369B; FRL-6057-7]', *US Federal Register* 66(139): 37772–817.

US EPA (2001b) 'Exemption from the requirement of a tolerance under the Federal Food, Drug, and Cosmetic Act for residues derived through conventional breeding from sexually compatible plants of plant-incorporated protectants (Formerly Plant-Pesticides). Final rule. 40 CFR Part 174 [OPP-300368B; FRL-6057-6]', *US Federal Register* 66(139): 37830–54.

US EPA (2001c) 'Exemption from the requirement of a tolerance under the Federal Food, Drug, and Cosmetic Act for residues of nucleic acids that are part of plant-incorporated protectants (Formerly Plant-Pesticides). Final rule. 40 CFR Part 147 [OPP-300371B; FRL-6057-5]', *US Federal Register* 66(139): 37817–30.

US EPA (2001d) 'Plant-incorporated protectants (Formerly Plant-Pesticides), Supplemental Proposal; Notice of data availability. 40 CFR Part 174 [OPP-300370B; FRL-6760-4]', *US Federal Register* 66(139): 37855–69.

US FDA (1992) 'Statement of policy: foods derived from new plant varieties; Notice', *US Federal Register* 57(104): 22984–3005.

US FDA (1994) 'Conference on scientific issues related to potential allergenicity in transgenic food crops', *US Federal Register* 59: 15415.

US FDA (1997) *Guidance on Consultation Procedures Foods Derived from New Plant Varieties*. US Food and Drug Administration, Center for Food Safety and Applied Nutrition, Office of Premarket Approval, and Center for Veterinary Medicine, Office of Surveillance and Compliance.

US FDA (1998) *Guidance for Industry: Use of Antibiotic Resistance Marker Genes In Transgenic Plants. Draft Guidance*. US Food and Drug Administration, Centre for Food Safety and Applied Nutrition, Office of Premarket Approval.

US FDA (2001a) 'Premarket notice concerning bioengineered foods. 21 CFR Parts 192 and 592', *US Federal Register* 66(12): 4706–38.

US FDA (2001b) 'Draft Guidance for Industry: Voluntary labelling indicating whether foods have or have not been developed using bioengineering; Availability. Notice', *US Federal Register* 66(12): 4839–42.

USDA (1987) 'Introduction of organisms and products altered or produced through genetic engineering which are plant pests or which there is reason to believe are plant pests. 7 CFR Part 340', *US Federal Register* 52: 22892.

USDA (1991) *User's Guide for Introducing Genetically Engineered Plants and Microorganisms*. United States Department of Agriculture, Animal and Plant Health Inspection Service. Technical Bulletin no. 1783. June.

USDA (1993) 'Genetically engineered organisms and products; Notification procedures for the introduction of certain regulated articles; and petition for nonregulated status. Final Rule. 7 CFR Part 340', *US Federal Register* 58(60): 17044–59.

USDA (1996) *Guide For Preparing and Submitting A Petition For Genetically Engineered Plants*. United States Department of Agriculture, Animal and Plant Health Inspection Service, November 5.

USDA (1997) *Notification for Interstate Movement, Importation, and Release Into The Environment. Section VI*. United States Department of Agriculture, Animal and Plant Health Inspection Service, May.

Vetter, J.L. (1996) *Food Laws and Regulations*. Manhattan, Kansas: American Institute of Baking (AIB).

WHO (1993) *Health Aspects of Marker Genes in Genetically Modified Plants: Report of a WHO Workshop*. Geneva: World Health Organization.

Chapter 2

The regulatory requirements for novel foods

A European perspective

Nick Tomlinson

Introduction

In Europe the sale of all novel foods, including genetically modified foods, is controlled by the EC Novel Foods Regulation (European Community, 1997a). The Regulation establishes a European-wide pre-market approval system for novel foods and novel food ingredients, that is foods which have not been used for human consumption in Europe before, including those containing or produced from genetically modified organisms. It also establishes specific labelling rules which apply in addition to general food labelling requirements. The main purpose of the legislation is to protect the consumer by requiring a rigorous safety assessment before a novel food can be approved for sale. A further important aspect of the regulatory framework is to ensure that consumers are able to make informed choices about foods that they eat.

Although the Novel Foods Regulation came into effect relatively recently, some European member states, notably France, the Netherlands and the UK, had operated approval systems for many years. Indeed much of the EC novel food approval process is based on experience gained from operating the UK system.

The previous UK voluntary approval process

The UK approval system for novel foods dates back to a system based on a voluntary arrangement with the food industry in 1980. Under the UK's previous voluntary arrangement for the safety assessment of novel foods, companies submitted applications for assessment by an independent advisory committee. Initially the Advisory Committee on Irradiation and Novel Foods was responsible for assessing applications, although in 1988 this committee was reconstituted into the Advisory Committee on Novel Foods and Processes (ACNFP).

To help companies identify the type of data that would be required to demonstrate that a novel food was safe, the committee developed a structured decision tree approach (Department of Health, 1991). The ACNFP updated this decision tree in 1994 (MAFF, 1994), using a series of linked questions which are designed to fully characterize the potential hazard of a novel food.

The opinions expressed are those of the author, they are not official statements of the organization for which the author works.

European legislation

The main piece of European Community legislation that applies to novel foods is the EC Novel Foods and Novel Food Ingredients Regulations (258/97) which came into effect on 15 May 1997. These regulations introduced a harmonized pre-market approval process for a wide range of novel foods, including genetically modified foods. The regulations describe procedures for assessing the safety of novel foods and also contain specific labelling rules. In addition, further EC Regulations no. 1139/98 (European Community, 1998), no. 49/2000 (European Community, 2000a) and no. 50/2000 (European Community, 2000b) set out detailed rules for the labelling of food containing ingredients derived from genetically modified soya and maize.

Before describing the current Novel Foods Regulation in detail it is worth first considering the legislative framework that governs the release of genetically modified organisms into the environment. This framework is closely linked to the Novel Foods Regulation.

Directive 90/220/EEC

In 1990 Europe adopted two directives which created a broad legislative framework covering the contained use, deliberate release and marketing of genetically modified organisms. Of these Directive 90/220/EEC (European Community, 1990a) controls the deliberate release into the environment of genetically modified organisms. The directive is intended to protect human health and the environment and to establish a single market for products containing genetically modified organisms. It defines a genetically modified organism as 'an organism in which the genetic material has been altered in a way that does not occur naturally by mating and/or natural recombination' and includes a non-exhaustive list of techniques that are regarded as genetic modification. The directive covers both the release for experimental purposes and the commercial sale of genetically modified organisms. However, Article 10 of the directive makes provision for product-specific regulations requiring a similar environmental risk assessment, such as the Novel Foods Regulation, to replace the need for a marketing consent under Directive 90/220/EEC.

Under the directive, all releases to the environment, whether for experimental purposes or for marketing, must be approved by the competent authority in advance. An applicant is required to provide all the necessary information set out in the directive, including information on the nature and stability of the genetically modified organism itself, on the potential receiving environment and on the interactions between the genetically modified organism and the environment. The authorities have the power to request further information as required. In addition an applicant is required to provide a statement assessing the risks that the genetically modified organism poses to human health and the environment. This risk assessment is then evaluated by the competent authority to which the application is made.

In the case of experimental releases, a consent is issued by the member state where the trial is to take place. A summary of the information provided, together with the decision, is circulated to other states of the European Community for information only. They cannot intervene in the decision but they may comment. This allows competent authorities to benefit from the experience of others, and to indicate any concerns if there is the likelihood of cross-border effects. In order to verify the assumptions of the risk assessment and to enable informed decisions to be made on future applications, a consent holder is required to provide a report of the release and details of any relevant information obtained from the release that relates to risks to human health or the environment.

Similar procedures apply in the case of a request to market a genetically modified organism. However, the consent, when issued, applies across the whole of the European Community. Accordingly once an application has been considered by the competent authority of the member state in which the product is first to be marketed, all other member states have the opportunity to comment or raise objections. If the initial assessment is favourable and no objections are raised the initial competent authority formally issues the marketing consent. If objections are raised, recent practice has been that they are referred by the European Commission to one of their scientific committees before the Commission submits a proposed decision to the 'Article 21' regulatory committee where decisions are taken by qualified majority voting. If the proposal is not adopted it is referred to Council where unanimity is required to amend the European Commission proposal.

In March 2001 Directive 2001/18/EC (European Community, 2001) was adopted. The existing Directive 90/220 will be repealed on 17 October 2002. Amongst the provisions in the new directive consents to market genetically modified organisms will be limited to a maximum of 10 years, during which time the notifier will be required to comply with a monitoring plan described in the consent. The purpose of the monitoring plan is to confirm that assumptions regarding the occurrence and impact of potential adverse effects of the genetically modified organism or its use made in the risk assessment are correct and to identify the occurrence of adverse effects of the genetically modified organism or its use on human health or the environment which were not anticipated in the original risk assessment.

The Novel Foods Regulation

What is a novel food?

The scope of the Novel Foods Regulation is defined in Articles 1 and 2 of the Regulation. In Europe a novel food is defined as a food or food ingredient which has not hitherto been used for human consumption to a significant degree within the Community and which falls within one of the following categories.

(a) foods and food ingredients containing or consisting of genetically modified organisms within the meaning of Directive 90/220/EEC;
(b) foods and food ingredients produced from, but not containing, genetically modified organisms;
(c) foods and food ingredients with a new or intentionally modified primary molecular structure;
(d) foods and food ingredients consisting of or isolated from microorganisms, fungi or algae;
(e) foods and food ingredients consisting of or isolated from plants and food ingredients isolated from animals, except for foods and food ingredients obtained

by traditional propagating or breeding practices and having a history of safe food use;
(f) foods and food ingredients to which has been applied a production process not currently used, where that process gives rise to significant changes in the composition or structure of the foods or food ingredients which affect their nutritional value, metabolism or level of undesirable substances.

In practice, any food or food ingredient, covered by any of these categories, that was not on sale in Europe before 15 May 1997 falls within the definition of a novel food. Accordingly products which had been approved under previous national procedures, but which had not been marketed are subject to the provisions of the regulation. Food additives, flavourings used in foodstuffs and extraction solvents used in the production of foodstuffs are excluded from these regulations on the basis that such products are already covered by existing community legislation. Indeed the Novel Foods Regulation specifically states that such exemptions only apply as long as such products are the subject of a comparable safety assessment.

The Novel Foods Regulation also makes provision for decisions to be taken on a case-by-case basis as to whether a specific food falls within the scope of the regulation. Member states have received many requests from potential applicants for clarification of the status of specific products. Most of these requests relate to dietary supplements, of which many are freely available in the USA, where under the US Dietary Supplements and Health Education Act 1994, they are exempt from any pre-market approval process. In Europe, if they do not have a significant history of consumption they require approval before they can be placed on the market.

Although the European Community definition of a novel food is somewhat general, there are several examples that help to clarify its scope. In April 1996 Monsanto's glyphosate-tolerant soya was approved for importation and processing to non-viable products following a rigorous safety assessment under Directive 90/220/EEC. Similarly, in February 1997 approval was issued for importation and cultivation of Ciba-Geigy's insect-resistant and glufosinate ammonium-tolerant maize. Both of these products, which were first grown commercially in the USA in 1996, had been consumed in significant quantities (greater than 1 million tonnes) by May 1997. Accordingly, member states ruled that they were outside the scope of the Novel Foods Regulation, and in any case they had both been fully assessed for food safety. Although, as will be discussed later, it was agreed that for labelling purposes they should be treated in the same way as products approved under the Novel Foods Regulation.

Other products that have been judged to fall outside the scope of the Regulation include the mycoprotein Quorn, first approved in the UK in 1983 and sold in most member states since then. Lactulose was also considered not to be novel on account of its use in biscuits in one member state. A more recent example which caused considerable debate was a margarine-based product containing plant sterols. This product, which had been sold in significant quantities in Finland since its introduction in 1995, was intended to reduce cholesterol absorption. Again member states ruled that the product was not a novel food. However, a similar product developed by another company was considered for approval under the Novel Foods Regulation because it had not been on sale before May 1997.

Based on the above examples it is possible to draw some conclusions as to what comes within the scope of the Regulation. Any product sold to consumers in one or more member states prior to May 1997 is not a novel food. If the food or food ingredient has not been sold in Europe, and is not a food additive, flavouring substance or extraction solvent, it is subject to the provisions of the Regulation. The foods generating most enquiries as to their status under the Novel Foods Regulation are dietary supplements. It is clear that the regulation will have a significant impact on this sector with companies now being required to generate the necessary safety data to support an application for approval under the Regulation.

Approval procedures for novel foods

Article 3 of the Regulation summarizes the procedures that apply to all foods covered by the Regulation. Before any novel food can be sold in Europe, member states must be satisfied on the basis of data provided in an application that the food does not:

- present a danger to the consumer
- mislead the consumer, or
- differ from foods or food ingredients which they are intended to replace to such an extent that their normal consumption would be nutritionally disadvantageous for the consumer.

Articles 4–9 of the Novel Foods Regulation describe the procedures that are followed in considering all applications, or in the case of Article 5, simplified procedure notifications. Each member state is required to establish a competent authority for the receipt and processing of applications. In the UK the competent authority is the Food Standards Agency. In addition, each member state is required to appoint a food assessment body to evaluate applications. In the UK this task is done by the independent Advisory Committee on Novel Foods and Processes, although this committee can seek advice from other expert committees such as the Food Advisory Committee, the Committee on Toxicity or the Advisory Committee on Releases to the Environment.

Article 4 also refers to the publication of guidelines which describe the information necessary to support an application, the presentation of such information and the format for assessment reports prepared by member states. These guidelines will be described later.

A company wishing to obtain approval for a novel food in Europe is required to submit two copies of their application; one copy goes to the member state where the product is intended to be marketed for the first time (the initial member state), the other copy is sent to the European Commission. The application consists of a summary document, copies of all study reports and any other supporting data and a proposal for the labelling of the product. Upon receipt of an application the initial member state first checks that the application is accompanied by all the necessary supporting data as identified by the guidelines. Once satisfied that an application is complete, the initial member state informs the European Commission and other member states that an application has been accepted. The date of the notification to the European Commission is taken as day 1 of the 90-day assessment period. At the same time the applicant's summary document is circulated to all member states. It is in the applicant's interests to ensure that their summary document lists all the supporting data and is sufficiently detailed to enable other member states to form an opinion in the absence of the supporting data. However, it is recognized that member states will have the option of requesting some or all of the supporting data if necessary.

Within 90 calendar days the initial member state is required to forward a copy of its assessment to the European Commission, The European Commission then forwards the initial assessment report without delay to all member states for their consideration. If during the initial assessment it becomes necessary to obtain more information, the applicant will be asked to provide the necessary additional information; during this period the 90-day clock is stopped. The initial member state is required to inform the Commission and other member states of the reason for the delay immediately. Similarly the European Commission and member states shall be informed when the clock is restarted. While recognizing that stopping the clock is necessary where further information is required, member states endeavour, as far as is possible with a quick scrutiny, to ensure that an application is complete before accepting it so as to keep the need for clock stopping to a minimum.

From the date the European Commission circulates the opinion of the initial member state the remaining member states have 60 calendar days in which to submit any comments or reasoned objections to the European Commission. There is no provision for stopping the clock within the 60-day period. Where member states raise objections they may seek to resolve them on a bilateral basis with the initial member state or the applicant.

If at the end of the 60-day period for comment the initial member state has received no reasoned objections, and no objections have been received by the European Commission, the applicant is immediately informed in writing by the initial member state that they may place their product on the market. If reasoned objections are received, which cannot be resolved on a bilateral basis, the initial member state shall write to the applicant informing them that an 'authorization decision' in accordance with Article 7 of the Regulation is required.

If an authorization decision is required, a decision on whether or not to issue an approval is taken centrally. The first step of the authorization procedure involves a meeting of the novel foods competent authorities where objections raised by member states are discussed. At this point member states decide that either the application requires a more detailed safety assessment, in which case the application is referred to the Scientific Committee for Food in accordance with Article 11 of the Regulation, or that the application does not raise any issues that have an effect on public health, in which case a decision as to whether the product should be allowed onto the market should be put to a vote at the next meeting of the Standing Committee for Foodstuffs, the body described in Article 13 of the Regulation.

If the novel foods competent authorities meeting decides that the Scientific Committee for Food needs to consider the safety of the product in more detail, they may wish to consider which aspects of the application the Scientific Committee for Food should be asked to address. Once the advice of the Scientific Committee for Food is available, the European Commission is required to put a proposal for an authorization decision to the Standing Committee for Foodstuffs for a vote. The proposal will specify the conditions of use of the food or food ingredient; the designation of the food or food ingredient, and its specification; and specific labelling requirements for the food.

If the Standing Committee for Foodstuffs decides by a qualified majority to issue an authorization decision, the European Commission informs the applicant in writing immediately. A copy of the decision is also published in the Official Journal of the European Communities. If no agreement is reached in the Standing Committee for Foodstuffs, the matter is referred to the Council of Ministers. The Council has three months in which to act, either adopting the European Commission proposal by qualified majority voting or amending it by unanimity. If it fails to act within the three months the European Commission proposal is adopted by default. The process of assessing applications under the regulations is summarized in Figure 2.1.

As of June 2001, 30 applications had been received, of which six had completed all stages of the approval process resulting in the approval of four products and the rejection of two products on the basis of incomplete supporting data. So far there have been 11 applications for foods derived from genetically modified crops, including processed tomato products, chicory, maize and soybeans. Other applications have included plant and dried leaves from *Stevia rebaudiana*, phospholipids from egg yolk, yellow fat spreads with added phytosterol-esters and the use of high pressure to treat fruit preparations.

Approval of foods containing or consisting of genetically modified organisms

As discussed earlier, Directive 90/220/EEC provides a European regulatory framework to protect human health and the environment from the introduction of genetically modified organisms. Article 9 of the Novel Foods Regulation removes the requirement to obtain a marketing consent, under Directive 90/220/EEC, for foods which contain or consist of genetically modified organisms. In support of an application for such a food, a company is required to submit copies of any consents already obtained under Directive 90/220/EEC for research and development purposes. In addition an applicant must provide all the relevant information that would have been required under that directive to ensure that all necessary measures are taken to safeguard both human health and the environment. An application under Directive 90/220/EEC must be made for those cases in which a genetically modified organism is intended for uses other than food, such as animal feed or cultivation within Europe, in order to secure scrutiny of human health and environmental effects.

Link with the EC Common Catalogue for seeds

Before crops can be grown for commercial purposes in Europe they have to be included on the National List of a member state or the EC Common Catalogue. Various European Community seeds directives require that varieties must be grown in official tests and trials, normally for two years, to establish that they are distinct, uniform and stable and have value for cultivation and use. Any seed marketed must be certified as to its varietal purity and germination standard. These requirements apply to any new variety, whether produced by conventional means or the use of genetic technology. In order to ensure that, once a final approval for food use of a

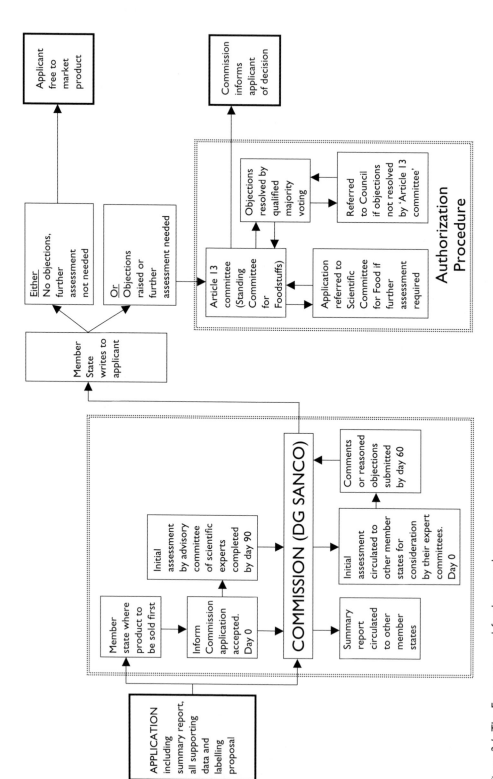

Figure 2.1 The European novel food approval process.

genetically modified crop is obtained the seed can be offered for sale to farmers without further delay, companies typically enter a variety into National List trials under an experimental trials consent issued under Directive 90/220/EEC. Addition to the EC Common Catalogue, which normally follows National Listing, allows the variety to be grown commercially anywhere in the European Community.

The intention of the link to the seeds directives in the Novel Foods Regulation was to enable a company to submit a single application for approval to grow a genetically modified crop commercially in Europe in accordance with Directive 90/220/EEC; to sell the food in accordance with the Novel Foods Regulation and to have the variety entered onto the EC Common Catalogue. In practice this linkage has never been used and is likely to be removed when the Novel Food Regulation is revised.

Simplified procedures

Articles 3 and 5 of the Regulation contain provisions for companies to notify the European Commission when they first place a product on the market which is considered to be 'substantially equivalent to existing foods or food ingredients in terms of composition, nutritional value, metabolism, intended use and level of undesirable substances'. Under the regulations companies have the option of submitting a notification based on the opinion of one competent authority that the particular food or food ingredient is 'substantially equivalent' to existing foods or food ingredients. Alternatively a company can submit a notification based on readily available and generally recognized scientific evidence such as papers from peer reviewed journals. Some companies have notified the European Commission that they have placed on the market various processed products derived from varieties of genetically modified maize and refined oil from herbicide-tolerant oilseed rape. The European Commission publishes a list of notifications in the C series of the Official Journal of the European Communities each year. A total of 11 notifications had been submitted under the simplified procedures by June 2001. However, particularly in the case of oilseed rape it is not clear whether the oil has in fact been marketed, given that at least one variety of oilseed rape has yet to be grown commercially. With the exception of oil from three varieties of oilseed rape, all the notifications submitted to date have been supported by a full safety assessment done prior to the novel foods regulation coming into force.

The simplified procedure has caused considerable confusion by use of the term 'substantially equivalent'. This has frequently been confused with the concept of substantial equivalence, which is used to structure the safety assessment process. Although member states recognize that the simplified procedure is an option available to companies, all member states agreed with the UK position that only highly processed, refined foods derived from genetically modified crops, such as hot pressed oil, white sugar and starch hydrolysates, would be suitable for consideration under the simplified procedures on the grounds that neither DNA nor protein resulting from the modification would be expected to be present. In submitting a notification under the simplified procedure a company has to demonstrate that no genetically modified material (protein or DNA) is present in the food ingredient. Nevertheless, this simplified procedure has not been well received in many countries and is likely to be phased out when the novel food regulations are revised.

Labelling

In reaching agreement on the final text of the Novel Foods Regulation, it was the issue of labelling which proved most difficult to reconcile from all viewpoints. A significant contribution to the debate at the time was made by the Group of Advisers on the Ethical Implications of Biotechnology of the European Commission when in May 1995 they published an opinion on 'Ethical aspects of the labelling of foods derived from modern biotechnology' (European Commission, 1995). Amongst the recommendations that this group made was that consumers must be provided with information which, for transparency should be:

- useful, adequate and informative;
- clear, understandable, non-technical;
- honest, not misleading or confusing, and which aims to prevent fraud;
- enforceable, i.e. possible to verify.

They also recommended that consumers have a right to be able to make informed choices about what they eat; so they can legitimately expect to receive a clear indication of where additional information can be obtained, especially when their choices include cultural and religious considerations.

Article 8 of the Novel Foods Regulation describes the specific labelling requirements that are required for novel foods in addition to the existing European Community law on the labelling of foodstuffs, the main source of which is contained in Directive 79/112/EEC (European Community, 1979). The main provisions of this directive, which applies to all foodstuffs placed on the market in the European Community, are to ensure that labelling does not mislead the consumer to a material degree as to the characteristics of the foodstuff, or by attributing properties to it which it does not have, or by suggesting that it has special characteristics when in fact all similar foods possess such characteristics. This directive applies to foods ready for delivery to the ultimate consumer or to mass caterers. The labelling requirements contained in the Novel Foods Regulation apply to food sold to the final consumer. They cover the following groups of foods:

(a) Foods and food ingredients obtained from genetically modified organisms when, on the basis of a scientific assessment, they are judged not to be equivalent to an existing food.
(b) Novel foods that contain material which is not present in an existing equivalent foodstuff and which may have implications for the health of some sections of the population. An example would be a protein from a known food allergen source such as peanuts.
(c) Novel foods that contain material which is not present in an existing equivalent foodstuff and which gives rise to ethical concerns.
(d) All foods that contain or consist of genetically modified organisms within the meaning of Directive 90/220/EEC.

In addition, if the novel food has no existing equivalent the Regulation contains powers for the adoption of provisions to ensure that consumers are adequately informed of the nature of the food ingredient concerned. An example of where such a provision might be required is in the context of low-calorie fat replacers, although under Directive 90/496/EEC on nutrition labelling (European Community, 1990b) the energy value for all fats is set at 9 kilo calories/gram (37 kJ/g) for the purpose of calculating energy values for the final foodstuff.

Although the Regulation makes provision for labelling where a food contains material which might give rise to health implications for groups of the population, it presupposes that such a product would be approved for sale. In the case of a food containing a major allergenic protein, it is unlikely that such a product would be approved.

In 1992 the then UK Minister of Agriculture, Fisheries and Food appointed a Committee under the chairmanship of the Reverend Dr J Polkinghorne 'to consider future trends in the production of transgenic organisms; to consider moral and ethical concerns (other than those related to food safety) that may arise from the use of such organisms and to make recommendations'. In its report (MAFF, 1993), one of the Committee's recommendations was that food should be labelled if it contains copies of ethically sensitive genes (human genes or genes of religious significance) or if a plant contains copies of an animal gene. This recommendation, and a similar recommendation from the European Commission's group of ethical advisers is reflected in the requirement to label if a food contains material which is not present in an existing equivalent foodstuff and which gives rise to ethical concerns.

In Europe, since the Novel Foods Regulation came into force the debate over the extent to which foods derived from genetically modified organisms should be labelled has been intense. The current labelling rules, which were agreed by all member states, reflect the view that where measurable differences in composition, compared with a non-modified counterpart exist, labelling should be required to enable consumers to make an informed choice. In this respect the labelling rules contained in the Novel Foods Regulation itself are not absolutely precise. However, the situation was clarified with the introduction of detailed rules for the labelling of genetically modified ingredients derived from Monsanto's soya and Ciba Geigy's maize contained in EC Regulation 1139/98. These rules require the labelling of foods sold to the final consumer containing ingredients derived from Monsanto's soya or Ciba Geigy's maize, in which either DNA or protein arising from the genetic modification is present. This approach was seen as setting a precedent for the labelling of all genetically modified foods approved under the Novel Foods Regulation. Nevertheless, since the introduction of Regulation 1139/98 no GM foods have been approved for sale in Europe. During negotiations on Regulation 1139/98 it was accepted that, although tests for DNA using the polymerase chain reaction were more sensitive than tests for proteins, labelling should be triggered by the presence of either protein or DNA resulting from a genetic modification. If a company or enforcement authority obtains a negative result with one method they would need to follow this up with the other method. However, if the first method used gave a positive result there would be no need for re-testing.

In the preamble to the Novel Foods Regulation itself there is a clause which would in effect enable companies selling bulk consignments of commodity crops to label the crop 'May contain genetically modified material'. During negotiations on the soya and maize labelling rules member states were determined to remove this option, which was seen as providing no meaningful information to the consumer. Accordingly, Regulation 1139/98 requires a definitive statement 'produced from genetically modified soya' or 'produced from genetically modified maize' whenever DNA or protein resulting from the modification is present.

Because products containing soya and maize were already on sale in Europe before labelling rules were agreed it was recognized that it would be impractical to require products already on sale to be re-labelled. Consequently the regulation only applies to foods manufactured and labelled after 1 September 1998. Furthermore, since European Community labelling policy is focused at the level of the ingredient rather than the whole food product, it was intended that enforcement would also be at the level of ingredients used in a foods production, thereby avoiding the need to develop detection methods capable of detecting protein or DNA in a finished food product.

In the preamble to Regulation 1139/98 there is a recognition that the labelling requirements should be no more burdensome than necessary but sufficiently detailed to supply consumers with the information they require. In order to avoid the need for enforcement authorities to sample those ingredients which, by virtue of the degree of refining during production, will not contain protein or DNA, the regulation makes provision for the establishment of a so-called 'negative list'. This was intended to be a list of those ingredients derived from soya and maize which have been clearly demonstrated to contain neither protein nor DNA; however, the list has never been developed.

Recognizing that there is a demand for supplies of non-genetically modified ingredients as a means of providing consumer choice, a number of European retailers have started to obtain supplies of soya and maize from identity preserved sources. The UK Government encouraged the provision of alternative supplies in 1998 when the then UK Ministry of Agriculture, Fisheries and Food published a list of suppliers offering non-genetically modified material to assist those manufacturers and retailers who wished to offer their customers a choice. Companies seeking to obtain supplies from identity preserved sources found that no company would offer an absolute guarantee that no genetically modified material had become accidentally mixed with the consignment during the supply chain. Regulation 1139/98 recognized this difficulty and proposed the development of a low-level *de minimis* threshold below which labelling would not be required.

In April 2000 EC Regulation 49/2000 came into effect. This Regulation amended EC Regulation 1139/98 by extending the requirements to foods sold to mass caterers and setting a *de minimis* threshold of 1% for the adventitious contamination of non-GM material. For such ingredients there is no need to label them as GM if they contain less than 1% GM material. The threshold applies only to ingredients obtained from non-GM sources; this flexibility does not apply to supplies obtained from sources of unknown origin. Companies also need to demonstrate that their ingredients are of non-GM origin, and it is possible that the use of documented and audited identity preservation systems could satisfy this requirement. Steps should also be taken to keep the level of adventitious contamination in non-GM supplies to a minimum. Although the level agreed for the threshold is 1%, in practice the need to provide proof that ingredients are of non-GM origin should ensure that actual levels are kept well below this figure. It is also important to realize that because the limit operates at the level of each individual ingredient, the actual amount in the final foodstuff will be much lower.

A consequence of the fact that Regulation 1139/98 was made under Article 4 of the food labelling Directive 79/112/EEC is that member states have the flexibility to exempt or otherwise from labelling provisions made under the directive those suppliers offering non-prepacked foods to the final consumer. In keeping with its policy of ensuring that all foods containing genetically modified material are clearly labelled, the UK Government introduced national regulations in 1999 requiring all businesses selling non-prepacked foods to the final consumer to ensure that information as to the presence of genetically modified soya or maize is provided where consumers request it.

In April 2000 EC Regulation 50/2000 also came into effect, requiring the labelling of foods containing additives and flavourings that had been obtained from genetically modified organisms where novel protein or DNA was present in the final food.

There is currently no requirement to label additives and flavouring substances obtained from genetically modified organisms sold as such to the final consumer (a very minor use), since this would require amendment of the additives and flavouring framework directives.

In December 2000, the European Commission announced that it was working on proposals for a regulation requiring the traceability of genetically modified organisms and food and feed ingredients derived from them. A second proposal, for a regulation for the approval of genetically modified food and animal feed, is expected to contain a requirement that all ingredients derived from genetically modified organisms are labelled, regardless of whether any genetically modified material can be detected in the final food. In taking these proposals forward, careful consideration will need to be given to their practicality, proportionality and enforceability.

Guidelines accompanying the Novel Foods Regulation

In order to identify the type of information that potential applicants would be required to submit in support of an application and to ensure a consistent approach to the assessment of an application by all member states, the European Commission published guidelines, developed by their Scientific Committee for Food, as a European Commission Recommendation (European Community, 1997b). These guidelines draw heavily on the structured approach to the safety assessment of novel foods developed by the ACNFP and first published in 1991 and revised in 1994. The ACNFP approach and the subsequent Scientific Committee for Food's guidelines use a series of linked questions to ensure that potential hazards of a novel food are fully characterized.

Underlying the structured questions there are a number of key issues which need to be addressed in assessing the safety of a novel food. First and foremost is the recognition that foods are frequently a complex mixture of many thousands of chemical entities. Furthermore, foods are generally ingested in much higher amounts than would be expected for discrete chemical entities such as food additives or pharmaceuticals.

For compounds such as pesticides, pharmaceuticals, industrial chemicals and food additives, animal studies represent a major element in the safety assessment. In such cases, the test substance is generally well characterized, of known purity, of no nutritional value and human exposure is generally low. It is therefore relatively straightforward to feed such compounds to animals at a range of doses, some several orders of magnitude greater than the expected human exposure levels, in order to identify any potential adverse effects of importance to humans. In this way it is possible, in most cases, to determine levels of exposure at which adverse effects are not present, and so set safe upper limits by the application of appropriate safety factors.

By contrast, foods are complex mixtures of compounds characterized by wide variation in composition and nutritional value. Due to their bulk and effect on satiety they can usually only be fed to animals at low multiples of the amounts that might be present in the human diet. In addition, a key factor to consider in conducting animal studies on foods is the nutritional value and balance of the diets used, to try to avoid the induction of adverse effects which are not related directly to the material itself. Picking up any potential adverse effects and relating these conclusively to an individual characteristic of the food can therefore be extremely difficult. It was very much in recognition of these difficulties that the concept of substantial equivalence was developed as a framework for the safety assessment of genetically modified foods in particular.

The guidelines also attach considerable importance to compositional analysis, particularly in respect of key nutrients and natural toxicants. However, the guidelines provide no real insight into how such compositional data should be generated to ensure that safety assessments are based on statistically valid data. The UK has issued a paper to prompt further discussion of the subject. Given the importance attached to the comparative approach to safety assessment of foods derived from genetically modified crops it is important to ensure that any real differences are not masked by natural variation. This natural variation may be large because of the wide variation in conditions, such as soil fertility, temperature, light or water availability, under which crops are grown. The issue may be compounded by other effects including agronomic treatments to the growing crop, such as herbicide applications to a herbicide-tolerant crop, or post-harvest stress during storage (e.g. for potatoes stored in clamps).

To minimize the possibility that any changes caused by a crop modification might be masked by natural variation in a crop, the UK has made a number of recommendations on the design of field trials. For crop varieties which are intended to be grown commercially in Europe, the location of trial sites should be representative of the range of environmental conditions under which the varieties would be expected to be grown. The number of trial sites should be sufficient to allow accurate assessment of agronomic and compositional characteristics over this range. Similarly, trials should be conducted over a sufficient number of years to allow adequate exposure to conditions met in nature. The UK has recommended that the minimum number of trial sites should be six and the minimum number of years should be two.

To reduce any effect from naturally occurring genotypic variation within a crop variety, and to minimize the impact of environmental effects at a given trial site, the minimum number of replicates used should be not less than three and will need to be appropriate to the crop species under trial. To allow effective comparison the UK also recommends that a genetically modified crop and its non-modified counterpart should be grown in adjacent plots at the same site and at the same time.

Information on likely intakes of a novel food is also of special importance in assessing its safety. In many instances the introduction of a novel food will not alter overall intakes of a particular dietary component. For example, the introduction of oil from genetically modified herbicide-tolerant oilseed rape is unlikely to alter the overall intake of oilseed rape oil. However, the introduction of a foodstuff fortified with plant sterols and sold on the basis of claims that it can help reduce cholesterol levels may result in a significant increase in the consumption of plant sterols. Where the introduction of a novel food is expected to give rise to significant shifts in consumption patterns the novel food guidelines recommend the introduction of a surveillance programme to accompany a products launch. In reaching conclusions on the overall safety of a novel food it is important to recognize that the assessment takes into account both food safety and nutritional safety implications. The latter can be of particular importance when considering the implications for groups such as infants, children, pregnant and lactating women, the elderly and those with chronic diseases that have an impact on nutritional status.

Two other aspects of the safety assessment which are considered elsewhere relate to the assessment of the allergic potential of a novel food and, in the case of genetically modified foods, the safety implications of any selective marker genes remaining in a crop as a result of the modification.

In designing a series of structured questions to cover all categories of novel foods whilst keeping the decision tree as simple as possible, the guidelines identify six classes of novel foods which present similar safety assessment issues. The six classes are:

1   Pure chemicals or simple mixtures from non-genetically modified sources. This class identifies data requirements for some foods falling within the third, fourth and fifth categories of the regulation and would include products such as low-calorie fat replacers.
2   Complex novel foods from non-genetically modified sources. This class identifies data requirements for foods derived from whole plants, animals or micoorganisms which fall within categories 4 and 5 of the Regulation. This class would include products such as stevia leaves.
3   Genetically modified plants and their products.
4   Genetically modified animals and their products.
5   Genetically modified microorganisms and their products. Classes 3–5 cover both viable genetically modified organisms and foods derived from them. These classes would include genetically modified chicory and food products derived from genetically modified processing tomatoes.
6   Foods produced using a novel process. This category covers foods covered by category 6 of the Novel Foods Regulation and would include processes such as high-pressure processing or new processes catalysed by enzymes.

The first five classes are then subdivided into:

(a) those foods for which there is a history of consumption of the source of the food, or in the case of genetically modified foods the host organism, and
(b) those foods for which there is no history of food use of the novel food or the host organism.

Having assigned a novel food to one of six classes, the guidelines identify the following information requirements which apply to some, or all, classes of novel foods:

•   Specification of novel food
•   Effect of production process applied to novel food

- History of source organism
- Effect of genetic modification on properties of host organism
- Genetic stability of the genetically modified organism
- Specificity of expression of novel gene
- Transfer of genetic material from genetically modified microorganisms
- Ability to survive in and colonize gut
- Anticipated intake/extent of use
- Information on previous human exposure
- Nutritional information
- Microbiological information
- Toxicological information.

Each information requirement consists of a series of structured questions that an applicant needs to address in order to provide all the necessary information to support their responses as part of the application package.

In submitting an application companies are advised to follow the recommended format contained in part II of the guidelines. In addition to providing details of the product, the applicant is advised to provide sufficient data to enable a food to be assigned to one of the six categories of novel foods set out in Article 1 of the Regulation. In working through the decision trees in the guidelines an applicant should provide all relevant data used to support the response to each question. All laboratory data should be generated in accordance with the principles of good laboratory practice. The application should consist of a summary document, full study reports for all supporting data and a proposal for the labelling of the food.

Transparency

Article 10 of the Novel Foods Regulation makes provision for detailed rules on data confidentiality to be adopted. To date, no such rules have been adopted and member states have introduced their own procedures. In the UK the ACNFP has continually sought to increase the transparency of its decision making. Since 1988 the ACNFP has published annual reports which have included copies of all the assessment reports prepared during that year. More recently the ACNFP has published agendas and full minutes of their meetings as well as annual reports.

In an effort to increase the transparency of the novel foods regulation, the ACNFP includes on its website details of all applications which have been submitted under the Novel Foods Regulation. In addition, in the case of applications submitted to the UK, the complete application dossier, except for any clearly commercially confidential information, is placed on the committee website as soon as the application is accepted for assessment. Members of the public have the opportunity to submit comments on the application. In addition, the Committee places a copy of its draft assessment report on the website and again invites comments.

Post-market monitoring

Article 14 of the Novel Foods Regulation requires the European Commission to monitor the application of the Regulation and its impact on health, consumer protection, consumer information and the functioning of the internal market. Whilst recognizing that novel foods are only approved for sale within the European Community following a rigorous safety assessment following the procedures described above, the UK has been considering proposals for the introduction of a post-market monitoring system for novel foods. Such a system would be very much an addition to the existing safety assessment procedures rather than an alternative. For post-market monitoring to be practical it is necessary to obtain data on consumption patterns and to relate these to any reported adverse health effects. The ACNFP has been looking at ways of integrating the two components necessary for such a system to operate. In terms of mechanisms for identifying patterns in health effects, the UK has an extensive array of health databases which enable disease patterns to be tracked over time and analysed for any regional differences. The other vital component is an ability to obtain details of what foods people have eaten and whether there have been any shifts in their consumption patterns. A number of sources already exist for food consumption data within the UK. Article 12 of the Novel Foods Regulation already contains a provision for action to be taken if new information suggests that a food approved under the Regulation might be endangering human health or the environment. If such a situation were to occur a member state may temporarily restrict trade in the food while the matter is considered by the Standing Committee for Foodstuffs.

Enforcement

Although both the Novel Foods Regulation and the regulation on the labelling of genetically modified soya and maize are directly applicable and legally binding in all member states, each member state is responsible for enforcement in its territory. Under the directive on the official control of foodstuffs (European Community, 1989) member states are required to submit an annual return to the European Commission giving details of the number and type of inspections carried out and details of any infringements found. In the UK local authorities are responsible for enforcing food law, enforcement powers have been provided in the form of regulations made under the UK Food Safety Act 1990.

Conclusions

Although the European Community Novel Foods Regulation only came into effect on 15 May 1997, it builds on many years' experience within some member states. In addition, in the case of genetically modified foods the regulation builds on experience gained from the approval for sale of genetically modified organisms under Directive 90/220/EEC. In addition to introducing a uniform approach to ensure that novel foods are rigorously assessed for safety to ensure consumer protection, the Regulation also makes special provision for labelling requirements to be included, where appropriate, as a condition of approval. This is particularly important in the case of genetically modified foods to ensure that consumers are able to make informed choices about foods that they eat. The Novel Foods Regulation has now been in force for four years, and it is clear that it is not operating effectively. Although initial problems with the operation of the Regulation were attributed to a lack of collective experience between member states, it soon became apparent that the problems were more deep seated. It is likely that the European Commission will shortly come forward with radically new proposals to improve the approval process for novel foods.

References

Department of Health (1991) *Report on Health and Social Subjects 38, Guidelines on the Assessment of Novel Foods and Processes*. Advisory Committee on Novel Foods and Processes. London: HMSO.

European Commission (1995) 'Ethical aspects of the labelling of foods derived from modern biotechnology', Opinion of the Group of Advisers on the Ethical Implications of Biotechnology of the European Commission, Opinion no. 5.

European Community (1979) 'Council Directive 79/112/EEC of 18 December 1978 on the approximation of the laws of the member states relating to the labelling, presentation and advertising of foodstuffs', *Official Journal of the European Communities* L33: 1–14.

European Community (1989) 'Council Directive of 14 June 1989 on the official control of foodstuffs 89/396/EEC', *Official Journal of the European Communities* L186.

European Community (1990a) 'Council Directive 90/220/EEC of 23 April 1990 on the deliberate release into the environment of genetically modified organisms', *Official Journal of the European Communities* L117.

European Community (1990b) 'Council Directive of 24 September 1990 on nutrition labelling of foodstuffs 90/496/EEC', *Official Journal of the European Communities* L276.

European Community (1997a) 'Regulation (EC) No. 258/97 of the European Parliament and of the Council of 27 January 1997 concerning novel foods and novel food ingredients', *Official Journal of the European Communities* L43: 1–6.

European Community (1997b) 'Commission Recommendation of 29 July 1997, concerning the scientific aspects and the presentation of information necessary to support applications for the placing on the market of novel foods and novel food ingredients and the preparation of initial assessment reports under Regulation (EC) No 258/97 of the European Parliament and of the Council', *Official Journal of the European Communities* L253.

European Community (1998) 'Council Regulation (EC) No. 1139/98 of 26 May 1998 concerning the compulsory indication on the labelling of certain foodstuffs produced from genetically modified organisms of particulars other than those provided for in Directive 79/112/EEC', *Official Journal of the European Communities* L159.

European Community (2000a) 'Commission Regulation (EC) No. 49/2000 of 10 January 2000 amending Council Regulation (EC) No. 1139/98 concerning the compulsory indication on the labelling of certain foodstuffs produced from genetically modified organisms of particulars other than those provided for in Directive 79/112/EEC', *Official Journal of the European Communities* L006: 13–14.

European Community (2000b) 'Commission Regulation (EC) No. 50/2000 of 10 January 2000 on the labelling of foodstuffs and food ingredients containing additives and flavourings that have been genetically modified or have been produced from genetically modified organisms', *Official Journal of the European Communities* L006: 15–17.

European Community (2001) 'Directive 2001/18/EC of the European Parliament and of the Council of 12 March 2001 on the deliberate release into the environment of genetically modified organisms and repealing Council Directive 1990/220/EEC', *Official Journal of the European Communities* L106.

MAFF (1993) *Report of the Committee on the Ethics of Genetic Modification and Food Use*. London: HMSO.

MAFF (1994) *Annual Report*. London: MAFF Publications.

Chapter 3

The concept of substantial equivalence

An overview

Paul R. Mayers, Peter Kearns,
Karen E. McIntyre and Jennifer A. Eastwood

Historical background

The concept of substantial equivalence was originally developed through discussions at the Organisation for Economic Co-operation and Development (OECD) (OECD, 1993a), though, to a large extent, these discussions built on previous work done by the World Health Organization (WHO) and the Food and Agriculture Organization (FAO) (WHO, 1991). In 1991, a Joint FAO/WHO Consultation had concluded that the evaluation of a food derived through modern biotechnology should consider both food safety and nutritional value using similar conventional food products as a standard and taking into account the processing of the food and its intended use (WHO, 1991).

In the 1980s, the OECD established a Group of National Experts on Safety in Biotechnology (GNE), which continued to work through to 1993. As an intergovernmental organization, OECD's GNE was attended by delegates nominated by the governments of the OECD member countries[1], who were, for the most part, representatives of those agencies and ministries with a responsibility for safety in biotechnology.

As a result of the work of the GNE, the OECD published a number of important documents relevant to the safety assessment of biotechnology-derived products during this period, which deal with a range of issues, including safety considerations for industrial, agricultural as well as environmental applications of organisms derived by recombinant DNA techniques (OECD, 1986). It was recognized at this stage that the safety assessment of an organism derived through recombinant DNA techniques would rely heavily on the knowledge of its parental organism, as well as on an analysis of how the new organism appears to differ from the parent (OECD, 1986). OECD recommended that the considerable data on environmental and human health effects of living organisms that exists should be used to guide the risk assessment.

By the early 1990s, there had already been large numbers of small-scale field trials of new crop varieties derived through recombinant DNA techniques (OECD, 1993b), and it became clear that new foods derived from these varieties would be marketed during the 1990s. In order to proactively address food safety issues related to these novel foods, in 1990, the GNE established a Working Group on Food Safety and

---

1 Current Member countries of the OECD are Australia, Austria, Belgium, Canada, Czech Republic, Denmark, Finland, France, Germany, Greece, Hungary, Iceland, Ireland, Italy, Japan, Korea, Luxembourg, Mexico, The Netherlands, New Zealand, Norway, Poland, Portugal, Slovak Republic, Spain, Sweden, Switzerland, Turkey, United Kingdom and the United States.

Biotechnology comprising experts, mainly from the ministries and agencies responsible for food safety issues in OECD member countries.

The main objective of the Working Group was to elaborate scientific principles and concepts to be used when evaluating the safety of new foods and food components of terrestrial microbial, plant or animal origin. The Working Group did not consider the safety assessment of food additives, contaminants, processing aids or packaging materials. Nor did it consider environmental safety issues which had been (or were being) addressed by other groups of the GNE.

Early in the life of the Working Group, an important recognition was that traditionally, the safety of food for human consumption had been based on the reasonable certainty that no harm will result from intended uses under the anticipated conditions of consumption. Foods prepared and used in traditional ways have usually been considered safe on the basis of long-term experience, even though they may have contained natural toxicants or anti-nutritional substances. Normally, new varieties of foods or crops have not been subjected to traditional toxicological testing. In fact, where toxicological testing has been applied to whole foods, the results have often been difficult to interpret (OECD, 1996).

Although the OECD's Working Group recognized that modern biotechnology might extend the scope of genetic changes that can be made – and might even broaden the range of the possible sources of foods – it was recognized that this would not inherently lead to foods that are less safe than those developed through conventional techniques. The Working Group therefore indicated that the evaluation of foods or food components derived through modern biotechnology does not require a fundamental change in established principles, nor does it require a different standard of safety.

Realizing the importance of using examples of new foods to identify and demonstrate the applicability of the proposed scientific principles, the Working Group organized a number of meetings and intergovernmental consultations which included case study presentations of novel foods (OECD, 1993a) such as enzymes, genetically modified bakers' yeast, mycoprotein and a number of genetically modified varieties of crop species. Although these case studies were not intended to be formal safety evaluations they were crucial in illustrating important concepts in the safety assessment of novel foods. Therefore, it was through their application in real examples that the concepts and principles were identified.

The Working Group proposed a scientific approach to the evaluation of foods derived through modern biotechnology, which is based on a comparison with traditional foods that have a safe history of use. One of the main concepts developed was that of substantial equivalence, and in describing this concept, the Group stressed that this was a principle that had been used in the past (perhaps intuitively) even if it had not been articulated as such.

Substantial equivalence: the concept

The concept of substantial equivalence is described as embodying the idea that existing organisms used as food, or as a source of food, can be used as the basis for comparison when assessing the safety of human consumption of a food or food component that has been modified or is new.

(OECD, 1993a)

As previously indicated, in elaborating the concept of substantial equivalence, the OECD Working Group noted that food safety is considered as a reasonable certainty, that no harm will result from intended uses under the anticipated conditions of consumption and that the most practical approach to the determination of safety is to consider whether a genetically modified organism (GMO)-derived food is comparable to an analogous conventional food product (OECD, 1993a). The concept of substantial equivalence therefore relies on the existing history of safe food use of a conventional food product as a useful consideration in framing the safety assessment of a GMO-derived food by permitting the identification of similarities and differences which can be considered in the assessment.

In 1996, participants at an expert FAO/WHO consultation recommended that safety assessment based upon the concept of substantial equivalence be applied in establishing the safety of foods and food components derived from genetically modified organisms (FAO, 1996). Assessing substantial equivalence was recognized as not being a safety assessment *per se* (FAO, 1996), but a process which establishes that the characteristics and composition of the new GMO-derived food are comparable to those of a familiar, conventional food which has a history of safe consumption.

A Joint FAO/WHO Expert Consultation on Foods Derived from Biotechnology was convened in 2000 to address food safety and nutritional questions regarding foods derived from GM plants, including a review of the scientific basis, application and limitations of the concept of substantial equivalence. It concluded, based on the current application of substantial equivalence and alternative strategies, that this concept contributed to a robust safety assessment framework. Moreover, the Consultation noted that substantial equivalence is a concept used to identify similarities and differences between GM food and a comparator with a history of safe food use which subsequently guides the safety assessment process (WHO, 2000).

While not a safety assessment *per se,* the substantial equivalence approach allows the structuring of the safety assessment through characterizing the similarities and identifying the differences which can then be the focus of further consideration. This approach has been referred to as a useful safety standard (The Royal Society of Canada, 2001). It therefore permits inference that the new food under consideration will be no less safe than the conventional food under conditions of similar exposure, consumption patterns and processing practices. Substantial equivalence is therefore clearly not intended to be a measure of absolute safety, but instead recognizes that while demonstrating absolute safety is an impractical goal, demonstrating that there is reasonable assurance that the GMO-derived product under consideration is no less safe than a conventional food product is an achievable goal.

Since it's development, the use of the substantial equivalence concept in the safety assessment of GMO-derived foods has been subject to misinterpretation and criticism. The comparative nature of the substantial equivalence concept without prescribing the extent of the phenotypic and compositional comparisons has led some to criticize the concept as not being measurable, and therefore inappropriate in safety assessment (Fagan, date unknown, Millstone *et al.*, 1999). This and other criticisms relate, in part, to the mistaken perception that the determination of substantial equivalence was the end point of a safety assessment rather than the starting point (FAO, 2000).

After several OECD countries had gained experience with safety assessment of GMO-derived foods, an OECD workshop examined the effectiveness of the application of substantial equivalence in safety assessment. This workshop concluded that the substantial equivalence approach provides equal or increased assurance of the safety of foods derived from genetically modified plants, as compared with foods derived through conventional methods. In 2000, the OECD Task Force for the Safety of Novel Foods and Feeds also reviewed the substantial equivalence concept, its interpretation and its application. While noting that the majority of international guidance documents addressing the safety assessment of genetically modified (GM) plants had interpreted substantial equivalence consistently, this Task Force reported that there were differences in how it was applied and that these differences needed to be resolved (OECD, 2000).

The Codex *Ad Hoc* Intergovernmental Task Force on Foods Derived from Biotechnology pursued the development of international guidance on the safety assessment of GMO-derived foods. To date, this Task Force has developed proposed draft principles and guidelines for the safety assessment of foods derived from modern biotechnology. These guidelines interpreted the concept of substantial equivalence as a way of structuring the safety assessment, consistent with the FAO/WHO interpretation. The 'Draft Principles for the Risk Analysis of Foods Derived from Modern Biotechnology' and 'Draft Guideline for the Conduct of Food Safety Assessment of Foods Derived from Recombinant-DNA Plants' have been forwarded by the Codex Task Force to the 25th Session of the Codex Alimentarius Commission for adoption at Step 8 of the Codex procedure (Codex *Ad Hoc* Intergovernmental Task Force on Foods Derived from Biotechnology, 2002). A proposed draft guideline for the food safety assessment of foods produced using recombinant-DNA organisms continues to be developed.

Substantial equivalence: the application

A conclusion from a 1995 WHO workshop on 'Application of the Principles of Substantial Equivalence to the Safety Evaluation of Foods or Food Components from Plants Derived by Modern Biotechnology' summarized the relationship between substantial equivalence and safety assessment as follows:

> Establishment of substantial equivalence is not a traditional safety assessment in itself, but a dynamic, analytical exercise in the assessment of the relative safety of a new food or food component to an existing food or food component.
>
> (WHO, 1995)

The application of the substantial equivalence concept is a key step in structuring the safety assessment of GMO-derived foods. Through appropriate analysis, the new food is compared phenotypically and compositionally to its conventional counterpart with a history of safe use. In this compositional comparison, the characteristics, including levels of key nutrients and toxicants, are considered relative to those of the conventional counterpart taking into account the natural variation for such characteristics. When a genetic modification results in the insertion of a specific trait, the comparative analysis of the new food will identify both the intended effect (inserted trait) and the potential unintended effects of the modification. Further assessment can then focus on new or altered characteristics for which no history of safe use can be established.

One of the important benefits of applying the substantial equivalence concept is that it provides flexibility which can be a powerful tool in terms of food safety assessment. The comparative approach to structuring the safety assessment can be applied at several potential levels along the food continuum (i.e. harvested primary food material or unprocessed food product, individual processed fractions, or final food product or ingredient). While from a practical point of view, the compositional comparison should typically be applied at the level of the unprocessed food product, the flexibility of the concept permits the determination to be targeted to the most appropriate level based upon the nature of the product under consideration.

Where multiple fractions from a single source are destined to different food products, the comparative approach might be targeted at the level of the unprocessed food product in order to permit the safety assessment to apply to all derived fractions (e.g. for soybean, where multiple fractions are used widely in foods, assessment at the level of the seed is appropriate). However, where a single fraction of a particular raw material is used as human food, then the comparison can focus at the level of the single fraction, thereby simplifying the safety assessment process (e.g. for canola, where the processed oil is the only fraction consumed by humans, safety assessment may appropriately be focused on comparison of the oil composition of the novel variety with traditional canola oil composition).

Application of the substantial equivalence concept in the safety assessment of a GMO-derived food depends on the identification of an appropriate comparator with an acceptable history of safe food use. It also requires that sufficient analytical data be available in the literature or be generated through analysis to permit an effective comparison. These requirements present a key limitation of the substantial equivalence concept since the consideration of similarities can only provide assurance of safety relative to those components assessed for the particular comparator. The choice of the comparator is therefore crucial to the effective application of substantial equivalence in establishing the safety of a GMO-derived food. An appropriate comparator must have a well-documented history of use. If adverse effects have been associated with the particular food type, specific components of the food which are considered to be causative of those adverse effects should be described and well characterized in order to permit effective comparison.

A joint FAO/WHO Expert Consultation on Biotechnology and Food Safety considered the application of substantial equivalence in the safety assessment of GMO-derived foods (FAO, 1996). The consultation recommended that applicacation of the substantial equivalence concept entail consideration of the molecular characterization of the new food source; phenotypic characterization of the new food source in comparison to an appropriate comparator already in the food supply; and the compositional analysis of the new food source or the specific food product in comparison to the selected comparator. The guidance further elaborates on compositional comparison by highlighting what information should provide sufficient information to permit effective comparison. The recommended focus of the compositional comparison is on analytical comparison of those components identified in the food source in question which are nutrients which provide a substantial impact in the overall diets (key nutrients) and toxicologically significant compounds known to be inherently present in the species (key toxicants). In addition, there is recognition that additional components might be identified for analysis based upon the molecular and phenotypic characterization and the nature of the genetic modification.

In addition to key nutrients and toxicants, additional parameters may be appropriate for assessing the potential for unintended effects of a genetic modification. When modifications are directed at metabolic pathways of key macro or micro nutrients, the possibility of an impact on nutritional value is increased, particularly if that food is a major dietary source of the nutrient affected. The potential for unintended effects would be determined, in part, by the nature of the intended alteration (e.g. consideration of fatty acid profile if an enzyme involved in fatty acid metabolism is introduced) and the data from molecular and phenotypic characterization.

The consideration of key nutrients and key toxicants in the comparison which is essential to applying a substantial equivalence in safety assessments also introduces another limitation of the concept. The nature of the comparative approach with respect to nutrients limits its universality since the relevance of nutrients in a particular crop are dependent on consumption patterns which might vary from region to region. Where differences in consumption exist for a particular crop, these must be considered in the identification of the key nutrients for assessment. This is particularly true for crops which form a significant portion in the diet in a particular region. The comparative approach to assessment can be applied in each region, but conclusions for one region will not automatically hold for another region if there are significant differences in consumption patterns and processing practices. Another potential regional limitation is related to the application of the concept of substantial equivalence as opposed to an inherent limitation of the concept *per se*. It may be difficult, particularly in developing countries, to apply the concept to assess the safety of foods where adequate nutritional databases are not available for a given population.

A safety assessment using the substantial equivalence approach does not demonstrate that a GMO-derived product is identical to its conventional comparator since the compositional comparison does not take into account all components. However, application of the guidance noted above provides assurance that the comparison has considered those components most likely to be relevant to the safety of the product as it is expected to be consumed in a particular region. Recognizing the importance of there being compositional data available for applying substantial equivalence, the OECD Task Force for the Safety Assessment of Novel Foods and Feeds has focused on the development of science-based consensus documents containing information on the nutrients, anti-nutrients or toxicants, product use and other data relevant to the assessment. The OECD have published such documents for potatoes, sugar beet, soybean and low erucic acid rapeseed (canola) (http://www.OECD.org/biotrack/). Such guidance will add to the understanding of appropriate parameters for a comparative assessment of composition while recognizing that additional parameters may be relevant to the safety assessment dependent on differences in consumption pattern.

The international guidance developed by the OECD and FAO/WHO has been practically applied to the safety assessment of GMO-derived food products in several countries. In order to facilitate such assessments, specific guidance documents which embrace the substantial equivalence concept have been published (Table 3.1). The majority of these currently address the safety assessment of genetically modified plants and have consistently interpreted the concept of substantial equivalence. Internationally, these guidance documents have been applied to the assessment of a significant number of GMO-derived plant products over a period of more than eight years, demonstrating that the concept of substantial equivalence can be applied effectively in the safety assessment of novel foods.

Table 3.1 Examples of international guidance on the safety assessment of GMO-derived foods

Country	Specific guidance documents	Year published
United Kingdom	Guidelines on the assessment of novel foods and processes – UK Advisory Committee on Novel Foods and Processes	1991
United States	FDA Statement of Policy: Foods derived from new plant varieties – United States Food and Drug Administration	1992
China	Safety Administration Regulation on Genetic Engineering – State Science and Technology Commission of the People's Republic of China	1993
Canada	Guidelines for the safety assessment of novel foods – Health Canada	1994
Japan	Guidelines for the safety assessment of foods and food additives produced by recombinant DNA techniques – Ministry of Health and Welfare Japan	1996
European Commission	EC Regulation 258/97 concerning novel foods and novel food ingredients – European Parliament and the Council of the European Union	1997
Australia/ New Zealand	Assessment guidelines for foods and food ingredients to be introduced in Standard A-18 – Food Derived from Gene Technology – Australia and New Zealand Food Authority	1997

Substantial equivalence – rapeseed oil (canola) as a case study

The comparative approach and hence the concept of substantial equivalence has been applied to the safety assessment of new and modified foods derived from plants developed using traditional breeding practices over the last 20 years and more recently to those products derived from recombinant DNA technology. Just as the OECD Working Group utilized case studies, the application of substantial equivalence in structuring the safety assessment of novel foods can be illustrated through discussion of the modifications made to rapeseed oil through traditional and recombinant DNA techniques.

Developments in rapeseed oil (canola)

Rapeseed breeding in Canada began soon after the crop was introduced during World War II. The initial goals of breeding were directed towards improving agronomic characteristics and oil content of rapeseed. Nutritional experiments conducted as early as 1949 indicated that consumption of large amounts of rapeseed oil with high levels of erucic acid could be detrimental to experimental animals (Boulter, 1983). Concerns about the nutritional safety of rapeseed oil and the potential impact on human health stimulated plant breeders to search for genetically controlled low levels of erucic acid in rapeseed oil. After 10 years of backcrossing and selection to transfer the low erucic acid trait into agronomically adapted cultivars, the first low erucic acid varieties, *Brassica napus* and *B. campestris* were released in 1968 and 1971, respectively (Eskin *et al.*, 1996).

Rapeseed meal is used exclusively in Canada as a high-protein feed supplement for livestock and poultry. Prior to the late 1970s, the use of this oilseed processing byproduct as an animal feed was limited by the presence of glucosinolates in the seed. The low palatability and the adverse effect of glucosinolates due to their anti-thyroid activity led to the development of varieties of rapeseed which have combined low levels of both glucosinolates and erucic acid (also known as 'double low' varieties).

Canola breeding programmes in the 1980s and 1990s have produced cultivars with higher yields, increased oil and protein contents, earlier maturity, yellow seeds, reduced green seed and improved disease, insect and herbicide resistance (Eskin *et al.*, 1996). (Note: The term canola has been registered and adopted in Canada to describe the oil (seeds, plants) obtained from the cultivars *B. napus* and *B. campestris*. In 1986 the definition of canola was amended to refer to *B. napus* and *B. campestris* lines containing <2% erucic acid in the oil and <30 mmol/g glucosinolates in the air-dried, oil-free meal.)

Low-erucic acid rapeseed oil – application of substantial equivalence

In 1987 low-erucic acid rapeseed oil (LEAR oil) was given generally recognized as safe (GRAS) status in the United States. Although this product was not produced as a result of recombinant DNA technology, it was considered to be a 'novel food'. The LEAR oil case study illustrated the application of the principles developed by the Working Group (established by the GNE) for assessing the safety of foods by applying the concept of substantial equivalence (OECD, 1993a).

The novel trait of LEAR oil was the low erucic acid content when compared with traditional rapeseed oil. In this case precise comparisons between LEAR oil and other vegetable oils could not be made because vegetable oils vary in composition depending upon the variety of plant and the growing conditions. The concept of substantial equivalence was applied to assess the safety of LEAR oil by comparing the individual fatty acid components to similar components present in other traditional oils including soy, corn, peanut, safflower, olive and sunflower. Except for the low levels of erucic acid, the individual fatty acid components of the LEAR oil were comparable to those similar fatty acids found in common vegetable oils.

Dietary exposure estimates for average and upper limit intakes for LEAR oil used by itself and as a component of blended oil products including shortening, margarine, salad oil and vegetable oil did not raise any safety concerns. The data considered in assessing the nutritional adequacy and digestibility of LEAR oil were similar to those that would be considered for any new oil (i.e. human and animal feeding studies). Due to concerns regarding the safety of the erucic acid component in both traditional rapeseed and LEAR oils, toxicological studies were an important consideration in the safety assessment. These studies included a review of an extensive toxicological database and a scientific rationale supported by the results of animal feeding studies in several species with a range of vegetable oils.

The development of LEAR/canola and the application of the comparative approach to directing its assessment is illustrative. We can further build on this illustration by considering the application of recombinant DNA technology to canola to develop herbicide-tolerant varieties. In these cases, considerations for the safety of the oil derived from these varieties follow the same approach. In this case the appropriate comparator is now the unmodified canola variety. The database used in establishing the safety of LEAR/canola oil provides an effective tool for the comparison of the composition of the oil derived from the herbicide-tolerant variety under consideration. Comparison of the oil from new canola varieties developed through recombinant DNA technology with the oil from unmodified canola with the same food use is an effective approach to determine its safety. Consideration is appropriately given to key nutrients (i.e. fatty acids) and key toxicants (i.e. ensuring that the level of erucic acid is sufficiently low for safe consumption).

The comparative approach can also be applied in the safety assessment of other compounds in plants such as canola. The canola meal, which may be used as an animal feed, would require a separate evaluation since measurable amounts of the introduced protein may be present in the meal, unlike in the refined oil. In the case of the meal, the safety of that protein in animal feeding may need to be demonstrated if the protein is new to the feed supply. If, based on the compositional comparison of the canola meal to an unmodified counterpart, the only difference is the presence of the introduced protein, then the next steps in the safety assessment would focus on determining if that new protein had the potential to be toxic or impact the nutritional quality for feed use.

The successful lowering of erucic acid led to continued interest in the compositional modification of canola oil. For example, plant breeders have used mutagenesis to genetically alter the plant's fatty acid biosynthetic pathways to obtain specialized fatty acid compositions. Canola oil has been developed with the linolenic acid content reduced from approximately 10% to less than 3%. Although high levels of linolenic acid are desirable from a nutritional point of view, they are undesirable in terms of chemical stability.

Other recent developments in canola oil compositional changes include the application of mutagenesis to produce high levels of oleic acid. The resulting high oleic acid-producing cultivar was then crossed to low linolenic cultivars to create a high oleic/low linolenic line.

Intentional modifications in the composition of canola oil may lead to the production of canola varieties that are not like other commercial varieties. For example, in the last 10 years the application of recombinant DNA technology has resulted in the production of increased levels of lauric and myristic acids in canola oil. Similarly, the application of mutagenesis has produced higher levels of oleic acid in canola oil (i.e. from 60% to 85% total fatty acid content). For these canola varieties the fatty acid profiles and levels will not fall within the ranges defined in the Codex standard for Edible Low Erucic Acid Rapeseed Oil (Codex Alimentarius Commission, 1992) or the Codex draft standard for Named Vegetable Oils (which includes low erucic acid rapeseed oil) (Codex Alimentarius Commission, 1997). In cases where the fatty acid composition of canola oil has been intentionally modified so that commercial canola varieties cannot be used as a comparator, the fatty acid component(s) will need to be considered on an individual basis using other commonly consumed oils as appropriate for comparison where such fatty acid component(s) are present in similar levels. Substantial equivalence could be applied at the component level to assess the safety of the oil produced since these fatty acids have a safe history of consumption as a significant component of other edible oils.

In addition to fatty acid profiles and levels, modifications may also result in alterations in the chemical structure (e.g. saturation, chain length and triglyceride structure) that may have nutritional consequences or result in changes in digestibility. Chemically altered fatty acids will need to be evaluated on their merit, which may involve a combination of nutritional and toxicological *in vivo* and *in vitro* testing.

Laurate canola provides an example of an oil that is not compositionally identical to any other food oils, although it shares many similar characteristics (WHO, 1995).

Unlike other commercial canola varieties which contain no detectable lauric acid (C12:0), laurate canola produces high levels of lauric acid (39.75% of total fatty acids). Laurate canola also produces lower levels of oleic acid (18:1) (32.55%) and higher levels of myristate (C14:0) (4.15%) compared with the Codex specifications for those fatty acids in low erucic acid rapeseed oil. Other fatty acids in laurate canola, including palmitic (C16:0), palmitoleic (C16:1), stearic (C18:0), linoleic (C18:2), linolenic (C18:3), gadoleic (C20:1), eicosadienoic (C20:2), behenic (C22:0) and lignoceric (C24:0), are similar to those levels found in commercial canola varieties. Levels of the naturally occurring toxicant erucic acid (C22:1) are very low. Substitution of laurate canola for coconut and palm kernel oils does not raise any safety concerns for the intended uses because the major components, laurate and myristate, are identical.

Conclusion

The safety assessment of novel foods based upon the concept of substantial equivalence relies on comparison with the conventional foods with a long history of safe use. The comparator can be the species itself, a product derived from that food source or a similar component from a different species (i.e. fatty acids in vegetable oils). The food safety issues for organisms that have been genetically modified are of the same nature as those that may occur through other ways of genetic modification, such as traditional breeding. These include potential food safety concerns such as toxicity or allergenicity. Once the safety assessment is completed it is reasonable to assume that the novel food does not pose a risk different to those of traditional foods that have been safely part of the diet for many years.

The term 'substantial equivalence' and its application in safety assessment has been criticized or challenged as lacking a clear definition and therefore being imprecise (Fagan, date unknown, Millstone *et al.*, 1999). However, when considered as a concept, the lack of a measurable definition does not restrict its application. Stated most simply, substantial equivalence encourages investigators to compare a product which they have to assess with one with which they are already familiar. The application of this concept to the assessment of foods derived from GMOs is therefore intended to permit the creation of a linkage between the new GMO-derived food and a familiar, conventional food in terms of their characteristics and composition.

Over the past eight years, the concept of substantial equivalence has been consistently interpreted and practically applied by numerous countries assessing the safety of foods derived from genetic modification. It has been demonstrated that the concept of substantial equivalence can be applied effectively in the safety assessment of those genetically modified foods developed for commercialization to date. It is noteworthy that the Codex *Ad Hoc* Intergovernmental Task Force on Foods Derived from Biotechnology has supported the comparative approach as a key element in structuring the safety assessment of GMO-derived foods, signifying that there is significant international consensus emerging on the application of substantial equivalence.[2]

2 While not yet a formal consensus, the agreement of the Codex Task Force to forward the 'Draft Principles for the Risk Analysis of Foods Derived from Modern Biotechnology' and 'Draft Guideline for the Conduct of Food safety Assessment of Foods Derived from Recombinant-DNA Plants' to the Codex Alimentarious Commission for adoption signifies emerging consensus (Codex *Ad Hoc* Intergovernmental Task Force on Foods Derived from Biotechnology, 2002)

Although the majority of the food products currently approved have resulted from relatively simple modifications involving the introduction of one or two novel proteins, it is expected that future foods will exhibit more complex modifications. In these cases, the application of substantial equivalence may not be appropriate. Therefore ongoing development of methodology for assessing the safety of genetically modified food is necessary.

Acknowledgements

The authors wish to acknowledge the contribution of Mr Brian Harrison, Health Canada for compiling the listing of examples of international guidance documents.

References

Boulter, G.S. (1983) 'The history and marketing of rapeseed oil in Canada', in J.K.G. Kramer, F.D. Sauer and W.J. Pidgen (eds) *High and Low Erucic Acid Rapeseed Oils. Production, Usage, Chemistry, and Toxicological Evaluation.* London: Academic Press.

Codex Alimentarius Commission (1992) *Codex Standard for Edible Low Erucic Acid Rapeseed Oil*, Vol. 8. Codex Stan 123–1981 (Rev. 1–1989). Rome.

Codex Alimentarius Commission (1997) *Codex Draft Standard for Named Vegetable Oils.* (At step 6 of the Procedure). ALINORM 97/17. Rome.

Codex *Ad Hoc* Intergovernmental Task Force on Foods Derived from Biotechnology (2002) *Report of the third session of the Codex Ad Hoc Interdepartmental Task Force on Foods Derived from Biotechnology.* Yokohama, Japan.

Eskin, N.A.M., McDonald, B.E., Przybylski, R. *et al.* (1996) 'Canola oil', in Y.H. Hui (ed.) *Bailey's Industrial Oil & Fat Products*, Vol. 2: *Edible Oil & Fat Products: Oils and Oil Seeds*, 5th Edition. New York: John Wiley & Sons.

Fagan, J. (date unknown) 'The failings of the principle of substantial equivalence in regulating transgenic crops', http://www.geocities.com/athens/1527/subequiv.html

FAO (1996) *Biotechnology and Food Safety: Report of a Joint FAO/WHO Consultation.* Rome: Food and Agriculture Organization of the United Nations.

Millstone, E., Brunner, E., Mayer, S. (1999) 'Beyond "substantial equivalence"', *Nature* **101**: 525–6.

OECD (1986) *Recombinant DNA Safety Considerations.* Paris: Organisation for Economic Co-operation and Development.

OECD (1993a) *Safety Evaluation of Foods Derived by Modern Biotechnology: Concepts and Principles.* Paris: OECD.

OECD (1993b) *Field Releases of Transgenic Plants, 1986–1992: An Analysis.* Paris: OECD.

OECD (1996) *Food Safety Evaluation.* Paris: OECD.

OECD (2000) *Report of the Task Force for the Safety of Novel Foods and Feeds.* Paris: OECD.

The Royal Society of Canada (2001) *Elements of Precaution: Recommendations for the Regulation of Food Biotechnology in Canada.* Ottawa.

WHO (1991) *Strategies for Assessing the Safety of Foods Produced by Biotechnology. Report of a Joint FAO/WHO Consultation.* Geneva: World Health Organization.

WHO (1995) *Application of the Principles of Substantial Equivalence to Safety Evaluation of Foods or Food Components from Plants Derived by Modern Biotechnology. Report of a WHO Workshop.* Geneva: World Health Organization, Food Safety Unit.

WHO (2000) *Safety Aspects of Genetically Modified Foods of Plant Origin. Report of a Joint FAO/WHO Expert Consultation on foods derived from biotechnology.* Geneva, World Health Organization.

Chapter 4

Strategies for analysing unintended effects in transgenic food crops

H.P.J.M. Noteborn, A.A.C.M. Peijnenburg and R. Zeleny

Introduction

The application of recombinant DNA technology has revolutionized plant breeding in recent years. Genetic transformation in plant breeding greatly increases the gene pool and broadens the scope of genetic changes that modern breeders may draw upon. The first generation of genetically modified (GM) crop plants hold great promise in carrying new or improved agronomical traits that require less intensive farming methods (Estruch *et al.*, 1997) or which have improved quality traits (Fromm *et al.*, 1993). These GM crop plants have been produced to improve farming production systems, aiming, for instance, to reduce negative impacts on the environment, such as fertilizers, herbicides and pesticides. In general they are of little direct interest to the consumer. However, the genetic manipulation of the primary and secondary metabolism of plants offers distinct possibilities in the development of nutritionally improved products, raw materials with added pharmaceutical value, and crops with real health benefits for the consumer. These are second-generation GM plants.

Recently, several prototypes in molecular plant breeding with novel traits for (non-) food and industrial applications have been introduced into the market, and many more GM crops have the prospect of commercial exploitation in the near future (Tsaftaris, 1996). The safety and wholesomeness of these GM crop plants raises questions beyond those posed by conventional foods. In this context, traditional plant breeding of most food crops has conformed to the standard 'generally recognized as safe' (GRAS). In some instances safety is assumed to be assured during performance and quality testing by breeders (e.g. glucosinolates and erucic acid in rape seed). However, the process is largely focused on the comparison of agronomic and phenotypic characteristics, such as colour, maturity, yield or disease resistance. Products with an unusual taste or that have harmful effects following ingestion (e.g. alkaloids in potatoes; Harvey *et al.*, 1985) or skin contact (e.g. cucurbitacin E in squash and zucchini, Coulston and Kolbye, 1990 and furanocoumarins in celery, Beier, 1990) have normally been rejected from the breeding programme (OECD, 1998). However, the safety of conventional plant varieties has usually been presumed from the assumption that prudent consumers will avoid those species once it is known that they cannot consume them without adverse health effects.

When the commercial production of GM plants started, it sparked a debate centred on issues associated with concerns on environmental, human and animal safety matters. Many people questioned whether the 'simple' GRAS approach, based on historical and empirical experiences, is sufficient to guarantee food safety of the GM crop plants. It is also asked whether the risks of chronic exposure of humans and animals to GM crops are adequately covered, and whether large-scale breeding of GM plants will have detrimental effects on ecosystems.

Food safety issues

Although the variety registration of existing crop plants has not resulted in adverse effects in humans, the conventional assessments carried out by plant breeders was not considered to be sufficient to ensure food safety of GM crop plants. As a result, Europe issued a regulation on novel foods and novel food ingredients that sets the legal framework for the market introduction of genetically modified organisms (GMOs). This came into force on 16 May 1997 (European Community, 1997). The Novel Foods Regulation establishes a system of mandatory pre-market approvals of novel foodstuffs, including GM plants. The accompanying EU guideline 97/618/EC gave a clear indication of the types of data that are needed to form the basis of an assessment. In particular, four central safety issues have been set forth related to food and feed safety:

1   the nutritional and toxicological consequences of inserted gene products;
2   the potential of pleiotropic (unintended) effects in the host organism due to the insertion event;
3   the allergenicity of expressed proteins and novel foodstuffs;
4   the potential of gene transfer to human and animal gut flora.

This means that in transgenic insect-resistant Bt tomatoes encoding the C-terminal truncated Bt2 gene derived from a *Bacillus thuringiensis* (Bt) strain, *IAb5*, the novel protein has to be evaluated on its own merit, while the safety and wholesomeness of the 'remaining' transgenic tomato is assessed separately (Noteborn *et al.*, 1995). Whereas, for example, in antisense RNA exogalactanase tomatoes (i.e. tEG), which have improved rheologic characteristics due to downregulation of the endogenous exogalactanase activity (De Silva and Verhoeyen, 1998), the safety assessment is primarily focused on the 'remaining' novel fruit (Noteborn *et al.*, 1998, 2002).

In general, there is basic agreement on the safety issues to be addressed. One particular area of concern is the possibility of unexpected or unintended metabolic perturbations due to genetic modification that may alter, for instance, levels of nutrients and health-influencing components. The use of rDNA techniques does not necessarily result in fundamental changes in crop plants compared to the food produced by conventional breeding methods. However, it should be emphasized that a uniform international agreement for the evaluation of GMOs is urgently needed. For example, an agreement on how to establish in practical terms the similarities or differences between the 'remaining' GM crop plant and the traditionally bred plant seems to be much more difficult to achieve (OECD, 1996, 1998).

This chapter is devoted to the description of advanced strategies for the evaluation of potential unintended effects in GM crop plants. Pleiotropic (unintended) alterations in agronomic traits or composition may arise from insertion mutagenesis or as a result of metabolic effects of the novel gene product(s) (Noteborn, 1998). The challenge is to gain the ability to discriminate between metabolic alterations due to somaclonal variations, unintended effects or natural diversity. Thereto, emphasis is given to a platform of technologies designed to identify changes in the molecular machinery of crop plants. As is the case for detection methods for GM crops and derived products, there are needs and criteria for sampling strategies, statistical models and manufacturing standard reference materials. This chapter will not attempt to catalogue these as they have been addressed by others recently (e.g. Nordic Council, 1998; ILSI, 1999).

Unintended effects

Strategies for assessing the food safety and wholesomeness of novel foodstuffs are currently in the exploration phase. Above all there is a need for uniform and harmonized quantitative methods for the testing of potential unintended effects in whole foods, such as GM crop plants are. The safety of, for example, synthetic chemicals and food additives is established by assessing separate elements of the compound in question. Although very successful in case of single chemicals, the evaluation of a whole food product may not be a simple addition of individual assessments of the constituents. Traditional testing protocols involving laboratory animal 90-day feeding trials with (whole) food products are far from ideal (Pauli and Takeguchi, 1986). This type of animal feeding study with complex food matrices is complicated by the likelihood of nutritional imbalances leading to dietary problems, confounding factors, an insensitivity for specific endpoints, and the impossibility of using large safety margins, if any (Noteborn and Kuiper, 1995; Noteborn *et al.*, 1995; Kuiper and Noteborn, 1996). Therefore, in a global context several regulatory, research and industry-based bodies have proposed alternative concepts for the safety evaluation of genetically engineered foods and food ingredients. In particular the Organisation for Economic Co-operation and Development (OECD), the Food and Agriculture Organization (FAO) and the World Health Organization (WHO) have developed concepts and principles for the safety evaluation of GMOs and derived products (WHO, 1991, 1995; FAO, 1996; OECD, 1993, 1996, 1997). Their reports concluded that risk assessments should be directed at demonstrating that a GM crop plant or derived food product is as safe as a traditional or previously authorized product. Accordingly the EU Competent Authority incorporated these general principles into their Novel Foods Regulation and accompanying guidelines. In this context, novel foods and food ingredients that are considered to be 'no longer equivalent' to traditionally bred plants are subject to labelling as defined in the Council Regulation 1139/98 (European Community, 1998).

Substantial equivalence

The term 'substantial equivalence' appears in the EU Novel Foods Regulation as a major principle in the risk assessment. It emphasizes that, within the context of the Regulation the evaluation of the safety and nutritional value of a GM plant should be focused on a comparative analysis with conventionally bred products. OECD's Group of National Experts on Safety in Biotechnology (OECD, 1993) enunciated this comparative approach towards the safety assessment of transgenic food plants as the concept of substantial equivalence. In their report, it is assumed that existing food organisms possess a long history of safe use (i.e. GRAS). Consequently, the most practical approach to the determination of safety is to consider whether the GM plants are substantially equivalent to analogous food product(s), if such exist. Once substantial equivalence of a novel foodstuff has been established, it provides assurance of safety that is equal to or better than that of its comparators. Further clarification of the concept of substantial equivalence originates from a WHO sponsored workshop in 1994 (WHO, 1995). It has been concluded that the establishment of substantial equivalence is not a safety assessment *per se*, but a dynamic, analytical exercise. Thus, it may mean that analysis of, for instance, gene expression patterns, global changes in protein expression and/or differences in metabolic capabilities (the metabolome) of the novel product in comparison with conventionally bred products should be performed in order to examine equivalence.

This strategy has worked satisfactorily for the assessment of the first-generation crop plants for which sufficient background knowledge is available. However, relatively less experience has been gained with the safety and wholesomeness evaluation of novel food plants for which no substantial equivalence can be established. Even in the case of single gene modification, potential alterations in metabolic pathways are difficult to identify and, consequently, the implications of the genetic modification process on the metabolism of plants are poorly understood (Nuccio *et al.*, 1999). Moreover, data on the mechanisms by which plants regulate, for instance, their response to environmental stress factors and other conditions are remarkably scanty (e.g. Degli-Agosti and Greppin, 1998; Lutts *et al.*, 1999).

Critical nutrients and key toxicants

Application of the concept of substantial equivalence appears in practice to lead to various interpretations. Different requirements for risk assessment with regards to the potential of unintended effects exist as a result of genetic modification. Actual approaches entail a consideration of the characteristics of host and donor organism, phenotypic properties and toxicological evaluation and include comparative determinations of any changes in critical (anti-)nutrients and key toxicants for the food source in question (OECD, 1998). In general, the analysis of an expanded spectrum of components is unnecessary, but should be considered if there is an indication from other traits that there may be an unintended secondary effect of genetic modification (FAO, 1996; OECD, 1996). However, compositional analyses based on single parameters using a list of crop-specific critical (anti-)nutrients and key toxicants as a minimum requirement of screening for possible variations in the content has its limitations. Limited data are typically available for the less important food components, as in less well-known crops most critical (anti-)nutrients and natural toxicants will be unknown. Moreover, the toxicants to be assessed may be partly influenced by the cellular function of the product encoded by the inserted gene. Furthermore, the natural variation of critical comparators may mask the evidence of secondary effects due to genetic modification. Thus, there is no criterion that determines whether a difference between a GM plant and its natural counterpart is outside the literature values for the ranges of parameters that express natural diversity. Relevant information on indicators of unexpected effects may not have been observed during the development of an edible transgenic crop species.

In the future other compounds could become of importance, such as anti-oxidative, oestrogenic and anti-carcinogenic compounds (Ames, 1983; Leiner, 1994). In determining critical nutrients and/or key toxicants, differences in consumption patterns and practices of processing and consumption in various geographical settings and cultures must also be recognized. It is our opinion that conclusions about relative safety and wholesomeness based on a list of selected single components may not be equally valid in a global context and in all times. Therefore, there is an urgent need to set up generic approaches using platform technologies to evaluate undesirable metabolic perturbations in GM food crops.

Unintended effects – a post-genome challenge

In our view the identification of unintended effects encompasses the ability to interpret and use innovations in molecular genetics research such as genomics, proteomics and metabolomics on a crop-by-crop basis. This will establish a whole data package directed towards a holistic view on possible unintended side effects due to genetic modification. In essence, the focus is shifting towards molecular characterization to understand functional activity. Vital to this approach will be informative profiles of molecules at different integration levels, e.g. mRNAs, proteins including post-translational modifications and molecules at the level of primary and secondary plant products and metabolites that might be useful benchmarks for the detection of unintended effects. The relative importance of these various 'profiling' techniques in establishing data sets for assessing unintended effects is not indicated by their order and will vary from species to species. Alterations in, for example, expression levels, post-transitional modification, interactions or chemical composition do not *per se* imply that the product is less safe or even unsafe.

Such a platform of technologies may determine whether unintended effects exist between the GM plant and its control(s) or whether there is substantial equivalence apart from certain well-defined expected differences. Measures of the relative activities of various molecular constituents associated with metabolic capabilities in non- and modified plant cells under different conditions will provide new insights into how metabolism and genetic modification are orchestrated. But it should be emphasized that the fields of plant genomics (Graves, 1999) and proteomics (Blackstock and Weir, 1999) are still in their infancy.

mRNA fingerprinting

A very promising strategy to evaluate the equivalence of genetically modified crops is gene expression profiling. Conventional methods for the analysis of differential gene expression include Northern blotting (Alwine *et al.*, 1977), S1 nuclease protection (Berk and Sharp, 1977), comparative expressed sequence tag (EST) sequencing (Adams *et al.*, 1991), differential display (Liang and Pardee, 1992) and serial analysis of gene expression (SAGE; Velculescu *et al.*, 1995). More recently, Kok *et al.* (1998) have designed a method of detecting altered gene expression by means of mRNA fingerprinting or reverse transcription polymerase chain reaction (RT-PCR). Based on the original concept of Liang and Pardee (1992) specific subsets of the mRNA population of a tomato plant were amplified. This enables the expression of different genes in the GM crop and genes in the parental line to be compared and any empirical information about significant alterations to be established. The method is still under development and requires further validation, but has been shown to be reproducible and able to detect important differences in gene expression. Some of the most significant remaining problems are that the method largely depends on individual skills based on experience and is time-consuming, even in its most simplified form.

cDNA microarrays

Over the past few years, the DNA microarray technology has emerged as a powerful high-throughput method for the analysis of gene expression (Schena *et al.*, 1995, 1996). At present, microarrays for gene expression studies generally consist of cDNAs, which are physically deposited onto small glass surfaces. The major advantage of DNA microarray technology over conventional gene-profiling techniques mentioned above, is that it allows small-scale analysis of expression of a large number of genes in a sensitive and quantitative manner. Furthermore, it allows comparison of gene expression profiles under a large number of different conditions. The cDNA microarray technology has already been successfully introduced in disciplines ranging from cell and developmental biology to drug development and pharmacogenomics. In the following a number of technical aspects of the technology as well as its potential value for the safety assessment of genetic modification of food plants will be discussed.

   The cDNA microarrays are produced by deposition onto a solid surface of purified PCR products corresponding to specific genes. Typically a few nanolitres of DNA solution (100–500 µg/ml) are spotted using a micro-dispensing robot. Essential information on robot construction and protocols can be found on the Internet (e.g. http://cmgm.stanford.edu/pbrown/mguide/index.html). In general, with respect to the spotting devices, a higher density of spots is accomplished when using a passive dispenser (>2500 DNA spots/cm^2). Thereto, glass slides covered with a positively charged layer (e.g. poly-L-lysine) or carrying reactive groups are most commonly used. To analyse the expression of genes in a plant cell or tissue sample, polyA+ RNA (typically 0.5–2.5 µg is required for each reaction) is purified and labelled by reverse transcription using an oligo-dT primer incorporating a fluorescently labelled nucleotide. The labelled cDNA pool is then hybridized to the microarray. The required hybridization conditions (i.e. sample concentration, stringency of hybridization) and parameters, such as detection level and the correlation between transcript concentration and hybridization have been determined for each experimental setting. The intensity of a hybridization signal, which is a measure of the relative expression level of each gene, can be read by charge-coupled device (CCD) cameras or confocal/non-confocal laser scanners. RNAs extracted from two different samples can also be simultaneously analysed with a microarray by labelling each with a different fluorescent label using either Cy-3 or Cy-5 (Shalon *et al.*, 1996).

   At RIKILT-DLO we are currently testing whether the analysis of differential gene expression using DNA microarrays is an informative strategy for the safety evaluation of GM plants. To address this question, cDNA libraries are constructed and enriched by a subtractive cloning approach (Diatchenko *et al.*, 1996) for cDNAs preferentially expressed in either green or red tomato fruit. These subtracted cDNAs, as well as control cDNAs representing known tomato genes (i.e. sequence information derived from public sequence databases), have been spotted on arrays and are used for comparison of gene expression profiling of control and genetically modified tomatoes. In the near future we will construct more defined arrays containing genes of toxicologically relevant pathways, such as those involved in the synthesis of natural plant toxins (e.g. the glycoalkaloids α-tomatine and α-solanine).

Proteome profiling

The introduction of Bt genes of bacterial origin (Noteborn and Kuiper, 1995; Noteborn *et al.*, 1995) may theoretically influence the nature of post-translation modification (e.g. glycosylation) in the transgenic tomato fruit. This is because as well as the impact of the protein moiety, for example, the type of glycosylation pattern may alter due to inactivation of glycosidases and glycosyltransferases in the host cell (Jenkins and Curling, 1994). In addition, glycosylation of the inserted Bt gene product in the plant might result in altered protein activity and stability. The importance of structurally elucidating these oligosaccharides, especially the *N*-glycans, can be attributed to their widespread functions in the cell. These functions include correct folding, biological activity and stability of proteins as well as involvement in plant development (Sturm, 1995). Moreover, *N*-glycans represent antigenic epitopes by themselves (Prenner *et al.*, 1992). Immunological as well as biochemical studies revealed the allergenic potency for the α1,3-fucosylation of the proximal *N*-acetylglucosamine residue (Tretter *et al.*, 1993; Wilson *et al.*, 1998). Additionally, also the β1,2-xylosylation of the β core mannose is suspected to be an allergenic epitope (Garcia-Casado *et al.*, 1996). Recently, there is increasing evidence that these *N*-linked oligosaccharides may have a substantial impact on the immunogenicity of glycoproteins, as *N*-glycans are suspected to play a certain role in the adverse food reactions of hypersensitized patients (Aalberse *et al.*, 1981; Foetisch *et al.*, 1998).

At the Institute for Agrobiotechnology, Austria, this issue was tackled by developing a generic technology for isolating, purifying and characterizing *N*-linked oligosaccharides in plant tissues. Our low-temperature acetone powder method appeared to be most suitable for the isolation of (glyco-)proteins from plant tissue (e.g. tomato fruit). The resulting (glyco-)protein fraction was proteolytically digested and corresponding glycopeptides purified by cation-exchange chromatography. The oligosaccharides (*N*-glycans including α1,3-fucosylated oligosaccharides) were released from the peptide backbones by *N*-glycosidase A (PNGase A) and fluorescently labelled with 2-aminopyridine. Subsequently, pyridylaminated glycans were separated and analysed structurally by repeated two-dimensional high-performance liquid chromatography (2D HPLC) using a size-fractionation and a reverse-phase column (Kubelka *et al.*, 1994) in combination with an exoglycosidase treatment. MALDI-TOF (matrix-assisted laser desorption ionization time-of-light) mass spectrometry was applied for further confirmation of the observed structures.

Sixteen different *N*-glycosidic structures could be detected in tomato fruit. The two most abundant glycans in all the tomato samples showed identical properties to those of the major *N*-linked oligosaccharides of horseradish peroxidase (MMXF: Man(3)Xyl-FucGlcNAc(2)) and pineapple stem bromelain (MOXF: Man(2)Xyl-FucGlcNAc(2)), respectively, and accounted for about 65–78% of the total glycan content. Oligomannosidic glycans occurred in only small quantities (3–9%). The majority of the *N*-glycans found was β1,2-xylosylated and carried an α1,3-fucose

Table 4.1 Relative abundance of pyridylaminated N-glycans in tomato fruit

| Structure carbohydrates | Relative abundance (%) | | | | | |
| | Variety | | State of ripening Moneymaker | | Bt-tomato San Marzano | |
	Notoro	San Marzano	Red	Green	GM line	Parent line
Oligomannosidic	7.9	6.1	3.0	7.8	6.0	6.1
MMXF[a]	49.7	56.1	58.7	54.0	55.4	56.0
MOXF[b]	18.0	19.2	19.6	22.1	20.1	19.3
α1,3-fucosylated	85.9	88.0	90.3	87.3	90.3	88.2

[a] $Man_3XylFucGlcNAc_2$.
[b] $Man_2XylFucGlcNAc_2$.

residue linked to terminal N-acetylglucosamine (Table 4.1). A structural element was identified and shown to be an IgE-reactive determinant, which contributes to cross-reactivities among non-related glycoproteins, and is present in a wide variety of plants extracts (Wilson *et al.*, 1998). It is proposed that this structural element might also be of importance in some cases of food and pollen allergy (Aalberse *et al.*, 1981; Vieths *et al.*, 1994; Staudacher *et al.*, 1995; Petersen *et al.*, 1996). The carbohydrate profiling of, for example, populations of green and red-ripe fruit of non-modified and modified Bt tomato plants (expressing the *cryIA5* gene) revealed no differences in the nature of N-glycans, nor were there significant differences in the relative amounts of glycan (Table 4.1). Interestingly, the relative amounts found were quite similar across seasons and comparable to the commercial variety Notoro used as an extended control. As far as the N-linked oligosaccharides were involved it was not possible to detect significant alterations in the α1,3-fucose content of Bt tomatoes due to genetic modification.

Profiling of the metabolome

Among other factors, identity of a complex GM crop plant with its traditionally bred parent entails metabolism and the level of (un-)desirable substances contained therein. Therefore, we would like to build up an overall picture of components associated with different metabolic capabilities in the (non-)modified plant tissue. A multi-compositional analysis of the so-called metabolome may overcome the test limitations of, for example, single critical nutrient or key toxicant analyses and problems related to their limits of natural diversity. An attractive alternative is the utility of high-resolution proton (^{1}H) nuclear magnetic resonance (NMR) spectroscopy (^{1}H-NMR) in combination with separation techniques (liquid chromatography (LC)) (Foxall *et al.*, 1996; Korhammer and Bernreuther, 1996; Lindon *et al.*, 1997). Recently, we have shown that the approach of chemical fingerprinting using off-line LC-NMR can obtain information on possible changes in complex plant matrices due to environmental effects as long as GLP-like conditions for sample handling, data acquisition and automation of data handling are maintained (Lommen *et al.*, 1999). In principle, it will be possible to screen all the different low molecular weight components (MW <10 kDa) present in the metabolome of the plant tissue.

Chemical fingerprinting

The principles for establishing a chemical fingerprint were directed towards the detection of alterations in compositions in ^{1}H-NMR spectra arriving from two populations, for example, non-modified counterpart(s) and an engineered one. In this way the unprocessed parts of the plant used for human consumption were fractionated and analysed, i.e. in tomato comparison on the fruit level (Noteborn *et al.*, 1998). Genetically modified tomato varieties, such as the antisense RNA exogalactanase fruit (De Silva and Verhoeyen, 1998), have been studied using this innovative technology. The performance of the chemical fingerprinting analysis is illustrated in Figure 4.1, showing a total of approximately 3000 amplitudes in five spectra that represent probably a couple of hundred plant cell constituents. The spread of measurements of the individual profiles was comparable for the GMO and its control(s). By subtraction of the ^{1}H-NMR spectra the differences of the means were calculated at 99% confidence intervals between the GM crop average ($n = 8$/batch per line) and its isogenic control average ($n = 8$/batch per line). The differences of the means was defined here as all differences of the means of the amplitudes of individual NMR signals obtained from spectra of various populations (Noteborn *et al.*, 2000). The differences of the means were used to quantify how the transgenic crop plant differed from its parental line. The normalized 99% confidence intervals of the normalized means of fractions A–E (Figure 4.1) after eightfold independent replicates demonstrated that differences in amplitudes exceeding at least 20% could normally be detected. This is an adequate sensitivity to allow statistical evaluation, and is in line with recommendations of the Nordic Council (1998). They proposed that if the average value of a parameter differs by more than 20% an explanation should be sought.

With the purpose of establishing those differences that may arise from metabolic effects linked to genetic modification the default statistical analysis used a mixed effects randomized block model (Sachs, 1984). Random block effects such as variations in the stage of cultivation, location, logistics or climate had a significant impact on the overall chemical fingerprints of cultivars. For example, chemical fingerprints varied much more between field sites and seasons than between replicates in one plot. A tomato background mean (i.e. crop mean value) was established by constructing an extended range of non-modified tomato references (WHO, 1995; FAO, 1996) for comparison with natural variability, and to assess the impact of external factors (i.e. variations in season, climate, etc.). The crop mean value was used to quantify how the differences of the means between the transgenic crop plant and its control differed from the natural diversity of commercial counterparts.

Figure 4.1 (opposite) Chemical fingerprinting analysis using off-line LC-NMR (vertical scale in arbitrary units). Representative examples of typical baseline correction and peak selection results obtained from the ^{1}H-NMR spectra of antisense-exogalactanase RNA tomato fruit (var. Moneymaker transformant p35tEG) using software as described in Lommen *et al.* (1999). A global compositional description per fraction is as follows: fraction A, polar compounds such as fruit acids, glucose and other monomeric and oligomeric sugars, amino acids and TCA-cycle substrates; fraction B, large variety of aromatic, aliphatic and sugar-like compounds; fraction C, primarily non-tomatine glycoalkaloids and aromatic compounds; fraction D, tomatines and aromatic/indole-like compounds and fraction E, primarily lipids and carotenoids. Typically the detection limits in the various fractions are estimated in the range of mg/kg wet weight tomato tissue.

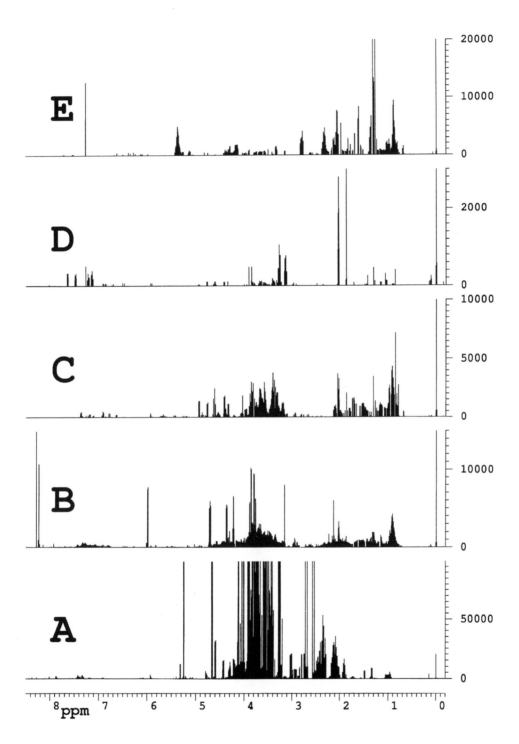

To differentiate between compositional changes either due to genetic modification, genetic variability or environmental variations the chemical fingerprints were studied in a hierarchical approach (Noteborn et al., 2002). Hence, the fingerprints of the GMO are compared to those of:

- the isogenic control line grown side-by-side under identical conditions in one plot;
- the isogenic control line grown side-by-side under identical conditions at multiple sites;
- an extended range of commercial varieties of that crop;
- the influence of downstream processing.

For example, a purely numerical analysis of statistical differences (p < 0.01) which showed up in comparing various populations is detailed in Table 4.2. It was recognized that 45% of all compounds of the control line harvested in different years varied considerably in concentration (Table 4.2, column C1/C2). Most differences were found in fractions A, C and D. A similar degree of impact on composition was observed in the case of processing temperatures (e.g. 100°C versus 70°C) on tomato juice (Table 4.2, column S70/S100). These findings were compatible with the protease

Table 4.2 Chemical fingerprinting analysis of the antisense-exogalactanase tomato line tEG

Distribution by degree of difference	Number of amplitudes that are significant different (p < 0.01)			
	C1/C2 S70/S100	tEG1/tEG2	C1/tEG	CON/tEG
>100	3			
20–100	3			
8–20	17			3
4–8	58	6	2	16
2–4	242	16	22	65
1.2–2	337	81	68	122
1–1.2			5	
1–1/1.2		14	41	9
1/1.2–1/2	223	15	74 10	285
1/2–1/4	242	6	37 11	84
1/4–1/8	55			28
1/8–1/20	10			14
1/20–1/100	2			2
<1/100				

The degree of difference was determined by calculating the individual peak amplitude ratios for all significantly different peaks over a total number of approximately 3000 amplitudes from 5 fractions (A–E). C1: Moneymaker-type control fruit (parent of p35StEG), tEG1: antisense RNA tomato (p35StEG), fruit of line 11 (plant 12+19+20), tEG2: some clone and sowing, fruit of line 11 (plant 3+9+11+15) and tEG: some clone and sowing, fruit of line 11 (plant 3+9+11+12+15+19+20) bred side-by-side and harvested at June 1997 in the greenhouse; C2: Moneymaker control fruit (parent of p35StEG) harvested at November 1997; CON: combined control tomato fruit C1 + C2 + batches of San Marzano-type control fruit (strain: TL0001) harvested at August 1996, June 1997 and November 1997 in the greenhouse; S70, tomato juice heat-treated at 70°C (PEF) and juice heat-treated at 100°C.

activity surviving the 70°C treatment (Lommen *et al.*, 1999). But in fruit harvested from various individual plants, divided into two populations, approximately 95% of all constituents were practically identical in concentration (Table 4.2, column tEG1/tEG2). Also the determination of the differences of the means between transformant tEG to its isogenic control bred side-by-side in the same plot showed that 249 out of 3000 amplitudes (8%) varied in concentration (Table 4.2, column C1/tEG). Assignments by NMR indicated that, for example, citric acid was present in fraction A of the control at 1.4 times the levels found in the transformant. However, in fraction A of the transformant glutamic acid and/or glutamine showed an increase of 1.3 times the levels found in the isogenic control.

On the other hand, one single aromatic compound, α-lycopene, appeared to be present in fraction E of transformant tEG at 2.5 times the concentrations found in the parental line (Noteborn *et al.*, 1998, 2002). No alterations were observed in β-carotene levels. These effects may be the result of the antisense downregulation of the *tEG1A* gene in tomato. In our view the natural diversity must be taken into consideration, however, when interpreting the biological relevance of any significant difference found in the differences of the means between the GMO and its parental line. As shown in Table 4.2 (column CON/tEG) the parental line was extended by entering non-isogenic commercial references (i.e. crop mean value). Although limited in the range of lines, the comparison of the crop mean value to the differences of the means showed that the contribution and magnitude of differences (e.g. citric acid, glutamic acid/glutamine, α-lycopene) diluted out. As listed in Table 4.2, the numbers of statistical differences approached approximately 1%.

It was recognized that in the agricultural practice the conditions of culturing, processing and storage of appropriate references will be mostly unknown and they may have a non-isogenic genotype. Notwithstanding these complicating random block effects we recommend that, for example, the chemical fingerprint should not be limited to the transgene and its isogenic parent bred under identical circumstances only, but should also include a sampling through multiple sites and a comparison with an extended range of controls of that crop. Otherwise it would be a too limited approach towards the establishment of the claim of substantial equivalence and it would neglect the ranges of natural variations in that crop. In future, if many more lines have to be compared with each other this approach may benefit from multivariate statistical methods.

Toxicological profiling

The significance of detected compositional alterations may need to be further explored by *in vitro* toxicity assays in order to screen for potential adverse effects and for mechanistic studies. It may even be that the latter approach leads to a redesign, refinement or even replacement of (sub-)chronic rodent feeding trials. *In vitro* systems derived from organs and tissues from animals and humans, and various types of cultured recombinant cell lines, are successfully used for screening for the toxic potential of single compounds. However, their ability to screen for the potential toxicity of whole foods or extracts thereof has been insufficiently explored up till now, and has been examined by us within the framework of a tiered safety evaluation of GM plants (Noteborn *et al.*, 1997, 1998). The validity of using *in vitro* test assays for comparative testing of whole food products, food ingredients and extracts thereof depends on the sensitivity and selectivity of the toxic response induced upon exposure. However, since these models are not yet validated, they should be considered as early warning or alert-systems to identify potential changes in toxicity of the engineered product only.

Assaying cytotoxicity and (anti-)genotoxicity endpoints

The mucosa of the gastrointestinal tract is clearly a primary, potential site of action for the effects of novel foods. Therefore effort has been concentrated on the development of a battery of intestinal cell systems (e.g. IEC-6, IEC-18, Caco-2 and INT407 cell lines) with different endpoints for initial cyto- and genotoxicity testing. The choice of a suitable method of sample preparation was considered of utmost importance in assaying vegetables. Freeze-dried tomato fruit was extracted (10% suspensions, w/v) in water, 0.9% saline or chloroform:methanol (2:1, v/v) immediately before incorporation into the culture medium. It was encouraging that aqueous and chloroform/methanol extracts of, for example, red-ripe tomatoes up to a maximum of 10% suspensions (w/v), had little or no general toxicity once the pH (>6.0) and osmotic pressure had been taken into account (Noteborn *et al.*, 1998). No toxicity to the intestinal cells Caco-2 and IEC-18 under short-term conditions was detected in red-ripe, non- and antisense RNA exogalactanase tomato using alpha-glutathione S-transferase (α-GST) lactate dehydroyrenase (LDH) leakage, reduction of 3-(4,5-dimethylthiazol-2-yl)-2,5-diphenyltetrazolium bromide (MTT conversion), neutral red (NR) uptake and total cellular protein as *in vitro* endpoints. On the other hand, green tomato fruit displayed a pronounced toxic and cell-damaging effect. No statistical significant differences were found between transformants and isogenic control lines in terms of their activity in the assays. Treatment of tomato tissue with pepsin (ratio pepsin to tomato protein 1:250, w/w) did not result in an increase of or loss of inherent cytotoxicity. Supplementary investigations with naturally occurring plant components, for example, quercetin, β-carotene, α-lycopene, genistein showed that only steroid glycoalkaloids, such as α-tomatine, exhibited a rapid toxic effect, whereas α-tomatidine was ineffective. It was noted that the *in vitro* models did show general matrix-related toxic effects. For example, α-tomatine diluted in a matrix of extracted red-ripe tomato tissue (final concentration of 5–10% w/v) showed an increase of dose–response-dependent cytotoxicity in human Caco-2 cells (e.g. IC_{50} 7.9 $\pm$ 2.1 mg/L versus IC_{50} 33.9 $\pm$ 5.3 mg/L of α-tomatine only). Using the COMET assay (Singh *et al.*, 1988), suspensions of freeze-dried tomato fruit up to a maximum of 10% (w/v) did not exhibit DNA-damaging effects in a range of gastrointestinal cells derived from rats (IEC-6 and IEC-18) and humans (Caco-2, INT407). Genetic modification did not result in increased genotoxicity, but fruit extracts were able to suppress the DNA-damaging effects of the known genotoxins H_2O_2 and N-methyl-N'-nitro-N-nitrosoguanidine (MNNG) (Noteborn *et al.*, 1997). Although these endpoints are important toxicological measures they do not indicate the molecular mechanisms of toxic damage. So far, pronounced adverse effects could only be observed with glycoalkaloids. The screening for subtle differences in toxic responses of other plant constituents appeared to be limited. However, therefore, the use of critical endpoints for such studies other than cytotoxicity should be clearly defined.

Assessing transcription factors

There is a demand for *in vitro* models that not only determine the (geno-)toxic potency of plant components or complex mixtures, but also provide information on the mechanism by which that compound exerts its toxic effect (Marshall, 1993). As an *in vitro* profiling system the eukaryotic stress gene assay (i.e. CAT-Tox(L)) was assessed. This utilizes human HepG2 cells stably transfected with chloramphenicol acetyltransferase (CAT) reporter constructs (Todd *et al.*, 1995). The aim of the CAT-Tox(L) assay is to collect and quantify molecular responses to cellular stress and toxicity that have a transcriptional component in their regulation. The assay employs an ELISA method to measure the transcriptional activity of the promoter or response element constructs. The advantage is that many of the stress responses occur before any measurable cytotoxicity, thus allowing monitoring of stress pathways at sublethal levels. Since 14 unique transcription responses can be measured simultaneously, the CAT-Tox(L) provides comprehensive cellular stress profiles. The assay generates specific stress gene induction profiles for a variety of stressful and toxic plant components such as genistein, diadzein and sinigrin. Aqueous extracts of red-ripe, non- and antisense RNA exogalactanase tomato fruit dosed up to 0.05 g per assay did not cause any cytotoxicity to the HepG2 transfectants nor were there molecular responses to cellular stress or toxicity. Extracts of non-modified and modified green tomato fruit showed cytotoxicity and induced the construct containing the xenobiotic response element (XRE). This gene-induction profile was not comparable to that of, for example, the reference molecules dioxin, 3-methylcholanthrene, benzo[a]pyrene or β-napthoflavone. These polycyclic aromatic hydrocarbons were characterized with a co-induction of cytochrome P450 1A1 and glutathione-*S*-transferase *Ya* subunit–CAT fusion constructs (Todd *et al.*, 1995). There was substantial evidence that the systemic fungicides fenarimol and bupirimate did not induce any of the genes in this assay.

There was no promotor in the assay that could be linked to universal cellular stress. As shown in Figure 4.2, α-tomatine caused severe cytotoxicity at high concentrations and yet did not induce any of the genes (i.e. causing membrane damage). However, it was observed that α-tomatine diluted in a matrix of red-ripe tomato tissue (final extract concentration 5% w/v) elicited an increased cytotoxicity concomitant with a dose-dependent response profile involving XHF (collagenase involved in inflammatory reactions sensitive to mitogens and cytokine IL–1). So far, the approach of the CAT-Tox(L) model shows potential but there is a need for validation based on molecular mechanisms of plant constituents.

Although the induction of these stress genes represents cellular perturbations, the mechanistic validations need to be extended before any conclusions can be drawn on the value of these findings. Certainly, the development of a database to explore relationships between a wide variety of compounds stressful to plants and meaningful stress gene profiles may serve as molecular fingerprints unique for a particular component.

Gene expression profiling

A second potential application of DNA microarray technology in safety evaluations concerns screening for biological effects of (fractions of) plant compounds that have

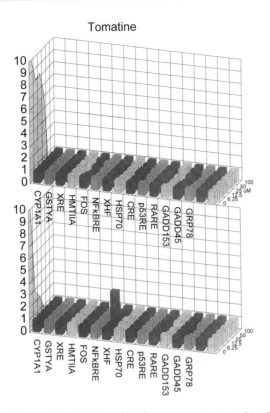

Figure 4.2   Typical profiles for α-tomatine tested in the CAT-Tox(L) assay. Upper panel, pure α-tomatine and lower panel, α-tomatine diluted in an aqueous extract of red-ripe tomato fruit. All recombinant cell lines in the CAT-Tox(L) assay were treated for 16–18 hours.

been found by other profiling techniques to be present or modified in the genetically modified crop plant when compared to the parental line and other references. At RIKILT-DLO we are testing the feasibility of this strategy by analysing defined food compounds for possible effects on human intestinal gene expression. To this end, human intestinal cell lines are exposed for different time periods to various amounts of crop plant components and the mRNA isolated. The labelled mRNA is hybridized to DNA microarrays containing unknown intestine-specific genes, as well as already identified control genes known to be intestine specific or to be functional indicators (for toxicological processes, apoptosis, etc.). As a source for unknown, intestine-specific genes we use cDNA libraries from human intestine biopsies and human intestinal cell lines which have been enriched by a subtractive cloning approach. Having established the usefulness of this approach for the identification of biological functions of defined plant compounds, these cDNA microarrays will be exploited to screen, for example, GM crop plants for possible effects on human health.

Summary and conclusions

The concept of substantial equivalence (OECD, 1993) is broadly accepted as a basis for the risk assessment of GM crops and derived novel foods. Data requirements for establishing substantial equivalence should be based on identifying unintended metabolic perturbations in the engineered crops that have been caused by genetic modification. A generic approach cannot be envisaged, however. In all cases the collation of data should be directed by the type of genetic modification and related consequences. It is recommended that unintended effects should be identified on a case-by-case basis, since such effects are dependent on the genotypic and phenotypic parameters of the new gene products involved.

The application of innovations in molecular genetics research will help to define the conditions under which the new food products can be marketed. The present approach has generated a sound scientific basis for the evaluation of the potential of unintended effects in GM crops and derived food products. The results may assist regulators and legislators to test and, if necessary, to improve the applicable regulations. Moreover, the technologies described and the results obtained may serve as a general framework for the risk assessment of GM crop plants, and may contribute to a better understanding by the public of recombinant DNA techniques in plant breeding and their implications. It is concluded that a parallel approach focusing on a comparative analysis using informative profiles of molecules at various integration levels (e.g. mRNA, protein, metabolites) including a toxicological profiling of relevant plant matrices, offers good prospects for the identification of hazards related to unintended effects. The choice of comparators and of external parameters to support a claim that a GM crop plant presents no unintended effects compared with its traditional counterpart should be based on sound scientific judgement. It is recommended that data are generated on the unprocessed part of the plant, such as the tomato fruit or potato tuber, used for human and/or animal consumption. Furthermore, data banks should be set up containing information on natural variations of essential plant or other product constituents (i.e. crop mean values).

The chemical fingerprinting technique, using a combination of off-line liquid chromatography and proton-NMR imaging, appears to be a powerful screening method for the detection of secondary effects in the metabolome that may be applied on a routine basis. The carbohydrate profiling technique is useful for the detection of unexpected changes at N-glycan levels; in particular it is applicable to the detection of post-translational modifications in newly expressed proteins as well as in the whole GM crop plant. DNA microarray technology is rapidly evolving and much effort is being put into improving spot density, reduced production time, and increased reproducibility and sensitivity. Improving the latter is critical because it makes it possible to use smaller amounts of starting material. It will depend on technical aspects such as the quality of the scanners and spotting machines, the development of fluorescent dyes with improved characteristics (e.g. narrow excitation and emission peaks; high level of photon emission; resistance to photo-bleaching), supports with reduced background and more target sequence binding capacity.

DNA microarrays generate a huge amount of complex hybridization data and major challenges are to develop more advanced computer software to help to find statistically significant correlations within, and between different experiments, and to link the data to sequence information and submitted expression profiles available in public databases. Application of these technologies in combination with other measurements of, for instance, the performance and quality may limit or even replace animal feeding studies aimed at the detection of unintended effects.

Together, these techniques, known as functional genomics, will enable us to answer questions about what happens to expression and protein composition when a plant cell changes metabolic state, for example, due to genetic modification or environmental factors. It is foreseen that this holistic view on the cellular machinery will be achieved soon, as the coincidence of genome sequencing with improvements in the analysis of expressed proteins is of growing importance and continues to be successful. Eventually, microarrays for genomic studies will probably also be used to evaluate the numerous constituents of GM crop plants. Some constituents might not cause obvious changes in cellular behaviour or morphologic appearance but could cause subtle metabolic alterations that would show up when the mRNA content was interrogated by an array. It is likely that microarrays might be used to evaluate the life cycle of a plant much more precisely and to understand the complex metabolic control systems of plants.

The ultimate test for GM crop plants will be acceptance by the public in large. It is therefore extremely important that industry, the scientific community and the regulatory authorities are able to respond to queries from the public by providing transparent information on the criteria for safety assessment. Improved safety assessment at the molecular level should refine and complement strategies by anticipating the potential risks and sources of unintended effects in GM plants.

References

Aalberse, R.C., Koshte, V. and Clemens, J.G.J. (1981) 'Immunoglobuline E antibodies that cross-react with vegetable foods, pollen and Hymenoptera venom', *Journal of Allergy and Clinical Immunology* 68: 356–64.

Adams, M.D., Kelley, J.M., Gocayne, J.D. *et al.* (1991) 'Complementary DNA sequencing: expressed sequence tags and human genome project', *Science* 252: 1651–56.

Altmann, F. (1998) 'Structures of the N-linked carbohydrate of ascorbic acid oxidase from zucchini', *Glycoconjugate Journal* 15: 79–82.

Alwine, J.C., Kemp, D.J. and Stark, G.R. (1977) 'Method for detection of specific RNAs in agarose gels by transfer to diazobenzyloxymethyl-paper and hybridization with DNA probes', *Proceedings of the National Academy of Sciences of the USA* 74: 5350–54.

Ames, B.N. (1983) 'Dietary carcinogens and anticarcinogens. Oxygen radicals and degenerative diseases', *Science* 221: 1256–64.

Beier, R.C. (1990) 'Natural pesticides and bioactive components in food', *Reviews of Environmental Contamination and Toxicology* 113: 47–137.

Berk, A.J. and Sharp, P.A. (1977) 'Sizing and mapping of early adenovirus mRNAs by gel electrophoresis of SI endonuclease-digested hybrids', *Cell* 12: 721–32.

Blackstock, W.P. and Weir, M.P. (1999) 'Proteomics: quantitative and physical mapping of cellular proteins', *TIBTECH* 17: 121–27.

Coulston, F. and Kolbye, A.C. (1990) 'Biotechnologies and food: assuring the safety of foods produced by genetic modification', *Regulatory Toxicology and Pharmacology* 12: S1–S196.

Degli-Agosti, R. and Greppin, H. (1998) 'Systemic stress effect on the sugar metabolism under photoperiodic constraint', *Archiv des Sciences* 51: 337–46.

De Silva, J. and Verhoeyen, M.E. (1998) 'Production and characterisation of antisense-exogalactanase tomatoes', in *Report of the Demonstration Programme on Food Safety Evaluation of Genetically Modified Foods as a Basis for Market Introduction.* The Hague: Information & Newssupply Department, Ministry of Economic Affairs, pp. 99–106.

Diatchenko, L., Lau, Y.F., Campbell, A.P. *et al.* (1996) 'Suppression subtractive hybridization: a method for generating differentially regulated or tissue-specific cDNA probes and libraries', *Proceedings of the National Academy of Sciences of the USA* **93**: 6025–30.

Estruch, J.J., Carozzi, N.B., Desai, N. *et al.* (1997) 'Transgenic plants: an emerging approach to pest control', *Nature Biotechnology* **15**: 137–41.

European Community (1997) 'Regulation (EC) No. 258/97 of the European Parliament and the Council', *Official Journal of the European Communities* **L43**: 1–7.

European Community (1998) Council Regulation 1139/98. *Official Journal of the European Communities* **L159**.

European Union guidelines 97/618/EC (1997) Commission Recommendation of 29 July 1997 concerning the scientific aspects and the presentation of information necessary to support applications for the placing on the market of novel foods and novel food ingredients and the preparation of initial assessment reports under Regulation (EC) No 258/97 of the European Parliament and of the Council. *Official Journal* **L253**.

FAO/WHO (1991) *Strategies for Assessing the Safety of Foods Produced by Biotechnology, Report of a Joint FAO/WHO Consultation.* Geneva: World Health Organization.

FAO (1996) *Biotechnology and Food Safety, FAO Food and Nutrition Paper 61, Report of a Consultation.* Rome: Food and Agriculture Organization of the United Nations.

Foetisch, K., Faeh, J., Wuethrich, B. *et al.* (1998) 'IgE antibodies specific for carbohydrates in a patient allergic to gum arabic (Acacia senegal)', *Allergy-Copenhagen* **53**: 1043–51.

Foxall, P.J.D., Lenz, E.M., Lindon, J.C. *et al.* (1996) 'Nuclear magnetic resonance and high-performance liquid chromatography nuclear magnetic resonance studies on the toxicity and metabolism of ifosamide', *Therapeutic Drug Monitoring* **18**: 498–505.

Fromm, M.E., Stark, D.M., Austin, G.D. *et al.* (1993) 'Improved agronomic and quality traits in transgenic crops: recent advances', *Philosophical Transactions of the Royal Society of London B Biological Sciences* **339**: 233–37.

Garcia-Casado, G., Sanchez-Monge, R., Chrispeels, M.J. *et al.* (1996) 'Role of complex asparagine-linked glycans in the allergenicity of plant glycoproteins', *Glycobiology* **6**: 471–77.

Graves, D.J. (1999) 'Powerful tools for genetic analysis come of age', *TIBTECH* **17**: 127–34.

Harvey, M.H., McMillan, M., Morgan, M.R. *et al.* (1985) 'Solanidine is present in sera of healthy individuals and in amounts dependent on their dietary potato consumption', *Human Toxicology* **4**: 187–94.

ILSI (1999) *Summary Report of a Workshop on Detection Methods for Novel Food derived from Genetically Modified Organisms.* Brussels: ILSI Europe Report Series, pp. 1–24.

Jenkins, N. and Curling, E.M.A. (1994) 'Glycosylation of recombinant proteins: problems and prospects', *Enzyme Microbial Technology* **16**: 354–64.

Kok, E.J., Keyer, J., Van Hoef, A.M.A. *et al.* (1998) 'mRNA fingerprinting of transgenic food crops', in *Report of the Demonstration Programme on Food Safety Evaluation of Genetically Modified Foods as a Basis for Market Introduction.* The Hague: Information & Newssupply Department, Ministry of Economic Affairs, pp. 37–49.

Korhammer, S.A. and Bernreuther, A. (1996) 'Hyphenation of high-performance liquid chromatography (HPLC) and other chromatographic techniques (SFC, GPC, GC, CE) with nuclear magnetic resonance (NMR): a review', *Fresenius' Journal of Analytical Chemistry* **354**: 131–35.

Kubelka, V., Altmann, F., Kornfeld, G. *et al.* (1994) 'Structures of the N-linked oligosaccharides of the membrane glycoproteins from three lepidopteran cell lines (Sf-21, IZD-Mb-0503, Bm-N)', *Archives of Biochemistry and Biophysics* **308**: 148–57.

Kuiper, H.A. and Noteborn, H.P.J.M. (1996) 'Food safety assessment of transgenic insect-resistant Bt tomatoes', in *Food Safety Evaluation*. Paris: OECD Documents, pp. 51–57.

Leiner, I.E. (1994) 'Implications of antinutritional components in soybean foods', *Critical Reviews in Food Science and Nutrition* 34: 31–67.

Liang, P. and Pardee, A.B. (1992) 'Differential display of eukaryotic messenger RNA by means of the polymerase chain reaction', *Science* 257: 967–71.

Lindon, J.C., Nicholson, J.K., Sidelmann, U.G. *et al.* (1997) 'Directly coupled HPLC-NMR and its application to drug metabolism', *Drug Metabolism Review* 29: 705–46.

Lommen, A., Weseman, J.W., Smith, G.O. *et al.* (1999) 'Automated comparison of complex matrices using NMR is a potentially powerful tool in detecting environmental influences', *Biodegradation* 9: 513–25.

Lutts, S., Majerus, V. and Kinet, J.M. (1999) 'NaCl effects on proline metabolism in rice (*Oryza sativa*) seedlings', *Physiologica-Plantarum* 105: 450–58.

Marshall, E. (1993) 'Toxicology goes molecular', *Science* 259: 1394–98.

Nordic Council (1998) *Safety Assessment of Novel Food Plants*. Copenhagen, Denmark: Nordic Council.

Noteborn, H.P.J.M. (1998) 'Chemical fingerprinting and in vitro toxicological profiling for the safety evaluation of transgenic food crops', in *Report of a Workshop on the Toxicological and Nutritional Testing of Novel Foods*. Paris: OECD Documents, pp. 32–33.

Noteborn, H.P.J.M. and Kuiper, H.A. (1995) 'Safety assessment of transgenic tomatoes expressing BT endotoxin', in *Application of the Principles of Substantial Equivalence to the Safety Evaluation of Plants Derived by Modern Biotechnology. Report of a WHO Workshop*. Geneva: WHO, Food Safety Unit, WHO/FNU/FOS/95.1, pp. 34–44.

Noteborn, H.P.J.M., Bienenmann-Ploum, M.E., van den Berg, J.H.J. *et al.* (1995) 'Safety assessment of the *Bacillus thuringiensis* insecticidal crystal protein CryIA(b) expressed in transgenic tomatoes', in K.-H. Engel, G.R. Takeoka and R. Teranishi (eds) *ACS Symposium Series 605*. Washington DC: ACS, pp. 134–47.

Noteborn, H.P.J.M., van der Jagt, R.C.M. and Rowland, I. (1997) 'Development of in vitro toxicological methods for the safety evaluation of transgenic food crops', *Developments in Animal and Veterinary Sciences* 27: 689–93.

Noteborn, H.P.J.M., Lommen, A., Weseman, J.M. *et al.* (1998) 'Chemical fingerprinting and in vitro toxicological profiling for the safety evaluation of transgenic food crops', in *Report of the Demonstration Programme on Food Safety Evaluation of Genetically Modified Foods as a Basis for Market Introduction*. The Hague: Information & Newssupply Department, Ministry of Economic Affairs, pp. 51–79.

Noteborn, H.P.J.M., Lommen, A., van der Jagt, R.C.M. *et al.* (2000) 'Chemical fingerprinting for the evaluation of unintended secondary metabolic changes in transgenic food crops', *Journal of Biotechnology* 77: 103–114.

Nuccio, M.L., Rhodes, D., McNeil, S.D. *et al.* (1999) 'Metabolic engineering of plants for osmotic stress resistance', *Current Opinion in Plant Biology* 2: 128–34.

OECD (1993) *Safety Evaluation of Foods Derived by Modern Biotechnology. Concepts and Principles*. Paris: Organisation for Economic Co-operation and Development.

OECD (1996) *Report of a Workshop on Food Safety Evaluation: Food Safety evaluation*. Paris: OECD.

OECD (1997) 'Draft Safety Assessment of New Foods: Results of an OECD Survey of Serum Banks for Allergenicity Testing and Use of Databases'. Paris: OECD.

OECD (1998) *Report of a Workshop on the Toxicological and Nutritional Testing of Novel Foods*. Paris: OECD.

Pauli, G.H. and Takeguchi, C.A. (1986) 'Irradiation of foods – an FDA perspective', *Foods Reviews International* 2: 79–107.

Petersen, A., Vieths, S., Aulepp, H. *et al.* (1996) 'Ubiquitous structures responsible for IgE cross-reactivity between tomato fruit and grass pollen allergens', *Journal of Allergy and Clinical Immunology* **98**: 805–15.

Prenner, C., Mach, L., Glössl, J. *et al.* (1992) 'The antigenicity of the carbohydrate moiety of an insect glycoprotein, honey-bee (Apis mellifera) venom phospholipase A-2: the role of alpha-1,3-fucosylation of the asparagine-bound N-acetylglucosamine', *Biochemical Journal* **284**: 377–80.

Sachs, L. (1984) *Angewandte Statistik: Anwendung statistischer Methoden*, 6th edition. Berlin, Heidelberg, New York: Springer-Verlag.

Schena, M., Shalon, D., Davis, R.W. *et al.* (1995) 'Quantitative monitoring of gene expression patterns with a complementary DNA microarray', *Science* **270**: 467–70.

Schena, M., Shalon, D., Heller, R. *et al.* (1996) 'Parallel human genome analysis: microarray-based expression monitoring of 1000 genes', *Proceedings of the National Academy of Sciences of the USA* **93**: 10614–19.

Shalon, D., Smith, S.J. and Brown, P.O. (1996) 'A DNA microarray system for analyzing complex DNA samples using two-color fluorescent probe hybridization', *Genome Research* **6**: 639–45.

Singh, N.P., McCoy, M.T., Tice, R.R. *et al.* (1988) 'A simple technique for quantitation of low levels of DNA damage in individual cells', *Experimental Cell Research* **175**: 184–91.

Staudacher, E., Dalik, T., Wawra, P. *et al.* (1995) 'Functional purification and characterization of a GDP-fucose: beta-N-acetylglucosamine (Fuc to Asn linked GlcNAc) alpha-1,3-fucosyltransferase from mung beans', *Glycoconjugate Journal* **12**: 780–86.

Sturm, A. (1995) 'N-glycosylation of plant proteins', in J. Montreuil, H. Schachter and J.F.G. Vliegenthart (eds), *Glycoproteins*. Amsterdam: Elsevier, p. 521.

Todd, M.D., Lee, M.J., Williams, J.L. *et al.* (1995) 'The Cat-Tox(L) assay: a sensitive and specific measure of stress-induced transcription in transformed human liver cells', *Fundamental and Applied Toxicology* **28**: 118–28.

Tretter, V., Altmann, F., Kubelka, V. *et al.* (1993) 'Fucose alpha-1,3-linked to the core region of glycoprotein N-glycans creates an important epitope for IgE from honeybee venom allergic individuals', *International Archives of Allergy and Immunology* **102**: 259–66.

Tsaftaris, A. (1996) 'The development of herbicide tolerant transgenic crops', *Field Crops Research* **45**: 115–23.

Velculescu, V.E., Zhang, L., Vogelstein, B. *et al.* (1995) 'Serial analysis of gene expression', *Science* **270**: 484–87.

Vieths, S., Schöning, B. and Petersen, A. (1994) 'Characterization of the 18-kDa apple allergen by two-dimensional immunoblotting and microsequencing', *International Archives of Allergy and Immunology* **104**: 399–404.

WHO (1991) '*Strategies for Assessing the Safety of Foods Produced by Biotechnology*'. Report of a Joint FAO/WHO Consultation. Geneva: WHO.

WHO (1995) *Application of the Principles of Substantial Equivalence to the Safety Evaluation of Plants Derived by Modern Biotechnology, Report of a WHO Workshop*. Geneva: WHO, Food Safety Unit, WHO/FNU/FOS/95.1.

Wilson, I.B.H., Harthill, J.E., Mullin, N.P. *et al.* (1998) 'Core alpha 1,3-fucose is a key part of the epitope recognized by antibodies reacting against plant N-linked oligosaccharides and is present in a wide variety of plant extracts', *Glycobiology* **8**: 651–61.

Chapter 5

Allergenicity of foods produced by genetic modification

Dean D. Metcalfe

Introduction

Immunologically mediated adverse reactions to components normally present in foods, and generally referred to as 'food allergens', are a significant health concern for both the pediatric and the adult population at risk. It is thus necessary to develop rational approaches to prevent or limit the spread of food allergens during a time when it has become technically possible to alter the genetic composition of foods derived from both plant and animal sources. To complete the argument that genetically modified foods must be considered for allergenic potential, refer to the identification of Brazil nut allergen in transgenic soybean (Nordlee *et al.*, 1996) or the ability of *Bacillus thuringiensis* pesticides to induce the production of specific IgE (Bernstein *et al.*, 1999).

The general pathway to protection appears clear: attempt to avoid the transfer of known significant allergens into other foods, i.e. try not to create new significantly allergenic foods. But are the issues this clear, and what are the difficulties and limitations in this approach? To understand the depth and breadth of this problem, it is first necessary to examine the spectrum and prevalence of food allergies, and where reasonable evidence exists as to the etiologic agents within the foods responsible. Given this database, it then becomes possible to examine what approaches may be used to reduce risk – and the limitations of these strategies.

Definitions

It is clear that a diversity of adverse food reactions exist – some of which are on an immunologic basis and are thus referred to as 'food allergies'. Following the classification of adverse food reactions as adopted by the European Academy of Allergy and Clinical Immunology (Bruijnzeel-Koomen *et al.*, 1995), such reactions may be divided into 'toxic' and 'non-toxic' groupings. Toxic reactions are defined as those that may develop in anyone who ingests a sufficient dose. Non-toxic reactions rely on person-to-person susceptibilities. These reactions may have at their basis either immune or non-immune mechanisms. Non-toxic reactions (Table 5.1) include those referred to as allergies or hypersensitivities. Adverse reactions from non-immune mechanisms, often referred to as intolerances, may be attributed to pharmacologic properties of the food and/or unique susceptibilities in the individuals affected.

Table 5.1 Non-toxic reactions to foods due to immune mechanisms

Disease	Target organs	Immune effector mechanism
I. Immediate		
Rhinoconjunctivitis	Eyes, upper respiratory tract	IgE: Basophils/mast cells
Oral allergy syndrome	Mouth	IgE: Mast cells
Urticaria/angioedema	Skin	IgE: Basophils/mast cells
Atopic dermatitis	Skin	IgE: Basophils/mast cells, eosinophils
Asthma	Lower respiratory tract	IgE: Mast cells, lymphocytes, eosinophils
Gastrointestinal reactions	GI mucosa	IgE: Mast cells, eosinophils
Systemic anaphylaxis	Skin, respiratory tract, GI tract, cardiovascular system	IgE: Basophils/mast cells
II. Delayed		
Allergic eosinophilic gastroenteritis	GI mucosa, submucosa	IgE: Mast cells, eosinophils lymphocytes
Food-induced colitis/enterocolitis	GI mucosa	IgA: Lymphocytes, mast cells
Celiac disease	GI mucosa	IgA?: Lymphocytes
Dermatitis herpetiformis	Skin	IgA,C3: Neutrophils

Demographics

Prevalence data relating to food allergies are somewhat limited. In a study of Danish infants, the prevalence of cow's milk allergy was found to be 2.2% (Host and Halken, 1990). In another study, newborns were followed through to their third birthday. Approximately 4% were thought to have IgE-mediated food allergies (Bock, 1987). Similar studies of food allergy are unusual in adult populations. One study in the Netherlands, which was based on questionnaires, clinical follow-up and double-blind placebo-controlled food challenge, estimated the prevalence of food allergy and intolerance together to be 2.4% (Nietizl-Jansen *et al.*, 1994). The prevalence of peanut and tree nut allergy in the US as determined in a random digit dial telephone survey was found to be 1.1% of the general population (Sicherer *et al.*, 1999). Such data gives credence to the conclusion that food allergies are a major health issue worldwide.

Types of reactions

Immediate reaction

IgE-mediated immediate reactions are the basis for the majority of allergic reactions to food, and may result in death from anaphylaxis. Thus, much of the concern that surrounds food allergies and their consequences relate to patients in this clinical category. These responses follow the release of chemical mediators of inflammation from mast cells and basophils following the interaction between food-specific IgE and a specific food allergen on the surfaces of these effector cells. Patients with food allergies appear to represent a subgroup of atopic individuals who have allergies in general and more difficulty regulating IgE levels in response to environmental antigens (Atkins *et al.*, 1985; Fiorini *et al.*, 1990).

Immediate hypersensitivity reactions to food antigens are evidenced by a spectrum of clinical findings from eczema to anaphylaxis (Table 5.1). The oropharynx is the initial site of exposure to food antigens. Edema and pruritus of the lips, oral mucosa and pharynx may be reported as the food contacts the mucosal surfaces. The term oral allergy syndrome is applied to the clinical situation dominated by oropharyngeal symptoms. This is most commonly associated with the ingestion of fruits or vegetables. It has been suggested that the observation that such foods as apple, which cross-reacts with birch, or watermelon and cantaloupe which cross-react with ragweed, relates to a relationship between the oral allergy syndrome and allergic rhinitis. Entry of food into the upper gastrointestinal tract may result in nausea, cramping, pain, abdominal distension, vomiting, flatulence and diarrhea. Symptoms of gastrointestinal involvement may be the only expression of food hypersensitivity, but this is unusual.

Food allergy is usually also expressed in one or more extra-intestinal target tissues. The skin is a common target organ. Reactions include acute urticaria, acute angioedema, atopic dermatitis and less frequently chronic urticaria. It has been estimated that clinically significant food hypersensitivity exists in approximately a third of children with atopic dermatitis (Burks et al., 1988). Asthma and rhinitis secondary to food hypersensitivity are more common in children (Bock et al., 1978) than in adults (Van Metre et al., 1968).

Systemic anaphylaxis associated with allergy to ingested foods generally occurs within 1–30 minutes after ingestion of the offending food. The first anaphylactic episode may be unexpected or may be preceded by prior symptoms, such as abdominal discomfort or urticaria, on previous exposure to the food. Anaphylaxis may include tongue itching and swelling, palatal itching, throat itching and tightness, wheezing and cyanosis, chest pain, urticaria, angioedema, abdominal pain, vomiting, diarrhea, hypotension and shock. Severe life-threatening reactions are most often associated with the ingestion of peanuts, nuts and seeds. Fatal reactions may progress rapidly or begin with mild symptoms and then evolve to cardiorespiratory arrest and shock over hours. Systemic anaphylaxis has also been reported after ingestion of food followed by exercise.

Laboratory procedures used to help diagnosis of immediate food hypersensitivity involve the identification of antigen-specific IgE to allergens within suspected foods. In the case of skin testing, the IgE examined is fixed to skin mast cells. In vitro tests identify antigen-specific IgE in serum. Allergic reactions to foods are unusual in the face of negative tests (false negatives). Patients should never be advised that they are allergic to certain foods solely on the basis of positive tests. This is because tests may also be positive to foods in the absence of symptomatic food allergy (false-positive tests). In cases of the oral allergy syndrome, the use of extracts of fresh fruits and vegetables is often necessary to exclude IgE-mediated food hypersensitivity.

The evaluation of an IgE-mediated food allergy utilizes the medical history, physical examination and relevant laboratory studies. The diagnosis in some instances must be confirmed by blinded food challenge (Bock et al., 1988; Metcalfe and Sampson 1990). When evaluating an adverse reaction to food, a number of other diseases, anatomic defects and reactions to additives, toxins and contaminants that may in some way mimic an allergic reaction must be considered and eliminated (Table 5.2).

Enzyme deficiencies both mimic or may complicate gastrointestinal inflammatory diseases. Abdominal cramping, bloating, and diarrhea accompany the ingestion of

Table 5.2 Considerations in the diagnosis of food allergy

I.   Enzyme deficiencies
     Disaccharidase deficiency (lactase etc.)
     Galactosemia
     Phenylketonuria
II. Gastrointestinal disease
     Developmental/structural (pyloric stenosis etc.)
     Peptic ulcer
     Gallbladder disease
     Post-surgical dumping syndrome
     Neoplasia
     Inflammatory bowel disease
     Pancreatic insufficiency
III. Endogenous food components
     A. Dyes
         Tartrazine
     B. Flavorings and preservatives
         Monosodium glutamate
         Sulfiting agents
         Nitrates and nitrites
     C. Pharmacologic substances
         Caffeine
         Tyramine
         Phenylethylamine
         Theobromine
         Tryptamine
         Alcohol
         Histamine
IV. Toxins
     A. Bacterial toxins
         Botulism
         Staphyloccoccal toxin
     B. Endogenous toxins
         Certain mushrooms (alpha-amanitine)
         Shellfish (saxitoxin)
         Ichthyotoxin
     C. Fungal
         Aflatoxin
         Ergot
V.   Psychological responses
     A. Bulimia
     B. Anorexia nervosa

milk and milk products in individuals with lactase deficiency. Deficiencies such as galactose-4-epimerase (galactosemia) may be diagnosed in infancy as a result of vomiting and diarrhea after milk ingestion. Cystic fibrosis may initially be confused with a food-induced malabsorption syndrome because of associated pancreatic enzyme deficiency. Chronic cough and wheezing secondary to aspiration may occur with hiatal hernia, pyloric stenosis and an H-type tracheoesophageal fistula. Overfeeding and chalasia are estimated to occur in up to 50% of newborns and result in vomiting associated with feeding. Abdominal pain following meals may be due to peptic ulcer disease or cholelithiasis.

Drugs, dyes, additives, bacteria and bacterial products may be present in foods. Tartrazine yellow (FD&C Yellow Dye No. 5) is a rare cause of hives. Sulfiting agents used to reduce spoilage of foods, to inhibit undesirable microorganisms during fermentation, to sanitize food containers and to prevent oxidative discoloration of foods induce problems reminiscent of allergic diseases including bronchospasm (Taylor et al., 1997). Monosodium glutamate in sufficient quantity (usually greater than 6 g) may lead to a transient syndrome consisting of a warmth or burning sensation over the head and shoulders, stiffness or tightness, extremity weakness, pressure, tingling, headache, light-headedness and gastric discomfort occurring approximately 15 minutes after ingestion.

Toxins may induce signs and symptoms resembling allergic reactions. In scrombroid poisoning, ingestion of fish containing high levels of histamine is followed by symptoms including diffuse erythema and headaches. Scrombroid fish commonly implicated include tuna, skipjack and mackerel (Saavedra-Delgado and Metcalfe, 1993). Ciguatera poisoning is seen especially in the Caribbean and Pacific islands. Symptoms include tingling of the lips, tongue and throat, followed by nausea, vomiting, diarrhea, headache, chills and myalgias. Paralytic shellfish poisoning is caused by ingestion of bivalve mollusks contaminated with dinoflagellates of the genus Gonyaulax which produce neurotoxins. Amnesic shellfish poisoning is an acute illness characterized by gastrointestinal symptoms, seizures, coma, disorientation and loss of memory following the ingestion of mussels. This disease is due to domoic acid, a potent neurotoxin produced by the bloom of the pennate phytoplanktonic diatom Nitzia pungens.

The only proven therapy for food allergy is strict elimination of the offending allergens. Severe elimination diets should only be instituted with nutritional guidance. Patients and parents must learn to read and understand food labels. No appropriately designed trial has demonstrated clear efficacy and practicality for the use of prophylactic medications, injection immunotherapy or oral desensitization in the prevention of allergic reactions to foods.

Food hypersensitivity reactions occurring in infants may be delayed, and in some cases avoided by breast feeding. Many infants eventually become tolerant to the foods provoking reactions. In prospective studies of adverse food reactions in infants, 80–87% of confirmed symptoms were no longer observed by three years of age (Bock, 1987). Older children (Sampson and Scanlon, 1989), and even some adults (Pastorello et al., 1989) may lose their sensitivity if the responsible food allergen is completely eliminated from the diet. Thus, after one or two years of allergen avoidance, up to one-third of children and adults lose their clinical sensitivity. Patients with an allergy to peanut, tree nut, fish or shellfish rarely lose their clinical reactivity (Bock and Atkins, 1989). Loss of sensitivity correlates with allergen avoidance, but whether reintroduction and repeated exposure to the same allergen will cause the sensitivity to reappear is unknown.

A patient may sometimes inadvertently consume a food to which he or she is sensitive. Treatment for a specific symptom which results from inadvertent exposure is the same as that employed when other factors provoke symptoms. Thus, laryngeal or pulmonary symptoms following an inadvertent food exposure should be treated immediately with epinephrine or bronchodilator therapy or both. The treatment of food-induced anaphylaxis is essentially the same as that for anaphylaxis due to a medication or insect sting. A patient with potential anaphylactic reactivity should be taught how to self-administer epinephrine, and have an epinephrine-containing syringe and an antihistamine available at all times. For children, daycare centers and schools should have a list of emergency numbers with backups to be called. It should be remembered that a patient may exhibit only mild symptoms in the first few minutes after ingesting a food to which he or she is allergic, but this may be followed 10–60 minutes later by hypotension and other severe problems. Following self-medication for systemic reactions, the patient should immediately seek medical attention. All patients with IgE-mediated food allergy should be warned about the possibility of developing a severe anaphylactic reaction and should be educated in the appropriate treatment measures to be taken in case of an accidental ingestion (Smith and Munoz-Furlong, 1997).

Delayed reaction

Food protein-induced gastroenteropathy

Food protein-mediated gastroenteropathy appears almost exclusively to be a disease of infants and children. Its basis is a hypersensitivity reaction to a food protein which results in damage to the intestinal mucosa. Food protein gastroenteropathy may be induced by cow's milk protein (Lake, 1997), soy protein, egg, fish, rice or chicken. The basis of food protein-mediated gastroenteropathy is unknown, although a cell-mediated delayed hypersensitivity mechanism has been suggested. Total serum IgE is often normal, and IgE specific to the inciting protein is usually absent in affected children.

The predominant symptoms of food protein-mediated gastroenteropathy are vomiting, diarrhea, malabsorption and gross or occult stool blood loss. Symptoms usually appear within weeks of introducing cow's milk or other proteins into the diet. Carbohydrate malabsorption may be present secondary to an intestinal mucosal injury. Eosinophilia may be present in some patients.

The diagnosis of protein hypersensitivity requires the demonstration of a relationship between ingestion of a given protein and symptomatology consistent with intestinal damage. In children exhibiting diarrhea and malabsorption, biopsy of the small intestine usually reveals villous atrophy, primarily in the jejunum. In some children only the colon is involved, manifesting as a colitis (Lake, 1997).

Treatment is dependent on removal of the sensitizing protein from the child's diet. An elemental formula or formulation or total parenteral nutritional may be required to allow for repair and return of normal function of the gastrointestinal mucosa. Reintroduction of the offending protein may be attempted, since protein-mediated gastroenteropathy typically resolves by 18–24 months of age.

Eosinophilic gastroenteritis

Eosinophilic gastroenteritis is characterized by eosinophilic infiltration of the gastrointestinal wall, peripheral eosinophilia and gastrointestinal symptoms. The sites of involvement include the esophagus, stomach, small intestine, colon and, rarely, extra-intestinal organs. Approximately half of the cases have allergic features and may be related to food allergy. It may occur at any age, although the peak age of onset is in the third decade. Most of those with food-dependent disease are under the age of 20 years (Thounce and Tanner, 1985).

The cause of food-induced eosinophilic gastroenteritis appears to be related to an IgE-dependent, mast cell-mediated mechanism. Atopic diseases, such as eczema, allergic rhinitis, bronchial asthma, urticaria and a positive family background for allergy are common. Many patients also have peripheral eosinophilia, an elevated serum IgE level and positive RASTs for specific IgE antibodies to food antigens.

All patients with eosinophilic gastroenteritis, however, are not atopic, and all cases cannot be explained by food allergy. Only approximately half of the patients with eosinophilic gastroenteritis have findings consistent with atopy. Many patients show no personal or family history of allergy, no positive skin tests for food allergens, no elevation in serum IgE, and no adverse reactions to foods. Even in patients who have suspected food allergies, sequential withdrawal of various food substances may fail to provide amelioration of symptoms, and there may be a poor correlation between results of skin tests to specific food antigens and the results of an elimination diet. In addition, most patients show no abnormality following extensive immunologic studies, including serum IgE, IgM and IgA levels; complement levels; lymphocyte quantification; and lymphocyte responses to non-specific mitogens.

Patients with eosinophilic gastroenteritis experience nausea with vomiting, abdominal pain, diarrhea, steatorrhea and either weight loss in the adult or growth failure in the child. Radiologic findings may reveal mucosal edema and nodularity of the folds of the small bowel. Muscular disease can precipitate small bowel obstruction.

Significant mucosal disease typically presents with evidence of iron-deficiency anemia, hypoalbuminemia and hypogammaglobulinemia secondary to a protein-losing gastroenteropathy, steatorrhea and an eosinophilic leukocytosis. The mucosal form of eosinophilic gastroenteritis may be confused with celiac disease, regional enteritis, neoplasms (lymphoma), polyarteritis nodosa, parasites, the hypereosinophilic syndrome or another protein-losing gastroenteropathy. The diagnosis of eosinophilic gastroenteritis is established with a gastrointestinal biopsy demonstrating an eosinophilic infiltration of the gastrointestinal wall.

Treatment of eosinophilic gastroenteritis is often unsatisfactory. Food hypersensitivity should be ruled out as the precipitating or exacerbating factor. A trial diet that eliminates multiple food antigens from the diet in a systematic order may be performed when no clear history or evidence of IgE-mediated hypersensitivity is elicited. In patients with eosinophilic gastroenteritis and food sensitivity, the number of foods involved may preclude the long-term use of an elimination diet for symptom control. Both patients who responded poorly to dietary restrictions and those without evidence of food hypersensitivity may require oral steroid therapy (Min and Metcalfe, 1997).

Gluten-sensitive enteropathy (celiac disease)

Gluten-sensitive enteropathy is a mucosal disease of the small intestine caused by the alcohol-soluble portion of gluten (gliadin) in susceptible individuals. The HLA-DR3DQ2 haplotype has the strongest disease associations (Sollid et al., 1989). Symptom onset typically occurs 6–12 months after introduction of gluten into the diet. Complaints include intermittent diarrhea, abdominal pain and irritability. Extensive mucosal injury may result in malabsorption with a clinical picture including stea-torrhea, peripheral edema from protein loss, anemia, bleeding diathesis, tetany and growth failure. An increase in the incidence of gastrointestinal lymphoma is reported. The acute reaction of the intestinal mucosa consists of edema, an increase in vascular permeability, and an eosinophil and neutrophil infiltration. Gliadin has been separated by gel electrophoresis into four proline and glutamine-rich fractions, each of which precipitates small bowel injury *in vitro*. Blunting of the mucosal surface, villous atrophy and a dense infiltration of the lamina propria with plasma cells, B cells and T cells is observed in chronic disease.

The diagnosis is dependent on demonstrating biopsy evidence of small intestinal mucosa injury upon gluten challenge. Treatment is directed at elimination of gluten from the diet. Wheat, barley, rye and oats contain gluten. Improvement in symptoms is seen as soon as two weeks after the institution of a gluten-free diet. Histologic improvement may take 2–3 months. Growth usually returns to normal once the small intestinal mucosa has healed. Strict gastrointestinal rest, and in some instances the use of steroids, is necessary to suppress diarrhea when inflammation is severe.

Dermatitis herpetiformis

Dermatitis herpetiformis is a chronic papulovesicular skin disorder, often associated with asymptomatic gluten-sensitive enteropathy. The histologic appearance of skin lesions is one of a granulocytic infiltration at the dermoepidermal junction associated with edema and blister formation. Granular IgA deposits with associated J chains are found in the papillary dermis. Complement-mediated injury is implicated. The histology of intestinal lesions is similar to celiac disease though usually less severe.

Skin lesions are symmetrically distributed on extensor surfaces of elbows, knees and buttocks. Most patients have little or no gastrointestinal complaints. The diagnosis rests on the typical appearance of skin lesions and histologic findings of IgA deposits in the perilesional or uninvolved skin. Fifteen per cent of patients will have a normal small intestinal mucosa on histologic evaluation. Treatment consists of the removal of gluten from the diet and the use of dapsone or sulfapyridine.

Respiratory disease

Asthma due to ingested foods and as part of a systemic IgE-mediated reaction is well recognized. Asthma induced by inhaled aerosolized food components, and with isolated organ involvement also occurs. It is encountered most frequently as 'occupational asthma'. Asthma defined as reversible airway obstruction must be separated from hypersensitivity pneumonitis (HP). HP is characterized by a diffuse, predominantly mononuclear cell infiltrate in the lung parenchyma, which may be followed by fibrosis. This disease also occurs after exposure to organic dusts, is principally an occupational disease, and is not IgE-mediated.

Similar foods (aerosolized during cooking or grinding) may cause both asthma or HP (Table 5.3) (O'Neil and Lehrer, 1997). Therapy is directed at identifying the etiologic agent and practicing avoidance. In some cases this is as direct as improving ventilation. Failure to identify HP in a timely fashion may result in permanent pulmonary function abnormalities.

Table 5.3 Examples of foods associated with occupational respiratory diseases

A.   Asthma
    Crustaceae
    Fish meal flour
    Dairy products
    Poultry/eggs
    Flour (wheat, rye, buckwheat, carob bean, soybean)
    Spices (garlic, coriander, ginger, paprika, etc.)
    Cinnamon
    Vegetables (beans, okra)
    Coffee/teas

B.   Hypersensitivity pneumonitis
    Poultry proteins
    Fish meal
    Tea plants
    Spices (paprika, curry, garlic, onions, mustard)
    Nuts (cashews)
    Seafoods (crustacea, oysters, fish)

Contact reactions

Contact dermatitis consists of a rash resulting from a substance touching the skin. Such reactions may be irritant in nature, result from IgE-dependent allergic mechanisms, or be related to photoallergic and delayed hypersensitivity (T cell-mediated eczematous disease). Contact urticaria is a term specifically related to skin disease due to IgE-dependent, immediate, mast cell-related mechanisms. Occupational eczema has been reported in association with a number of foods including fish, meat, vegetables, crustaceae, celery and spices. Inciting substances are identified by history and patch testing. Treatment is avoidance.

Allergens

The specificity of IgE-mediated reactions as defined by the identification of allergen-specific IgE, the demonstration that allergens interacting with allergen-specific IgEs on the surface of mast cells and basophils initiate inflammatory reactions, and the association of these observations provoked *in vivo* in those with immediate reactions to foods to clinical disease, focused attempts at risk protection on such reactions. The evidence is thus persuasive that 'food allergens' as defined by specific IgEs induce the immediate clinical reactions to foods including anaphylaxis, asthma, hives, angioedema and some contact reactions. Indeed, this approach covers all of the immediate reactions in Table 5.1, the immediate component of allergic eosinophilic gastroenteritis, and occupational asthma (Tables 5.1 and 5.3).

Unfortunately, the components in foods responsible for food-induced enterocolitic syndromes are poorly characterized, and thus the strategy of preventing spread of known antigens cannot be relied on to protect children with these disorders, although they constitute a small subset of those that experience adverse reactions to foods. Fortunately, the components within certain foods that provoke celiac disease and dermatitis herpetiformis (gliadins) are known. Advisory groups have uniformly recommended that gliadins not be transferred during genetic modification of foods (FAO, 1995; FAO/WHO, 1996), seeming to place this issue at rest.

Thus, the focus in protection has been on allergens and their characteristics. And while the general focus has been on food allergens, there is general consensus that inhalant allergens, usually not consumed to any degree, should also be considered in protection approaches, even though they are often not in the edible portion of the plant, and knowing that allergens in fruits and vegetables causing the oral allergy syndrome may cross-react with certain inhalant allergens.

Food allergens in general are water-soluble glycoproteins that range in size from 10 to 60 kDa. For poorly understood reasons, they tend to be stable to treatment with heat, acid and proteases (Astwood *et al.*, 1996). It has been suggested that this is indirect evidence for the conclusion that the allergenic potential of food allergens as defined by IgE binding (B cell epitope) lies in linear, rather than conformational (continuous or discontinuous) epitopes. This is in apposition to the general thought that T cell epitopes are linear, whereas B cell epitopes are conformational (Lehrer *et al.*, 1996).

Allergenic foods for the most part contain multiple significant allergens. Major allergens are usually defined by the observation that more than 50% of patients sensitive to that food react with these proteins. While virtually any food may cause an allergic reaction in someone, somewhere, and extensive surveys of such occurrences have been published (Hefle *et al.*, 1996), a relatively small number of foods appear to cause over 90% of all reported immediate reactions in both children and adults. These foods are listed in Table 5.4, along with examples of major allergens

Table 5.4 Major allergenic foods and examples of major food allergens

Food	Allergen source	Allergen
Milk	*Bos domesticus* (cattle/milk)	Bos d 4; α-lactalbumin
		Bos d 5; β-lactoglobulin
		Bos d 6; serum albumin
		Bos d 7; immunoglobulin
		Bos d 8; caseins
Soy	*Glycine max* (soybean)	Coly m1A; HPS
		Coly m1B; HPS
Peanuts	*Arachis hypogaea* (peanut)	Arah 1 vicilin
		Arah 2 conglutinin
Tree nuts	*Bertholletia excelsa* (Brazil nut)	
	Juglans regia (walnut)	
	Pistacia vera (pistachio)	
Crustaceae	*Penaeus indicus* (Indian shrimp)	Pen i 1; tropomyosin
Fish	*Gadus callarias* (cod)	Gad c 1; allergen M
	Salmo salar (salmon)	Sal s 1; parvalbumin
Egg	*Gallus domesticus* (hen)	Gal d 1; ovomucoid
		Gal d 2; ovalbumin
		Gal d 3; conalbumin
		Gal d 4; lysozyme

within these foods. These allergens are named according to the accepted taxonomic designation, using the first three letters of the genus and the first letter of the species with an Arabic number assigned in the order of their designation (the same number is used to designate homologous antigens). In infants less than six months of age, the majority of allergic reactions are due to milk or soy. In adults, the most common food allergens are peanuts, tree nuts, crustaceae (Shanti *et al.*, 1993), fish and egg (FAO, 1995). It has been recommended that sesame seed and wheat be added to this list (Bousquet *et al.*, 1998). It has also been suggested that members of the Prunoideae subfamily (peach, plum, apricot, cherry, almond), celery and rice may have to be added to the list (Bousquet *et al.*, 1998).

Risk protection

Given that protection strategies focus on the avoidance of the creation of new and significant allergenic foods as defined by alterations or additions to the genome of a plant or animal that would result in the synthesis by that organism of a newly expressed allergen, there are at least three major considerations. First, how to avoid transferring a known major allergen from one source to another. Second, what strategies may be employed to identify whether a gene selected for transfer codes for a protein with significant allergenic potential where there is no previous significant population exposure to that protein? Third, how to monitor modified foods to see if modifications led to the expression of new allergens, or upregulated the expression of existing allergens.

Individuals and groups reviewing these issues have generally identified four technical approaches that may be used to determine if a modified food may be or is enhanced in its allergenic potential (Table 5.5) (FAO/WHO, 1991; Miller 1993; Metcalfe *et al.*, 1996; FAO/WHO, 1996; Lehrer *et al.*, 1996; Kimber *et al.*, 1997; Van Dam and de Vriend, 1999). It is also clear that each approach has limitations (Wal and Pascal, 1998). However, without use of available technology to help prevent the marketing of modified foods with substantial allergenic potential, consideration for food labeling (Miller, 1999) may be appropriate. That is, labeling in the US is required 'if a food derived from a new plant variety differs from its traditional counterpart such that the common or usual name no longer applies, or if a safety or usage issue exists to which consumers must be alerted' (Federal Register, 1992). These issues are not lost on the general public, where allergenicity of modified foods is one concern among many worldwide (Enserink, 1999; Ferber, 1999; Gaskell *et al.*, 1999; Gavaghan, 1999) dating back a number of years (Gannes, 1993).

An examination of the amino acid sequence coded for by a transferred gene is recommended both in the situation where the gene transferred is from a known allergenic source and for instances where the gene transferred codes for a protein whose allergenic potential is largely unknown (Table 5.5). This latter instance would be expected if the gene was obtained from an animal or plant source rarely consumed by the general population. The search for sequence similarity should include a review of similarity of allergens from both plant- and animal-derived foods, and inhalant allergens such as pollens, fungal spores, insect venoms and allergens which provoke immediate contact reactions. The chief difficulty in this approach is to define the number of contiguous amino acids required to create concern. Because the optimal

Table 5.5 Technical approaches to address major questions concerning the possible allergenicity of modified foods

Questions	Technical approaches			
	Sequence similarity[1]	Reaction to specific IgE[2]	Stability to digestion[3]	Allergenicity in an animal model[4]
I. Does the gene transferred code for a known allergen? (source is a common allergenic food)	X	X		
II. Does the gene transferred code for a protein that may be allergenic?	X		X	X
III. Did the genetic manipulation lead to the expression or upregulation of an allergen?				X

[1] For both T cell and B cell epitopes.
[2] Serum is obtained from individuals allergic to the food which is the source of the genetic material transferred (if applicable).
[3] Includes stability to acid and proteases.
[4] Animal model must have predictive value.

peptide length for binding appears to be between 8 and 12 amino acids for T cell epitopes and even longer for B cell epitopes, some have suggested that the sequence identity requires a match of at least 8 contiguous identical amino acids. There are certainly limitations in this approach. For example, not all food allergens have been sequenced, and this strategy will not identify confirmational or non-contiguous epitopes (Metcalfe *et al.*, 1996).

For a gene transferred from a known allergenic food or an allergenic food of concern, sequence similarity studies are strengthened by the determination as to whether a serum from individuals sensitive to the food of origin contains allergen-specific IgE which will bind with the gene product. These assays are usually performed as solid-phase immunoassays such as the RAST or RAST inhibition assay or the enzyme-linked immunosorbent assay (ELISA). As a general procedure, this screening strategy for the transfer of a gene coding for an allergen gains strength when the number of sera from individuals sensitive to the food of origin is increased in number. Because *in vivo* skin prick tests have been held to be somewhat more sensitive than *in vitro* test results, some have suggested that if the *in vitro* tests are negative, skin testing should be performed. Finally, double-blind placebo-controlled food challenges could be performed in controlled clinical conditions with patients sensitive to the food in question if the *in vivo* and *in vitro* tests as above are negative or equivocal. This latter consideration is one which must address a number of ethical issues, including the possibility of inducing anaphylaxis in test subjects, the availability of appropriate procedures, the availability of clinical safety data, utilization of institutional review board procedures, instructions to the subject to be tested, and what steps may be used to decrease risks to participants.

If the gene transferred is derived from a source where there is no general history of its use in the population and thus its allergenic potential cannot be assessed, the reactivity to specific IgEs from sensitive individuals cannot be generally applied. Thus in this situation, in addition to sequence similarity, stability to digestion may be employed and the issue of the use of an animal model to determine allergenicity becomes an issue (Table 5.5). As has been discussed previously and as a general prin- cipal, food allergens appear to be resistant to acids and proteases. It has thus been argued that stability of food allergens to digestion may be used to help identify the products of transferred genes as to their possible allergenicity (Astwood *et al.*, 1996). However, insufficient information is available on possible differences in susceptib- ility to acid-denaturation and susceptibility to proteases between various allergenic food proteins and with proteins that possess weak or no allergenic potential to use this test in any other way but a relative indication of allergenic potential (Houben *et al.*, 1997).

Any discussion of technical approaches to address the possible allergenicity of modified foods must consider animal models and their use in the determination of food protein allergenicity. Such an approach would help identify as to whether a gene product transferred from a protein where there is no history of exposure of that protein in the general population, is allergenic (Table 5.5). Perhaps more import- antly, a screening test to determine the allergenic potential of a modified food using an animal model would be ideal to address the question as to whether the genetic manipulation led to the expression or upregulation of an allergen in the recipient organism. It is, of course, possible to make animals of various species allergic to specific proteins. The usual approach is to inject the protein in an adjuvant using the intraperitoneal route. For instance, the most commonly applied mouse model of asthma employs ovalbumin as its antigen, and the protein is administered intraperitoneally in alum. The difficulty arises in using sensitization by this means as a test for allergenicity, since it may be that both allergenic and non-allergenic proteins applied in this manner will provoke antigen-specific IgE, as determined by pulmonary or intradermal challenge, or using specific *in vitro* diagnostic assays for antigen-specific IgE. Indeed, it may be too much to expect any one animal model to predict allergenicity in humans of a specific food. The differences between immune responses in an inbred population and in an outbred population are obvious. How- ever, it may be reasonable to expect, with careful selection of the study animal, and the route and frequency of administration of the antigen, that an animal model might be developed that provides some ranking of allergenicity. That is, in such a model it would be expected that injection of an allergen such as Pen i 1 (Table 5.4) would provoke a stronger antigen-specific IgE response than would injection of a protein with low allergenic potential. Reviews of the application of animal models to the identification of food proteins with allergenic potential (Houben *et al.*, 1997) look to the time when a specific animal model may prove of value in the evaluation of allergenicity. In the absence of a widely accepted animal model, it may be possible on a case by case basis to examine allergenicity of genetically modified foods, comparing the response in a specific animal model with the response that the animal exhibits to a non-allergenic food.

Summary

In the genetic modification of foods directed at increasing crop production, modifying foods to increase their health benefits (Zeisel, 1999), decreasing a specific food's allergenicity, or increasing palatability, caution must be taken to avoid creating a new and significant allergenic food. Those with food allergies mediated by IgE and immediate in onset should reasonably expect that application of technical approaches to monitor allergenicity will be employed in an attempt to identify and prevent the marketing of modified foods with significant allergenic potential. However, because almost any food may be allergenic in one or a very few individuals (see above) it is not reasonable to expect that modified foods will be absolutely and consistently without allergenic potential in everyone. It is reasonable to expect that the technical approaches available (Table 5.5) will help prevent the marketing of a modified food with significant allergenic potential. Clearly, the modification of foods offers great promise in addressing worldwide issues such as increasing world grain production (Mann, 1997), adding specific nutrients to the diet (Friedrich, 1999), in producing edible vaccines (Haq et al., 1995; Arntzen, 1998; Tacket et al., 1998; Mason et al., 2002) and in reducing the allergenicity of existing allergenic foods (Astwood et al., 1997). These beneficial aspects of technology applied to food modification should not necessarily be limited by concerns over allergenicity of modified foods, given the technical approaches available to monitor and help prevent the creation of new major allergenic foods.

References

Arntzen, C.J. (1998) 'Pharmaceutical foodstuffs – oral immunization with transgenic plants', *Nature Medicine* 4: 502–3.

Astwood, J.D., Leach, J.N. and Fuchs, R.L. (1996) 'Stability of food allergens to digestion in vitro', *Nature Biotechnology* 14: 1269–73.

Astwood, J.D., Fuchs, R.L. and Lavrik, P.B. (1997) 'Food biotechnology and genetic engineering', in D.D. Metcalfe, H.A. Sampson and R.A. Simon (eds) *Food Allergy: Adverse Reactions to Foods and Food Additives*, 2nd edition. Cambridge: Blackwell Science, pp. 65–92.

Atkins, F.M., Steinberg, S.S. and Metcalfe, D.D. (1985) 'Evaluation of immediate adverse reactions to foods in adults. I. Correlation of demographic, laboratory, and prick skin test data with response to controlled oral food challenge', *Journal of Allergy and Clinical Immunology* 69: 348–55.

Bernstein, I.L., Bernstein, J.A., Miller, M. *et al.* (1999) 'Immune responses in farm workers after exposure to bacillus thuringiensis pesticides', *Environmental Health Perspectives* 107: 575–82.

Bock, S.A. (1987) 'Prospective appraisal of complaints of adverse reactions to foods in children during the first 3 years of life', *Pediatrics* 79: 863–88.

Bock, S.A. and Atkins, F.M. (1989) 'The natural history of peanut sensitivity', *Journal of Allergy and Clinical Immunology* 83: 900.

Bock, S.A., Lee, Y., Remigio, L.K. *et al.* (1978) 'Studies of hypersensitivity reactions to foods in infants and children', *Journal of Allergy and Clinical Immunology* 62: 327–34.

Bock, S.A., Sampson, H.A., Atkins, F.M. *et al.* (1988) 'Double-blind placebo-controlled food challenge as an office procedure: A manual', *Journal of Allergy and Clinical Immunology* 82: 986–97.

Bousquet, J., Bjorksten, B., Bruijnzeel-Koomen, C.A. *et al.* (1998) 'Scientific criteria and the selection of allergenic foods for labelling', *Allergy* 53: 3–21.

Bruijnzeel-Koomen, C., Ortolani, C., Aas, K. *et al.* (1995) 'Adverse reactions to food', *Allergy* 50: 623–35.

Burks, A.W., Mallory, S.B., Williams, L.W. *et al.* (1988) 'Atopic dermatitis: clinical relevance of food hypersensitivity reactions', *Journal of Pediatrics* 113: 447–51.

Enserink, M. (1999) 'Ag biotech moves to mollify its critics', *Science* 286: 1666–67.

FAO (1995) *Report of the FAO Technical Consultation on Food Allergies, Rome, Italy, 13 November–14 November.* Rome: Food and Agriculture Organization.

FAO/WHO (1991) *Strategies for Assessing the Safety of Foods Produced by Biotechnology: Report of a Joint FAO/WHO Consultation.* Geneva: World Health Organization.

FAO/WHO (1996) *Biotechnology and Food Safety Report of a Joint FAO/WHO Consultation, Rome, Italy, 30 September–4 October.* FAO Food and Nutrition Paper 61. Rome: Food and Agriculture Organization.

Federal Register (1992) 'Statement of policy: foods derived from new plant varieties' 57, 22984 (29 May 1992).

Ferber, D. (1999) 'GM crops in the cross hairs', *Science* 286: 1662–66.

Fiorini, G., Rinaldi, G., Bigi, G. *et al.* (1990) 'Symptoms of respiratory allergies are worse in subjects with coexisting foods sensitization', *Clinical and Experimental Allergy* 20: 689–92.

Friedrich, M.J. (1999) 'Genetically enhanced rice to help fight malnutrition', *Journal of the American Medical Association* 282: 1508–509.

Gannes, S. (1993) 'Take a tomato, add a gene of flounder. What do you get? Future food', *Self* March, 152.

Gaskell, G., Bauer, M.W., Durant, J. *et al.* (1999) 'Worlds apart? The reception of genetically modified foods in Europe and the US', *Science* 285: 384–87.

Gavaghan, H. (1999) 'Britain struggles to turn anti-GM tide', *Science* 284: 1442–44.

Haq, T.A., Mason, H.S., Clements, J.D. *et al.* (1995) 'Oral immunization with a recombinant bacterial antigen produced in transgenic plants', *Science* 268: 714–16.

Hefle, S.L., Nordlee, J.A. and Taylor, S.L. (1996) 'Allergenic foods', *Critical Reviews in Food Science and Nutrition* 36: 569–89.

Host, A. and Halken, S.A. (1990) 'A prospective study of cow milk allergy in Danish infants during the first 3 years of life', *Allergy* 45: 587–96.

Houben, G.F., Knippels, L.M. and Penninks, A.H. (1997) 'Food allergy: predictive testing of food products', *Environmental Toxicology and Pharmacology* 4: 127–35.

Kimber, I., Lumley, C.E. and Metcalfe, D.D. (1997) 'Allergenicity of proteins', *Human Experimental Toxicology* 16: 516–18.

Lake, A.M. (1997) 'Food protein-induced colitis and gastroenteropathy in infants and children', in D.D. Metcalfe, H.A. Sampson and R.A. Simon (eds) *Food Allergy: Adverse Reactions to Foods and Food Additives*, 2nd edition. Oxford: Blackwell Scientific, pp. 277–86.

Lehrer, S.B., Horner, W.E. and Reese, G. (1996) 'Why are some proteins allergenic? Implications for biotechnology', *Critical Reviews in Food Science and Nutrition* 36: 553–64.

Mann, C. (1997) 'Reseeding the green revolution', *Science* 277: 1038–43.

Mason, H.S., Haq, T.A., Clements, J.D. *et al.* (2002) 'Edible vaccine protects mice against *E. coli* heat-labile enterotoxin (LT): potatoes expressing a synthetic LT-B gene', *Vaccine* (in press).

Metcalfe, D.D. and Sampson, H.A. (1990) 'Workshop on experimental methodology for clinical studies of adverse reactions to foods and food additives', *Journal of Allergy and Clinical Immunology* 86: 421–42.

Metcalfe, D.D., Astwood, J.D., Townsend, R. *et al.* (1996) 'Assessment of the allergenic potential of foods derived from genetically engineered crop plants', *Critical Reviews in Food Science and Nutrition* 36: S165–86.

Miller, H.I. (1993) 'Foods of the future. The new biotechnology and FDA regulation', *Journal of the American Medical Association* 269: 910–11.

Miller, H.I. (1999) 'A rational approach to labeling biotech-derived foods', *Science* **284**: 1471–72.

Min, K.-U. and Metcalfe, D.D. (1997) 'Eosinophilic gastroenteritis', in J.A. Anderson (ed.) *Food Allergy. Immunology and Allergy Clinics of North America.* Philadelphia: W.B. Saunders, Vol. II, pp. 799–813.

Nietizl-Jansen, J.J., Kardinaal, A.F.M., Huijbers, G.H. *et al.* (1994) 'Prevalence of food allergy and intolerance in the adult Dutch population', *Journal of Allergy and Clinical Immunology* **93**: 446–56.

Nordlee, J.A., Taylor, S.L., Townsend, J.A. *et al.* (1996) 'Identification of a brazil-nut allergen in transgenic soybeans', *New England Journal of Medicine* **334**: 688–92.

O'Neil, C.F. and Lehrer, S.B. (1997) 'Occupational reactions to food allergens', in D.D. Metcalfe, H.A. Sampson and R.A. Simon (eds) *Food Allergy: Adverse Reactions to Foods and Food Additives*, 2nd edition. Cambridge: Blackwell Science, pp. 311–35.

Pastorello, E.A., Stocchi, L., Pravettoni, V. *et al.* (1989) 'Role of the elimination diet in adults with food allergy', *Journal of Allergy and Clinical Immunology* **84**: 475.

Saaverda-Delgado, A.-M. and Metcalfe, D.D. (1993) 'Seafood toxins', *Clinical Reviews in Allergy* **11**: 241–60.

Sampson, H.A. and Scanlon, S.M. (1989) 'Natural history of food hypersensitivity in children with atopic dermatitis', *Journal of Pediatrics* **115**: 23.

Shanti, K.N., Martin, B.M., Nagpal, S. *et al.* (1993) 'Identification of tropomyosin as the major shrimp allergen and characterization of its IgE-binding epitopes', *Journal of Immunology* **151**: 5354–63.

Sicherer, S.H., Munoz-Furlong, A., Burks, A.W. *et al.* (1999) 'Prevalence of peanut and tree nut allergy in the US determined by a random digit dial telephone survey', *Journal of Allergy and Clinical Immunology* **102**: 559–62.

Smith, L.J. and Munoz-Furlong, A. (1997) 'The management of food allergy', in D.D. Metcalfe, H.A. Sampson and R.A. Simon (eds) *Food Allergy: Adverse Reactions to Foods and Food Additives*, 2nd edition. Oxford: Blackwell Scientific, pp. 431–44.

Sollid, L.M., Markussen, G., Ek, J. *et al.* (1989) 'Evidence for a primary association of celiac disease to a particular HLA-DQ α/β heterodimer', *Journal of Experimental Medicine* **169**: 345–49.

Tacket, C.O., Mason, H.S., Losonsky, G. *et al.* (1998) 'Immunogenicity in humans of a recombinant bacterial antigen delivered in a transgenic potato', *Nature Medicine* **4**: 607–609.

Taylor, S.L. (1997) 'Food from genetically modified organisms and potential for food allergy', *Environmental Toxicology and Pharmacology* **4**: 127–35.

Taylor, S.L., Bush, R.K. and Nordless, J.A. (1997) 'Sulfites', in D.D. Metcalfe, H.A. Sampson and R.A. Simon (eds) *Food Allergy: Adverse Reactions to Foods and Food Additives*, 2nd edition. Oxford: Blackwell Scientific, pp. 339–57.

Thounce, J.Q. and Tanner, M.S. (1985) 'Eosinophilic gastroenteritis', *Archives of Disease in Childhood* **60**: 1186–88.

Van Dam, F. and de Vriend, H. (1999) 'Genetic engineering and food allergy: friend or foe', *C and B Bulletin* **2**: 1–12.

Van Metre, T.E., Anderson, S.A., Barnard, J.H. *et al.* (1968) 'A controlled study of the effects on manifestations of chronic asthma of a rigid elimination diet based on Rowe's cereal-free diet 1,2,3', *Journal of Allergy* **41**: 195–208.

Wal, J.M. and Pascal, G. (1998) 'Benefits and limits of different approaches for assessing the allergenic potential of novel foods', *Allergy* **53**: 98–101.

Zeisel, S.H. (1999) 'Regulation of nutraceuticals', *Science* **285**: 1853–55.

Chapter 6

Biosafety of marker genes

The possibility of DNA transfer from genetically modified organisms to the human gut microflora

Kieran M. Tuohy, Ian R. Rowland and Paul C. Rumsby

Introduction

The first food products containing genetically modified material are now available on supermarket shelves throughout the world. Much public debate has arisen concerning the safety of such products and indeed the need for genetically modified foodstuffs in the well-stocked larders of the Western World. However, because genetic engineering offers such technical advantages to the food industry in the mass production of cheap processed food of predictable consistency and quality, there will be increased commercial pressure to broaden the range of genetically modified foodstuffs available in the marketplace. Similarly, as our confidence in the safety of this new technology grows and our ability to apply genetic engineering to products which offer real benefits to the consumer (e.g. safer, more wholesome food) and in the development of 'functional' foods which may well play an important role in human health and disease prevention, consumer acceptance and demand for such products may grow (Verrips and van den Berg, 1996). However, consumer confidence will rely on rigorous assessment of the various potential risks involved and appropriate safety testing. This chapter outlines what is at present known about one of these potential risks: the transfer of DNA from genetically modified organisms to the human gut microflora.

Microorganisms, particularly lactic acid bacteria, have been used in the production of fermented foods for millennia. Recent advances in genetic engineering allow, for the first time, accurate identification of microorganisms traditionally used in food fermentation and the design of novel strains with improved characteristics (McKay and Baldwin, 1990; Castellanos *et al.*, 1996; Verrips and van den Berg, 1996). Age-old production problems such as instability of industrially important traits and failure of starter cultures due to bacteriophage attack may now be tackled at the molecular level (McKay and Baldwin, 1990; Klaenhammer and Fitzgerald, 1992). It has long been proposed that certain lactic acid bacteria or 'probiotics' contribute greatly to intestinal health and well-being (Fuller, 1997; Fuller and Gibson, 1997). Similarly, genetically modified lactic acid bacteria have also been proposed for use as oral vaccines (Wells *et al.*, 1996). Recent advances in molecular microbial ecology allow the scientific basis of such claims to be determined and genetic engineering will enable the design of probiotic bacteria with specific health-promoting properties. No food products containing live genetically modified microorganisms (GMMs) are available in the marketplace at present. However, commercial pressure on the food industry and the consumer benefits promised by probiotic strains designed with scientifically proven health-promoting capabilities will encourage their use in fermented foods in the near future. Clearly the biosafety of such products must be rigorously investigated before they become commercially available, particularly in view of the public back-lash towards genetically modified plant material used in the food products in Europe. Of particular concern when considering the release of live genetically modified microorganisms in food is the possibility of recombinant DNA transfer from genetically modified organisms (GMOs) to members of the human gut microflora (Verrips and van den Berg, 1996).

Transfer of DNA between bacteria occurs naturally in the environment and offers prokaryotes a unique means of evolution and adaptation in response to changing environmental conditions (Veal *et al.*, 1992; Salyers, 1993; Yin and Stotzky, 1997). Interest in DNA transfer between bacteria in the mammalian gastrointestinal tract stems from three main areas:

1   The possibility of DNA transfer from GMOs which may be ingested in food to members of the human gut microflora.
2   The spread of antibiotic resistance amongst bacteria as a result of gene transfer.
3   The emergence of novel human pathogens as a result of transfer of virulence factors or antibiotic resistance determinants between bacteria.

The focus of this review is to discuss the ability of bacteria to undergo DNA transfer in the human gastrointestinal tract with particular reference to the risks posed by genetically modified microorganisms, which may be used in the production of fermented foods such as bread, beer, cheese and yoghurt. We will look at the existing procedures available to monitor DNA transfer in the human gut microflora and investigate some of the factors governing frequencies of such transfer events. These studies will provide information relevant to our understanding of the possibility of the transfer of marker genes used in the construction of genetically modified crops to the bacteria present in the mammalian gastrointestinal tract. First let us look at the mechanisms of DNA transfer available to bacteria in natural environments.

Transformation

Transformation is the process by which a naked piece of DNA from the environment binds to the surface of a competent bacterial cell and is taken up by the bacterium. DNA may then be incorporated into the host genome, depending on the recombinational abilities of the host and the 'foreign' DNA. The ability to translocate DNA across the cell boundary is called competence (Stewart and Carlson, 1986; Lorenz and Wackernagel, 1994). Competence is a specific physiological state of a bacterial cell (genetically encoded in some cases, e.g. *Bacillus subtilis*), which occurs transiently and is restricted to certain stages of the growth cycle (Mazodier and Davies, 1991; Stewart, 1992; Yin and Stotzky, 1997). Natural competence has been observed in bacteria from a variety of genera, including *Haemophilus*, *Neisseria*, *Streptococcus*, *Bacillus*, *Acinetobacter* and *Pseudomonas* spp. and *Helicobacter pylori* (Trevors *et al.*, 1986; Mazodier and Davies, 1991; Tsuda *et al.*, 1993; Claverys *et al.*, 1997). Recalcitrant species, such as *Escherichia coli*, may be rendered competent using a variety of chemical, enzymatic and physical procedures, e.g. $CaCl_2$ treatment and electroporation (Saunders and Saunders, 1988). Such physiochemical conditions may sometimes be prevalent in the local environment of recalcitrant bacteria. For example, Ca^{2+} concentrations in drinking water sometimes approach the levels used *in vitro* to induce a state of competence in *E. coli* (Baur *et al.*, 1996).

Competence is not the only bacterial-encoded parameter shown to play a role in natural transformation. In some cases, requirements for specific lengths of DNA, DNA states (double or single stranded) and the presence of specific DNA sequences have been observed (Stewart and Carlson, 1986; Hirsch, 1990; Mazodier and Davies, 1991; Lorenz and Wackernagel, 1994; Yin and Stotzky, 1997). Competent *Bacillus* and *Streptococcus* spp. may take up any piece of DNA but only homologous DNA will be maintained in the bacterial genome. *Haemophilus* spp., on the other hand, require the presence of specific 11-bp sequences before DNA uptake occurs. These 11-bp recognition sequences occur at a number of locations on the *Haemophilus* genome (Kahn and Smith, 1984). It has also been reported that a diffusible factor may play a role in the induction of competence in *Streptococcus* spp. (Goodgal, 1982). Here, induction of competence was found to be dependent on cell density and the pH of the surrounding environment.

Factors affecting transformation in natural environments

The presence of naked DNA has been demonstrated in a number of natural environments. Naked DNA may arise in a given habitat via a number of routes. Cell lysis as a result of cell death or the activities of bacteriophage will release bacterial DNA and DNA may be released from actively growing bacteria during certain stages of the growth cycle (Lorenz and Wackernagel, 1994). The uptake of DNA from the environment will not only depend on a state of competence in the bacterium but also on the persistence of naked DNA in a given environment. In this respect, different microhabitats will vary greatly in their abilities to either protect naked DNA by the presence of favourable salt concentrations or absorption onto solid supports (e.g. the surface of soil particles or food particles in the gut) or to degrade DNA, for example, by active nucleases present in the microenvironment (Mazodier and Davies, 1991).

Despite the longevity of DNA in the soil and the presence of bacteria with known competence for transformation, gene transfer from plant to soil bacteria appears to be an extremely rare event: for example, transfer of an ampicillin resistance gene from a transgenic potato line to the plant pathogenic bacterium *Erwinia chrysanthemi* was at a calculated frequency of 2×10^{-17} (Schlütter *et al.*, 1995) and from transgenic plants to the soil bacterium *Acinetobacter calcoacetinus* at a frequency lower than 10^{-13} (Neilson *et al.*, 1997). However, the uptake and integration of transgenic plant DNA via natural transformation has been shown for *Acinetobacter* sp., strain BD413 by *in vitro* marker rescue using DNA from various transgenic plants containing the bacterial kanamycin-resistance gene *nptII* (Gerhard and Smalla, 1998; De Vries and Wackernagel, 1998).

Naked DNA enters the human gastrointestinal tract from a number of sources. These include ingested food and foreign microorganisms as well as members of the human gut microflora. Very little is known about the ability of naked DNA to persist in the gut and evade the activities of mammalian and bacterial nucleases. Factors which may affect the persistence of naked DNA, and thus the incidence of transformation in the human gastrointestinal tract, include pH, salt concentrations, cell densities, local nuclease activities and protection afforded by absorption onto surfaces such as food particles or mucosal surfaces (Mercer *et al.*, 1999). Such factors would act at the level of the microhabitat and as such would vary greatly even within specific regions of the gut. It has been shown that the *CRY1A* protein and DNA from genetically modified maize are digested by simulated gastric juices *in vitro* (EU Scientific Committee for Food, 1997; Estruch *et al.*, 1997). However using an *in vitro* model of the intestinal tract (Minekus *et al.*, 1995), van der Vossen *et al.* (1998) showed that 6% of transgenic tomato DNA survived the stomach and small intestine and concluded that the presence of raw mashed tomato helped to preserve the DNA. Free chromosomal DNA of *Bacillus subtilis* persists for weeks in milk and dairy produce. Such bacteria develop natural competence and can be transformed with free chromosomal or plasmid DNA in such produce (Bräutigam *et al.*, 1997; Zenz *et al.*, 1998).

Schubbert *et al.* (1997) showed that the wall of the gastrointestinal tract is exposed to a variety of DNA fragments and remains exposed to DNA fragments of dietary origin for hours after ingestion of the food. The authors found that upon feeding mice 50 µg M13mp18 DNA, approximately 95% of ingested DNA was lost during passage through the stomach. However, phage DNA could be detected by PCR and fluorescent *in situ* hybridization in peripheral leukocytes, spleen and liver cells as well as the contents of mouse small intestine, caecum, large intestine and faeces. Ingested phage DNA was detected for up to 18 hours after ingestion in caecal contents, for up to 8 hours in DNA from the peripheral blood cells and for up to 24 hours but not 48 hours in DNA from spleen and liver.

Mercer *et al.* (1999) monitored the survival of recombinant plasmid DNA (pVACMC1) in fresh human saliva. The fraction of naked DNA remaining amplifiable in saliva ranged from between 40 and 65% after 10 minutes and 6–25% after 60 minutes. Amplifiable plasmid DNA was still present after 24 hours incubation in fresh saliva. The authors also found that plasmid DNA, which had been exposed to degradation by saliva, was capable of transforming naturally competent *Streptococcus gordonii* DL1 in filtered saliva. Transformation activity decreased rapidly with the extent of plasmid DNA degradation. Such studies suggest that DNA released from food or bacteria ingested with food may undergo transformation in not only the oral cavity but other regions of the human gastrointestinal tract. Clearly, much more research is needed on the ability of naked DNA to persist in the lower regions of the human gastrointestinal tract and the extent to which transformation contributes to DNA transfer in the human gut microflora.

Transduction

Transduction is the mechanism by which DNA may be transferred between bacteria by bacteriophage. In essence, the bacteriophage act as microbial couriers, picking up DNA from one bacterial chromosome and delivering the heterologous DNA to another bacterial chromosome (Mazodier and Davies, 1991; Veal *et al.*, 1992; Yin and Stotzky, 1997). Where degradation of naked DNA is one of the chief factors limiting DNA transfer by bacterial transformation in natural environments, bacteriophage protect DNA in natural environments via their proteinaceous capsid (Saye *et al.*, 1990; Veal *et al.*, 1992). Bacteriophage, on the whole, contribute to gene transfer between bacteria by two main mechanisms, namely specialized and generalized transduction.

Specialized transduction

Specialized transduction involves the incorporation of a lysogenic phage into the bacterial chromosome. Lysogeny is favoured by conditions of environmental stress such as the limitation of nutrients and probably aids the survival of the phage and host bacterium (Ogunseitan *et al.*, 1990; Saye *et al.*, 1990). Upon excision from the chromosome, elements of host DNA adjacent to the prophage DNA may become excised along with the phage DNA. The host DNA may then become incorporated into the bacteriophage genome and packaged along with the phage DNA. Thus, when infection of a recipient bacterium occurs the heterologous DNA may become integrated into the recipients chromosome during lysogeny (Veal *et al.*, 1992).

Generalized transduction

Generalized transduction on the other hand, occurs when host DNA is packaged into phage particles instead of the phage genome (Kokjohn, 1989). This mechanism of transduction has been observed in the lytic phage, P1 and P2 of the enterobacteraceae, for example. Upon transduction to a novel recipient, the transferred DNA may either be incorporated into the host chromosome via recombination or where it possesses the means of self-replication, it may replicate autosomally (Mazodier and Davies, 1991; Veal *et al.*, 1992; Yin and Stotzky, 1997).

Factors limiting DNA transfer by transduction

Transduction is greatly dependent on the host range of the transducing bacteriophage, which is generally narrow, since bacteriophage infection is dependent on the phage recognizing specific receptor sites on the bacterial cell surface. Some bacteriophage that are able to mediate transduction between different species of bacteria have been described, e.g. P1 and Mu (Wilkins, 1988). The frequency of transduction in nature may be much higher than previously recognized, since the number of bacteriophage particles in many environments appears to be much higher than first thought (Kokjohn and Miller, 1992; Veal *et al.*, 1992; Yin and Stotzky, 1997). Whether the transferred DNA is maintained in the new host is greatly dependent on the ability of the bacteriophage to insert itself and the heterologous DNA into the host genome and evade the bacterial restriction modification system (Yin and Stotzky, 1997). Bacterial restriction enzymes may recognize specific sequences on heterologous or phage DNA. It has been proposed that restriction modification systems may reduce infection of unmodified phage DNA by 2–3 orders of magnitude (Hirsch, 1990). The amount of DNA that may be transferred by transduction is also limited, being about equivalent in size to the phage genome itself (Saye *et al.*, 1990; Mazodier and Davies, 1991).

Transduction in the human gastrointestinal tract

Despite the fact that the bacteriophage are ubiquitous members of natural microbial ecosystems, little is known about their distribution or activity in the human gut microflora. To date no reports of transduction in the human gastrointestinal tract have been presented.

However, where sufficient effort has been made to isolate bacteriophage from natural environments and examine their activity *in situ*, they have been found to play a significant role in microbial ecology (Kokjohn *et al*., 1991). The fact that about 90% of all bacteriophage isolated from natural environments are temperate suggests that transduction may be more important in microbial genetic plasticity than previously appreciated (Yin and Stotzky, 1997). Thus specifically designed studies involving *in vitro* or *in vivo* models of the human gastrointestinal microflora may well provide evidence of transduction in this ecosystem and elucidate some of the ecological factors governing transduction. Bacteriophage specific for major groups of bacteria present in the gut microflora, e.g. *Bacteroides* spp., bifidobacteria, lactobacilli, methanogens, clostridia, enterobacteriaceae, streptococci and staphylococci have been identified. However, transduction has not been observed in many of these bacterial groups e.g. *Bacteroides* spp. or *Clostridium* spp. (Wells and Alison, 1995). Bacteriophage of the lactic acid bacteria play a major role in the dairy industry both as destructive agents, causing the failure of starter cultures in cheese and yoghurt production, and as valuable genetic tools in the development of genetically altered industrial strains (Klaenhammer and Fitzgerald, 1992; Gasson, 1996).

The variety of bacteriophage associated with the lactobacilli suggest that transduction may be a significant means of gene transfer within this genera. Tohyama *et al*. (1971) demonstrated that the *Lactobacillus salivarius* temperate phage PLS-1 mediates generalized transduction of auxotrophic markers (lysine, proline and serine) and lactose metabolism at frequencies of 10^{-7} to 10^{-8} transductants per CFU *in vitro*. Later, Raya *et al*. (1989) showed that phage ϕadh replicates in a lytic cycle, establishes lysogeny, confers superinfection immunity on the host and mediates plasmid DNA transduction in *L. acidophilus* ADH. It was also observed that plaque formation on cell lawns of *L. acidophilus* NCK102 was pH dependent, with the optimal pH for plaque formation being pH 5.5. Transfer of antibiotic resistance determinants have been reported between strains of *Desulfovibrio desulfricans* via phage Dd1 mediated transduction at frequencies of 10^{-5} to 10^{-6} transductants per recipient (Rapp and Wall, 1987). Transduction has also been observed in methanogens, although not strains found in the human gut microflora (Bertani and Baresi, 1989; Leisinger and Meile, 1990).

Despite demonstrations that transduction does occur under laboratory conditions, little is known about the significance of transduction-mediated gene transfer in the natural and animal-associated environments. Important questions remain unanswered regarding the prevalence and survival of bacteriophage in different environments, the limitations imposed on transduction by the host specificity of the transducing bacteriophage and the frequencies at which transduction occurs during phage replication (Veal *et al*., 1992; Yin and Stotzky, 1997).

Bacterial conjugation

Plasmid DNA, extrachromosomal, self-replicating genetic elements, may mediate DNA transfer between bacteria via conjugation. Conjugation involves cell-to-cell contact between a donor (plasmid-bearing) and recipient (plasmid-free) cell (Mazodier and Davies, 1991; Veal *et al*., 1992; Yin and Stotzky, 1997). Not all plasmids are conjugative but many conjugative plasmids carry environmentally important traits such as antibiotic resistance determinants, virulence factors, novel degradative pathways (Clewell, 1990; Simonsen, 1991). Different bacteria or groups of bacteria display different mechanisms of conjugation, a comprehensive discussion of which is beyond the scope of this review. However, a short, generalized description of some of the conjugative mechanisms employed by different groups of bacteria may serve to highlight both the mechanistic diversity of bacterial conjugation and that DNA transfer via conjugation-like mechanisms has been observed in bacteria from a wide phylogenetic background.

Conjugation in Gram-negative bacteria

In Gram-negative bacteria, cell-to-cell contact between a donor and recipient cell is accomplished via the formation of a pilus (which can be short and rigid or long and flexible) by the donor cell (Bradley, 1980). This pilus, a proteinaceous tube, forms a cytoplasmic bridge between donor and recipient, along which a single-stranded copy of the conjugative plasmid is transferred to the recipient cell. DNA replication then leads to generation of the complementary strand of plasmid DNA in both recipient and donor (Day and Fry, 1992; Clewell, 1993a; Salyers and Shoemaker, 1994). Certain plasmids originally found in Gram-negative species display an extremely broad host range (Guiney and Lanka, 1989; Thomas, 1989). Such plasmids (e.g. the R plasmids) may carry a number of antibiotic-resistance determinants and have been shown to transfer to a variety of bacterial species in different natural environments (Abdul and Venables, 1986). Some conjugative plasmids may mediate transfer of chromosomal DNA between donor and recipient cells upon integration into the chromosome via homologous recombination and low-fidelity excision from the chromosome, leading to incorporation of segments of bacterial DNA into the circularized plasmid before transfer (Yin and Stotzky, 1997). Transfer of Gram-negative plasmids has been observed in many natural environments, including soil, rhizosphere, salt and fresh water and effluent (Mazodier and Davies, 1991; Veal *et al.*, 1992; Yin and Sotozky, 1997).

Conjugation in Gram-positive bacteria

Conjugation of plasmids in Gram-positive bacteria differs considerably from those in Gram-negative bacteria. Two main mechanisms have been described.

Certain plasmids found in *Enterococcus* spp. and *Lactococcus* spp. have been shown to initiate conjugation in response to a small peptide signal released extracellularly by plasmid-free recipient cells (Gasson and Davies, 1980; Clewell, 1993b; Valdivia *et al.*, 1996). These extracellular peptides act as sex pheromones, and lead to clumping of donor and recipient cell to form a conjugative aggregate. This increases cell-to-cell contact between donor and recipient cells and results in the transfer of pheromone-induced plasmids at high frequencies. Plasmid pAD1 of *Enterococcus faecalis* is the best characterized example and is transferred at high frequencies in conjugative aggregates formed in response to pheromones released by plasmid-free recipient cells (Clewell *et al.*, 1982; De Freire Bastos *et al.*, 1997).

The second type of conjugation elucidated in Gram-positive bacteria, especially *Streptococcus* spp. and *Staphylococcus* spp., depends on the direct contact between donor and recipient cell imposed by growth on a solid surface (Clewell, 1993a; Zatyka and Thomas, 1998). Plasmids employing this form of DNA transfer are not induced to conjugate by sex pheromones and the mechanism of their transfer remains unclear. Such plasmids often display a remarkable broad host range, i.e. they can be transferred to a wide spectrum of both Gram-positive and Gram-negative bacteria, including *Streptococcus* spp., *Staphylococcus* spp., *Clostridium* spp., *Lactobacillus* spp., *Listeria* spp., *Pediococcus* spp. and *Bacteroides* spp. (Yin and Stotzky, 1997). Many of these plasmids encode one or more antibiotic-resistance determinants and are thought to have played a role in the dissemination of antibiotic-resistance determinants among bacteria of clinical importance. The best characterized examples include pAMβ1 and pIP501 (Zatyka and Thomas, 1998).

Conjugation in Archaea

Little is known about the molecular biology of Archaea because of the difficulties in cultivating such organisms routinely under laboratory conditions and the major differences that exist between Archaea and Bacteria. However, a conjugation-like mechanism of DNA transfer has been observed in the halophilic Archaea *Halobacterium volcanii* (Rosenshine *et al.*, 1989). Here, a 'cytoplasmic' bridge forms between parental cells along which chromosomal DNA has been shown to transfer. The mechanism of DNA transfer displays characteristics of both bacterial conjugation and eukaryotic cell fusion. However, until we gain a greater understanding of the molecular biology and indeed microbiology of the Archaea, few conclusions can be drawn about the frequencies or mechanisms of DNA transfer in Archaea.

Conjugative transposon

Transposons, genetic elements borne by microbial genomes and possessing the ability to 'transpose' between different locations on the same genome, have long been known to play a role in microbial genetic plasticity. Some such elements not only possess the machinery required for transposition but are also capable of mediating their own transfer between the genomes of different bacteria via conjugation. Such conjugative transposons have been found in a wide array of both Gram-positive and Gram-negative bacteria. They are thought to play a significant role in the dissemination of antibiotic-resistance determinants among a wide range of bacteria and the emergence of multiple antibiotic-resistant pathogenic strains (Clewell and Gawron-Burke, 1985; Salyers and Shoemaker, 1995; Salyers *et al.*, 1995). Conjugative transposons generally excise from the bacterial chromosome to form a covalently closed circular double-stranded DNA transposition intermediate. The transposition intermediate can either reinsert itself into the bacterial chromosome or onto a plasmid in the same cell, or it can mediate its own conjugation to a second bacterial cell and integrate into the recipient chromosome (Scott, 1992; Salyers and Shoemaker, 1995). Conjugative transposons have a wide host range and have been shown to transfer between bacteria at high frequencies typically 10^{-5} to 10^{-4} per recipient (Yin and Stotzky, 1997).

Mobilization

Some conjugative plasmids may 'mobilize' plasmids co-resident in a bacterial cell. Mobilization results in the transfer of a non-self transmissible plasmid from one bacterial cell to another and is mediated by the conjugative plasmid or conjugative transposon (Yin and Stotzky, 1997). Mobilization may occur as a result of the non-conjugative plasmid 'hijacking' the conjugative machinery of the conjugative plasmid or the two plasmids may form a co-integrate plasmid. Plasmids with *ori*T sequences may be mobilized upon formation of efficient cell-to-cell contact between donor and recipient bacteria mediated by the *trans* acting products of *tra* and *mob* genes borne on self-conjugative plasmids co-resident in the same cell. Plasmids devoid of *ori*T sequences may be mobilized by formation of a co-integrate with a self-conjugative plasmid co-resident in the same cell (Crisona *et al.*, 1980; Bennett *et al.*, 1986). Such co-integrate formation is characteristic of bacterial transposable elements (Klecner, 1981). The co-integrate once transferred to a recipient strain via conjugation, may then resolve into its constituent plasmids. Mobilization of genetically modified plasmids has been demonstrated from introduced genetically modified donor strains to bacteria present in different environments (Henschke and Schmidt, 1990; Sandt and Herson, 1991; Hoffmann *et al.*, 1998). Such observations have implications for the release of genetically modified microorganisms into natural environments, even when steps have been taken to limit the transfer potential of recombinant DNA. Many broad host range plasmids are also mobilizable, which aids their dissemination within microbial consortia as well as the dissemination of antibiotic resistance often borne on such plasmids. Some conjugative plasmids can also mobilize chromosomal DNA between different bacteria. Conjugative transposons have also been shown to mediate mobilization of non-conjugative plasmids or chromosomal genetic elements between bacteria (Klecner, 1981).

Retrotransfer

Transfer of plasmid DNA between bacteria is usually unidirectional, i.e. from donor to recipient. However, Mergeay *et al.* (1987) demonstrated the transfer of genetic markers from a recipient strain to the donor strain during conjugation experiments. This form of plasmid transfer has been called retrotransfer. The mechanism involved has not been fully elucidated, but is thought to involve multiple donor and recipient cells. Retrotransfer has been demonstrated in a number of bacterial species (Mazodier and Davies, 1991; Perkins *et al.*, 1994).

Factors governing DNA transfer by conjugation in natural environments

Plasmid transfer by conjugation is determined by characteristics of the plasmid itself, e.g. tra^+, mob^+ and oriT$^+$ genotypes, and the donor and recipient bacteria involved (Veal *et al.*, 1992; Wilkins, 1992; Yin and Stotzky, 1997; Zatyka and Thomas, 1998). Environmental parameters which affect pilus formation or the rate of plasmid curing, such as SDS, organic solvents, proteases, high temperature and acridine orange, will affect the frequency of conjugation (Viljanen and Boratynski, 1991). The metabolic character of the host bacterium, e.g. growth rate or stress, may also affect the frequency of conjugation (MacDonald *et al.*, 1992; Klingmuller, 1994; Schäfer *et al.*, 1994). The frequency of conjugation may be enhanced by the presence of surfaces and bacterial aggregation (e.g. mediated by bile salts or sex pheromones) (Bradley, 1980; Gasson and Davies, 1980; Walsh and MacKay, 1981; Doig *et al.*, 1996; Heaton *et al.*, 1996). The presence of solid surfaces, e.g. soil particles, food particles or walls of mucosa, in an environment may also affect the frequency of plasmid transfer depending on the mode of conjugation employed, by increasing cell-to-cell contact between donor and recipient cells. Substrate availability may also affect conjugation with increased frequencies of conjugation sometimes observed with a plentiful supply of bacterial substrates (Sun *et al.*, 1993; Goodman *et al.*, 1994; Björklöf *et al.*, 1995).

Physiological conditions extant in the environment, such as pH, temperature, substrate availability and bacterial cell density, may all play a role in governing the frequency of conjugation in a plasmid- and bacterium-specific manner (Fernandez-Astorga *et al.*, 1992; MacDonald *et al.*, 1992; Schäfer *et al.*, 1994). Plasmid mainten-ance by bacteria exacts a metabolic cost on the cell, i.e. the energy needed for the plasmid to replicate and express plasmid encoded proteins. This 'metabolic burden' may be counteracted by adaptational advantages endowed by plasmid maintenance (Simonsen, 1991; Kado, 1998). As mentioned previously, plasmids often encode envir-onmentally important traits such as resistance to antibiotics and bacteriophage, unusual metabolic degradative pathways, virulence factors and genes encoding anti-bacterial compounds. Such traits may enable a plasmid-bearing cell to out-compete a plasmid-free cell in response to altering environmental conditions, e.g. presence of antibiotics, unusual substrates or bacteriophage. Similarly, environmental factors affecting the competitiveness or growth rate of a bacterial cell will affect the ability of the cell to both undergo conjugation and maintain newly acquired plasmids.

Properties of the plasmids themselves may limit the extent to which conjugation occurs in the environment. For example plasmid incompatibility, host range, surface exclusion, evasion of host restriction modification systems and segregational stability may all play a role in the extent to which a particular plasmid may spread and per-sist in natural microbial communities (Wilkins, 1992; Yin and Stotzky, 1997). Thus it appears that the frequency of plasmid transfer in a given environment will be a function of the interaction between environmental factors and the bacterial populations involved, characteristics of the plasmid itself and the mode of conjugation employed by the plasmid.

Monitoring transfer of DNA from live GMOs to members of the human gut microflora

There is potential for the presence of live GMOs in foodstuffs such as yoghurt and fermented sausage produced with starter cultures containing genetically modified organisms or perishable foods treated with protective cultures of GM organisms. The environment within living food may permit gene transfer by conjugation or transduction and this has been seen at high frequency in complex food matrices (Gabin-Gauthier *et al.*, 1991; Heller *et al.*, 1995; Hertel *et al.*, 1995). Free recombinant DNA may also be present in food products resulting from the killing of GMMs during food processing.

The human gut microflora is extremely complex, being made up of more than 500 different species of bacteria and climax microbial populations reaching up to 10^{11} CFU/g luminal contents in the colon (Conway, 1995b; Steer *et al.*, 2000). The high density and species diversity of bacteria present in the colon provides ideal conditions for the transfer of DNA between bacteria. However, the diversity of the human gut microflora also presents significant problems in monitoring the DNA transfer events between bacteria and, in particular, monitoring the fate of recombinant DNA, which may be ingested in food. It has been determined that only a fraction of the bacteria present in the human gut microflora may be cultured under laboratory conditions and characterized (Langendijk *et al.*, 1995; Wilson and Blitchington, 1996).

Many species found in the human gut microflora defy culture by standard microbiological techniques, and selective agars and growth conditions exist for only a small proportion of the predominant species present. Much progress as been made in recent years in the development of culture-independent methods for identifying bacteria present in natural environments. Using the phylogenetic information present in 16S rRNA and fluorescent *in situ* hybridization techniques, it is now possible to identify bacteria *in situ* in natural environmental samples including human faeces (Amann *et al.*, 1995). However, such techniques tell us little about the genetic capabilities of bacteria and by themselves do not allow us to monitor the distribution or indeed dissemination of specific genetic determinants amongst natural assemblages of bacteria. Thus, studies on the transfer of DNA between members of the human gut microflora and from GMOs to members of the human gut microflora have employed specific marker systems borne on the mobile genetic elements or on recombinant DNA (Saunders *et al.*, 1990; Prosser, 1994).

Traditional marker systems used in pure culture microbiology such as the *lac* operon or X-gal are of little use when employed to monitor DNA transfer events in complex assemblages of gut microorganisms, because of the widespread occurrence of such genes amongst the mammalian gut microflora. Antibiotic-resistance determinants, when shown to be absent from background bacteria, have been widely employed in laboratory-based studies. However, such marker systems cannot be used in studies with human volunteers because of the emergence of antibiotic-resistance determinants in pathogenic strains. Novel marker systems such as green fluorescent protein are now being employed in laboratory studies to monitor the persistence of GMOs in the human gut microflora and DNA transfer events between bacteria (Dahlberg *et al.*, 1998; Scott *et al.*, 1998). Another hurdle to monitoring the dissemination of recombinant DNA in complex microbial consortia and investigating the factors governing DNA transfer in natural environments is the non-expression of marker genes in novel bacterial hosts (Saunders *et al.*, 1990; Prosser, 1994). Marker genes transferred between bacteria may not always be expressed in the recipient cell. Thus not all DNA transfer events will be detected by existing marker systems. One way around this problem is to use a specific plasmid transfer system comprising a donor and recipient bacterium in which the marker gene is known to be expressed, and then examining the factors governing DNA transfer between these two strains in models of the ecosystem under investigation.

Recent advances in molecular biology offer some hope that more reliable means of monitoring the dissemination of genetic determinants amongst diverse bacteria may be developed. Fluorescent *in situ* hybridization is now being employed to enumerate bacteria in natural assemblages by hybridizing 16S rRNA targets to fluorescently labelled DNA probes and visualization with fluorescent microscopy (Amann *et al.*, 1995; Langendijk *et al.*, 1995; Harmsen *et al.*, 1998; Zoetendal *et al.*, 1998). Similarly, it is now possible to visualize PCR products within bacterial cells fixed on microscope slides (Hodson *et al.*, 1995; Tani *et al.*, 1998). By combining these techniques it may soon be possible to visualize concomitantly, fluorescently labelled PCR products of single-copy genes and labelled 16S rRNA sequences within a single cell in environmental samples.

To enable us to ask specific questions about the human gut microflora, *in vitro* and *in vivo* models of the human gut have been developed (Gibson *et al.*, 1988; Rumney and Rowland, 1992; Conway, 1995a). *In vitro* models generally employ anaerobic chemostats of varying degrees of complexity or short-term batch cultures of human faeces or luminal samples extracted from animals. They allow generation of data on the effect of specific biotic and abiotic environmental parameters on the gut microflora, which may otherwise be unavailable due to inaccessibility of the target environment in the gastrointestinal tract or for ethical reasons (Macfarlane *et al.*, 1998; Steer *et al.*, 2000). Freter *et al.* (1983) studied the transfer of plasmid DNA between *E. coli* strains using a single-stage continuous flow culture model of the mouse gut microflora. This *in vitro* model of the mouse caecal microflora simulated microbial interactions observed in the mouse gut. Data generated from the *in vitro* model also allowed validation of mathematical models of plasmid transfer between *E. coli* strains in the presence of the complex mouse gut microflora. Rang *et al.* (1996) employed short-term batch cultures of luminal extracts from the mouse gastrointestinal tract to monitor transfer of plasmid RP1 between *E. coli* strains. More complex *in vitro* models of the human gastrointestinal tract developed recently have yet to be employed to monitoring DNA transfer events in the human gut microflora.

In vitro models allow us to examine interactions between specific members of the gut microflora or between the whole human faecal microflora and introduced GMOs and non-GMOs under the physiological conditions of the human gut. They provide a vital first stage in the biosafety assessment of GMOs intended for use in human food (Conway, 1995a). However, such models do not take into consideration the biological complexity of the mammalian gut ecosystem. *In vitro* models may not simulate the biological surfaces or the input of endogenous substrates and secretions of the human immune system, which may play a vital role both in the maintenance and establishment of the gut microflora, and the frequency of DNA transfer events.

Gnotobiotic and germ-free animals have been employed as *in vivo* models of the human gastrointestinal ecosystem for some years. Gnotobiotic technology provides a useful tool in studying the interactions between specific members of the human gut microflora, e.g. the role played by specific members of the microflora in colonization, resistance to pathogenic bacteria, bacterial pathogenesis, metabolism of dietary constituents or xenobiotic compounds (Rumney and Rowland, 1992). Transfer of plasmid DNA between specific members of the human gut microflora has also been studied using di- or poly-associated animals. Conventional animals may not provide information of direct relevance to the human gut ecosystem because of the major differences in the composition of the gut microflora between animals and man (Drasar, 1988). To circumvent these problems, germ-free animals have been associated with groups of bacteria from the human gut microflora or the whole human gut microflora

(Raibaud *et al.*, 1980; Hentges *et al.*, 1995). Such human flora-associated (HFA) animals provide an essential tool in studies of human gut microbiology in that they reflect the complexity of the human gut ecosystem in terms of microbial species and numbers, and they provide some of the biotic parameters provided by the mammalian mucosa, immune system and digestive functions. Such *in vivo* models provide an extremely useful tool in the generation of data on the safety of GMOs in food until a greater degree of confidence in the safety of such products is attained and human volunteer studies can ethically be considered (Hentges *et al.*, 1995). Selected examples of DNA transfer events monitored using *in vitro* and *in vivo* models of the mammalian gastrointestinal tract are shown in Table 6.1.

Factors affecting transfer of plasmid DNA from bacteria in vivo

Very little is known about the extent to which transformation and transduction contribute to DNA transfer between bacteria in the human gastrointestinal microflora. Only one study on bacterial transformation in fresh human saliva samples has been presented in the literature (Mercer *et al.*, 1999). No reports of transformation or transduction in lower regions of the human gastrointestinal tract were found after exhaustive search of the literature. This is probably due to the fact that conjugation has been viewed as the dominant contributor to DNA transfer between bacteria in natural environments, because of the ubiquitous occurrence of conjugative elements amongst diverse bacterial species. Thus most studies looking at DNA transfer in the gut have concentrated on conjugation.

Earlier, it was proposed that the frequency of plasmid transfer in a given environment will be a function of the interaction between environmental factors and the bacterial populations involved, characteristics of the plasmid itself and the mode of conjugation employed by the plasmid. Some of these parameters have been studied *in vivo* in animal models.

Effect of antibiotics on plasmid transfer

A number of early investigators observed the transfer of plasmid DNA between bacteria in the gastrointestinal tract of human volunteers (Smith, 1969; Farrar *et al.*, 1972; Anderson *et al.*, 1973; Neu *et al.*, 1973; Anderson, 1975; Petrocheilou *et al.*, 1976; Williams, 1977; Marshall *et al.*, 1981; Levine *et al.*, 1983). However, many of these studies were conducted with antibiotic-resistance determinants in *E. coli* strains and in the presence of antibiotic selective pressure. In the light of the recent emergence of pathogenic bacteria with multiple antibiotic resistance determinants it is no longer considered ethically justifiable to carry out such investigations in human volunteers. Ingestion of antibiotics leads to perturbation of the natural homeostasis of the human gut microflora and presents a selective pressure for the evolution of antibiotic resistance strains through mutation and transfer of resistance determinants (Rowland, 1988; Conway, 1995b; Amyes and Gemmell, 1997).

The human gut microflora is commonly exposed to antibiotics, both through clinical practice and use of antibiotics in animal husbandry. Similarly, antibiotic-resistant bacteria have been isolated from a wide range of human foodstuffs (Corpet, 1998).

Table 6.1 Selected examples of DNA transfer between bacteria using *in vitro* and *in vivo* models of the human gastrointestinal tract

Genetic element	Mechanism	Bacteria involved	Model	References
In vitro models of the gut microflora				
Plasmid DNA	Conjugation	*E. coli* strains	CFC of mouse gut microflora	Freter *et al.*, 1983
Plasmid RP1	Conjugation	*E. coli* strains	Batch culture of mouse gut luminal extracts	Rang *et al.*, 1996
Plasmid pVACMC1	Transformation	*Streptococcus gordonii* DL1	Sterilized and fresh human saliva	Mercer *et al.*, 1999
In vivo models of the gut microflora				
R plasmid	Conjugation	*Serratia liquifaciens* to *E. coli.*	HFA mice	Duval-Iflah *et al.*, 1980
Plasmid pAMβ1	Conjugation	*Lactobacillus reuteri* to *Enterococcus faecalis*	Gnotobiotic mice	Morelli *et al.*, 1988
Plasmid pIL205	Conjugation	*Lactococcus lactis* to *Ent. faecalis*	Gnotobiotic mice	Gruzza *et al.*, 1990
Plasmid pAMβ1	Conjugation	*L. lactis* strains	Gnotobiotic rats	Schlundt *et al.*, 1994
Plasmid pIL205	Conjugation	*L. lactis* to *Ent. faecalis*	Gnotobiotic and HFA mice	Gruzza *et al.*, 1994
Plasmid pIL253	Mobilization	*L. lactis* to *Ent. faecalis*	Gnotobiotic and HFA mice	Gruzza *et al.*, 1994
pAMβ1	Conjugation	*L. lactis* strains	Gnotobiotic rats	Brockman *et al.*, 1996
pAMβ1	Conjugation	*L. lactis* to *Ent. faecalis*	HFA rats	Tuohy *et al.*, 1997
PBR derived plasmids	Conjugation Mobilization	*E. coli* K12 to *E. coli* of human origin	Gnotobiotic mice	Duval-Iflah *et al.*, 1994
Plasmids R388, pCE325 & pUB2380	Conjugation Mobilization	*E. coli* K12 to *E. coli* PG1	Gnotobiotic mice	Duval-Iflah *et al.*, 1998

Duval-Iflah *et al.* (1980) observed the transfer of an R plasmid from *Serratia liquifaciens* to an *E. coli* recipient strain, naturally present in the human microflora of the HFA mice. The donor *S. liquifaciens* strain became established in the HFA mice at between 10^6 and 10^7 CFU/g faeces. Transfer was observed 12 days after inoculation with the donor strain and their numbers ranged from between 10^3 and 10^5 CFU/g faeces. The number of transconjugants greatly increased after addition of selective antibiotic in drinking water. Morelli *et al.* (1988) showed that a subtherapeutic dose of selective antibiotic (erythromycin at 10 µg/ml drinking water) increased the level of pAMβ1 transfer from *L. reuteri* to *Ent. faecalis* from 1×10^{-7} to 1×10^{-4} transconjugants per donor in anexic mice (i.e. gnotobiotic mice associated with donor and recipient strains). Although selective antibiotic pressure previously has been employed to demonstrate transfer of genetic elements between bacteria in animal models and in humans, the observation that subtherapeutic doses of selective antibiotic increase the rate of plasmid transfer has important implications for both the design in GMOs and the use or misuse of antibiotics in clinical practice and agriculture.

Persistence of GMOs and their recombinant DNA in vivo

The ability of GMOs, which may be ingested in a viable form in fermented foods, to persist in the human gastrointestinal tract will greatly determine the extent to which recombinant DNA may be transferred to the human gut microflora. Similarly the stability of recombinant DNA within GMOs *in vivo* is of major interest when designing GMOs for use in food or in the human gut itself. Stable maintenance of recombinant DNA will greatly determine the ability of a GMO to effectively carry out the task for which it was designed. On the other hand, a certain degree of recombinant DNA instability may limit the risks posed by recombinant DNA persisting in non-target ecosystems.

Gruzza *et al.* (1990) investigated the colonization potential and recombinant DNA stability of genetically modified *L. lactis* strains in germ-free mice. All *L. lactis* constructs reverted to plasmid-free derivatives in the digestive tract of anexic mice and plasmid-free derivatives became dominant over plasmid-bearing parental strains. Two high copy number, non-self transmissible plasmids exerted strong ecological disadvantage on plasmid-bearing *L. lactis* strains and were quickly lost from the digestive tracts of the anexic mice. Transfer of the conjugative plasmid pIL205 was observed between an *Ent. faecalis* strain colonizing the digestive tract of anexic mice and an *L. lactis* donor strain. Both the *L. lactis* donor strain and *Ent. faecalis* (pIL205) transconjugants were lost from the anexic mice 10 days after dosing with the donor strain. Thus it appears that carriage of recombinant DNA on non-conjugative plasmids and indeed conjugative plasmid DNA, imposed considerable ecological disadvantages on plasmid-bearing strains in anexic animals.

Schlundt *et al.* (1994) investigated transfer of pAMβ1 between *L. lactis* donor and recipient strains in gnotobiotic and conventional rats. In the conventional animals, both donor and recipient strain were quickly lost from the system and no transconjugants were recovered. In gnotobiotic animals both donor and recipient strain colonized the gastrointestinal tract. Tranconjugants were recovered within a few days of dosing and maintained at between 10^3 and 10^5 CFU less than the recipient strain in faecal samples. Upon sampling of luminal contents of the jejunum, caecum and colon it was observed that carriage of the plasmid seemed to endow a competitive advantage to transconjugants in the small intestine but not in the caecum. The population of transconjugants was maintained at about 10^4 CFU/g throughout the intestine but numbers of recipient strain varied from between 10^4 and 10^5 CFU/g jejunum contents to 10^8–10^9 CFU/g caecal and colonic contents. Clearly the ability of plasmid DNA either to impose ecological advantage or disadvantage on the bacterial host will depend on the plasmid and bacterium involved, and may very well vary considerably between different environmental habitats or microhabitats.

Effect of microbial interactions on DNA transfer

The human gut microflora exerts considerable colonization resistance on extraneous microorganisms entering the colonic environment (Hentges, 1992). Limiting the persistence of a GMO in a given environment will also limit its ability to undergo DNA transfer with the microflora. Gruzza et al. (1994) investigated the ability of members of the human gut microflora to limit the colonization ability of and DNA transfer from, genetically modified *Lactococcus lactis* strains, similar to those likely to be used in the food industry. *L. lactis* strains bearing the conjugative plasmid pIL205 and/or a non-conjugative but a pIL205 mobilizable plasmid pIL253 were used. Gnotobiotic mice were associated with four strictly anaerobic bacteria commonly found in the human gut microflora, i.e. *Bacteroides* sp., *Bifidobacterium* sp., *Peptostreptococcus* sp. and *Ent. faecalis*. No transfer of plasmid DNA was observed from *L. lactis* donor strain to *Bacteroides* sp., *Bifidobacterium* sp. or the *Peptostreptococcus* sp. However, upon addition of *Ent. faecalis* to the polyassociated mice, *Ent. faecalis* (pIL205) transconjugants were observed despite the competitive exclusion imposed on the *L. lactis* donor strain by the colonizing *Ent. faecalis* strain. Gene transfer from the *L. lactis* strains was also investigated in HFA mice. HFA mice were dosed with *L. lactis* donor strain, bearing pIL205 and pIL253, seven days after introduction of the human faecal microflora. From an initial inoculum of 8×10^8 CFU, no *L. lactis* donor cells were recovered from the faecal samples 24 hours after dosing (detection limit 10^2 CFU/g faeces). However, 21 hours after introduction of the donor strain, transconjugant facultatively anaerobic streptococci were recovered resistant to antibiotic markers borne on each of the two plasmids. The number of transconjugants increased until day 15 after dosing. Thus, although rather quickly eliminated from the HFA mice, the *L. lactis* donor strain was able to transfer its plasmid DNA (both the conjugative pIL205 and the non-conjugative but mobilizable pIL253) to members of the human gut microflora in HFA mice.

Brockmann et al. (1996) investigated the transfer of plasmid pAMβ1 and pLMP1 (bearing a genetically modified proteinase gene) from *L. lactis* donor strains to a proteinase-deficient recipient strain in gnotobiotic and conventional rats. No transfer of the proteinase gene to the recipient *L. lactis* strain was observed in anexic or conventional rats. Plasmid pAMβ1 was transferred between *L. lactis* donor and recipient strains in anexic animals but not in conventional rats where both donor and recipient were quickly lost (within 48 hours) from faecal samples. Transfer of pAMβ1 from the *L. lactis* donor strain to *Ent. faecalis* (or a close relative) was observed in one conventional rat. Using an *in vitro* single stage continuous flow culture (CFC) of human faecal microflora and *in vivo* human flora-associated (HFA) rat model of the human colonic microflora we have shown that the human gut microflora exerts a

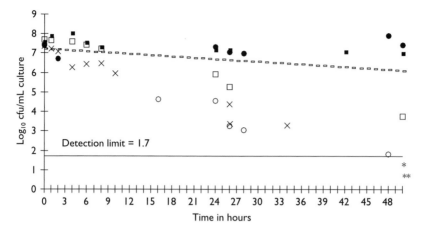

Figure 6.1 Persistence of *L. lactis* MG1614 (□, ○ and ×) in three human faecal microflora anaerobic semi-continuous flow chemostats (sCFC I, II and III) over a 50-h period. Population levels of *Enterococcus* spp. (● and ■) are shown for comparison. Data are shown in $\log_{10}$ CFU/mL culture fluid. The dotted line represents the loss of non-growing cells from the sCFC at the dilution rate used (0.05 h^{-1}). The detection limit was 1.7 $\log_{10}$ CFU/mL sCFC liquor. *L. lactis* MG1614 numbers in sCFC I and II fell below the detection limit by 50 h (* and **).

considerable barrier to persistence of *L. lactis* strains in the colonic ecosystem (Tuohy, 2000). A plasmid-free *L. lactis* strain MG1614 was washed out from the *in vitro* model at a rate greater than expected for non-growing cells (Figure 6.1). Similarly, *L. lactis* MG1614 introduced into HFA rats was eliminated from the HFA rats within eight days. By continuously dosing HFA rats with *L. lactis* MG1614 in the drinking water we succeeded in maintaining a stable population of *L. lactis* MG1614 in the rats. It was hoped that this strain implanted in the microflora of HFA rats would act as a recipient in plasmid transfer studies with a closely related *L. lactis* strain.

Transconjugants were maintained at between 10^4 and 10^5 CFU/g faeces for 27 days (Figure 6.2). Thus despite the fact the human gut microflora exerts such a drastic barrier effect on the persistence of *L. lactis* strains in models of the human colonic microflora, transfer of plasmid DNA from an ingested *L. lactis* donor strain to *Enterococcus* spp. present in the microflora was observed.

Design of GMOs and colonization of gut by GMOs

GMOs considered for release into the environment in a viable form will employ several measures to limit the ability of recombinant DNA to be transferred to indigenous bacteria in the target environment (Verrips and van den Berg, 1996). Recombinant DNA may be borne on non-conjugative vectors, on the chromosome of the host strain or by the use of suicidal genetic elements (Molin *et al.*, 1993; Klijn *et al.*, 1995). However, inbuilt safeguards which prevent the transfer of recombinant DNA under laboratory conditions cannot be relied upon to prevent transfer of recombinant DNA in the natural environment. Thus studies employing *in vivo* models of the human gastrointestinal tract have been conducted with such GMOs in order to determine their ability to persist in the digestive tract and transfer DNA.

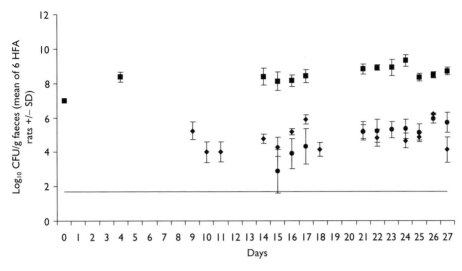

Figure 6.2 Population levels of *Enterococcus* spp. (■), *L. lactis* MG1614 (♦) and *Enterococcus* spp. (pAMβ1) transconjugants (closed circles) in faecal samples taken from HFA rats over a 27-day period. Bacterial counts are expressed as $\log_{10}$ CFU/g faeces ± standard deviation (SD). *L. lactis* MG1614 was maintained in HFA rats at between 4.1 and 6.2 $\log_{10}$ CFU/g faeces by continuous dosing via the drinking water at 9.3 $\log_{10}$ CFU/mL drinking water. HFA rats were dosed with *L. lactis* HU514 at 10 $\log_{10}$ CFU/mL on day 14. *Enterococcus* spp. (pAMβ1) transconjugants were recovered in faecal samples within 24 hours of dosing with donor strain.

Gruzza *et al.* (1992) investigated the ability of genetically modified *L. lactis* strains to colonize gnotobiotic mice. The genetically modified *L. lactis* strains were constructed with antibiotic-resistance marker genes inserted on different replicons, chloramphenicol resistance was inserted into a self-transmissible plasmid pIL205, erythromycin resistance was inserted into non-self transmissible plasmids pIL252 (low copy number) and pIL253 (high copy number) and a second erythromycin-resistance determinant was implanted into the *L. lactis* chromosome. All strains carried a naturally occurring pIL9 plasmid encoding the machinery for lactose fermentation. It was observed that anexic mice were colonized with the recombinant strains or their plasmid-free derivatives, with plasmid-free derivatives becoming dominant in all cases. Plasmids pIL9 and pIL205 were lost from donor strains but plasmid-bearing parental strains and plasmid-free derivatives were maintained at co-dominant levels, suggesting an equilibrium between plasmid loss and conjugation in the mouse gastrointestinal tract.

L. *lactis* strains bearing the low-copy number non-self transmissible plasmid pIL252 rapidly lost their plasmid complement in the gut of anexic mice, probably due to high segregational instability of this plasmid observed *in vitro*. *L. lactis* strains bearing pIL253, the high-copy number, non-self transmissible plasmid colonized the digestive tract of anexic mice but only became dominant when plasmid pIL205 was not present in the same cell, indicating a degree of incompatibility between the two plasmids (both plasmids belong to the same incompatibility group *inc* 18 (Brantl *et al.*, 1990). The erythromycin-resistance determinant borne chromosomally was maintained stably in the digestive tract of anexic mice. The authors concluded that *in vitro* determinations of recombinant DNA stability may highlight if a particular insert is extremely unstable but does not give reliable data on the *in vivo* stability of the construct. This, highlights the need for *in vivo* studies both on the stability of recombinant DNA for efficient GMO function and the effect instability has on the persistence of recombinant DNA in the digestive tract, which is a major factor governing the extent to which DNA transfer between ingested GMOs and members of the human gut microflora occurs.

Duval-Iflah *et al.* (1994) investigated the mobilization of pBR-derived recombinant plasmids (*mob*[-], *tra*[-] but *ori*T[+/-]) from donor *E. coli* K12 strains to *E. coli* strains of human origin *in vitro* and in the digestive tract of anexic and gnotobiotic mice associated with the human gut microflora. The genetically modified plasmids used were designed to be non-conjugative (*mob*[-], *tra*[-]) and either mobilizable or non-mobilizable (*ori* T[+/-]). *E. coli* strains isolated from human faeces were screened for their ability to mobilize the *ori*T[-] and *ori*T[+] plasmids. In anexic mice, associated with donor and recipient *E. coli* strains, mobilization of an *ori*T[+] pBR derivative by trans-complementation with the gene products of *tra* (encoded by conjugative plasmid R388) and *mob* (encoded by plasmid pUB2380) was observed. Such mobilization was only observed sporadically with one *E. coli* strain of human origin in di-associated mice and not at all in HFA mice. It was determined that about 50% of *E. coli* strains of human origin were able to mobilize *ori*T[-] pBR derivatives in *in vitro* matings but not in anexic mice or HFA animals.

Dietary factors affecting DNA transfer in the gut and the effect of donor cell concentration

The ecological significance of the metabolic burden imposed on a bacterium by plasmid carriage will be greatly determined by the availability of substrates and nutrients to the cell. Similarly, densities of donor and recipient cells have been shown to affect conjugation frequencies *in vitro* (Fernandez-Astorga *et al.*, 1992). In our studies, transfer of plasmid pAMβ1 from an *L. lactis* donor strain to an *Enterococcus* spp. recipient strain naturally present in the microflora of HFA rats was found to be strongly affected by dose of donor strain employed. Significant numbers of *Enterococcus* spp. (pAMβ1) transconjugants were recovered only at a donor cell concentration of 10^9 CFU/ml or above. The diet fed to the HFA rats also had an effect on transfer of pAMβ1 from the *L. lactis* donor strain to the *Enterococcus* spp. *in vivo*. A high-fat, low-fibre synthetic diet (similar to human high-fat diets) reduced the numbers of *Enterococcus* spp. (pAMβ1) transconjugants recovered from the faeces of HFA rats compared with the commercial rodent chow (higher levels of complex carbohydrate and fibre). The presence of increased substrate availability in the caecum of HFA rats fed the rodent chow or increased surfaces provided by the higher fibre content of the rodent chow may be involved in the higher numbers of *Enterococcus* spp. (pAMβ1) transconjugants recovered from HFA rats fed this diet.

Duval-Iflah *et al.* (1998) looked at the effect of milk fermented with *Lactobacillus bulgaricus* and *Streptococcus thermophilus* on the transfer and mobilization of plasmids between *E. coli* strains in the gut of anexic mice. Plasmid transfer studies were conducted in gnotobiotic mice associated with *E. coli* K12 donor strain (bearing plasmids R388, conjugative; pCE325 and pUB2380, mobilizable) and an *E. coli* recipient strain PG1 of human origin. The affect of milk fermented with *L. bulgaricus*, *S. thermophilus* or with both lactic acid bacteria (LAB; S85) on transfer of plasmid between the *E. coli* strains was determined. Fermented milk consumption had little effect on the populations of R388 or pUB2380 transconjugants in the anexic mice but the milk fermented with both LAB resulted in a slight increase in *E. coli* PG1 (R388) transconjugants. Long-term consumption of milk fermented with both LAB strains resulted in inhibition of occurrence and maintenance of *E. coli* PG1 (pCE325) transconjugants in the anexic mice. *L. bulgaricus*-fermented milk resulted in a reversible reduction in the numbers of *E. coli* PG1 (pCE325) transconjugants in mouse faecal samples. When *L. bulgaricus* and *S. thermophilus* were grown in BHI broth, the transconjugant lowering capabilities observed with the LAB plus fermented milk were reversed. Administration of *L. bulgaricus* and the combination of both strains grown in BHI broth increased the population of *E. coli* PG1 (pCE325) transconjugants recovered from mouse faecal samples. Thus, the authors concluded that probiotic LAB have an effect on the formation and subsequent maintenance of transconjugants in the gastrointestinal tract of anexic mice, and that this effect was dependent on whether the LAB were administered in fermented milk or laboratory growth media, indicating that bacterial metabolic end products as well as viable LAB may be involved. Clearly much work remains to be done to elucidate the role played by specific environmental parameters, such as diet, ratio of donor and recipient cells, activity of donor and recipient and antagonistic effects of the human gut microflora on plasmid transfer and persistence of GMOs which may be ingested in a viable form in fermented foods.

Conclusion

Genetic engineering is now being employed by the food industry to improve the performance of microorganisms used in the production of fermented foods. Concern has been expressed over the exposure of consumers to genetically modified microorganisms which may be ingested in a viable form in future food products. Of chief concern is the risk of recombinant DNA transfer from ingested GMOs to members of the human gut microflora. A range of natural DNA transfer mechanisms are available to microorganisms in the environment. Chief among these are conjugation, transformation and transduction. Apart from one case of transformation in the oral cavity, most studies examining DNA transfer in the human gastrointestinal microflora have been concerned with conjugation. *In vitro* and *in vivo* models of varying degrees of complexity have been employed to simulate the prevailing ecological conditions in the human gastrointestinal microflora. Such studies have demonstrated the transfer of plasmid DNA, both naturally occurring and genetically modified, in the digestive tract. Some of the factors thought to affect conjugation between bacteria in the gut microflora have also been studied *in vitro* and *in vivo*, e.g. metabolic burden of plasmid maintenance, recombinant DNA stability, colonisation ability of GMOs and antagonistic activities of the microflora, presence or absence of antibiotics, cell concentrations of bacteria involved in conjugation and the effect of dietary supplements on DNA transfer events in the human gut microflora. In order to provide information of direct relevance to risk assessment of the use of genetically modified microorganisms in food, further scientific investigations employing *in vivo* and *in vitro* models of the human gut microflora must be conducted. Particular attention needs to be focused on the frequency of DNA transfer events involving GMOs similar to those likely to be used in the food industry and the environmental factors governing such events. The extent to which transformation and transduction contribute to microbial genetic plasticity in the gastrointestinal ecosystem also remains to be determined.

References

Abdul, P. and Venables, W.A. (1986) 'Occurrence of R-plasmids in porcine faecal waste and comparison of their transfer rates in laboratory mating systems and anaerobic digesters', *Applied Microbiology and Biotechnology* **24**: 149–52.

Amann, R.I., Ludwig, W. and Schleifer, K.-H. (1995) 'Phylogenetic identification and *in situ* detection of individual microbial cells without cultivation', *Microbiological Reviews* **59**: 143–69.

Anderson, E.S. (1975) 'Viability of, and transfer of a plasmid from, *E. coli* K-12 in the human intestine', *Nature* **255**: 502–504.

Anderson, J.D., Gillespie, W.A. and Richmond, M.H. (1973) 'Chemotherapy and antibiotic resistance transfer between Enterobacteria in the human gastro-intestinal tract', *Journal of Medical Microbiology* **6**: 475–73.

Amyes, S.G.B. and Gemmell, C.G. (1997) 'Antibiotic resistance', *Journal of Medical Microbiology* **46**: 436–70.

Baur, B., Hanselmann, K., Schlimne, W. *et al.* (1996) 'Genetic transformation in freshwater: Escherichia coli is able to develop natural competence', *Applied Environmental Microbiology* **62**: 3673–78.

Bertani, B.J. and Baresi, L. (1989) 'Genetic transformation in the methanogen *Methanococcus voltae* PS', *Journal of Bacteriology* **169**: 2730.

Bennett, P.M., Heritage, J., Comanducci, A. *et al.* (1986) 'Evolution of R plasmids by replicon fusion', *Journal of Antimicrobial Chemotherapy* **18**: 103–11.

Björklöf, K., Suoniemi, A., Haahtela, K. *et al.* (1995) 'High frequency of conjugation versus plasmid segregation of RP1 in epiphytic *Pseudomonas syringae* populations', *Microbiology* **141**: 2719–27.

Bradley, D.E. (1980) 'Morphological and serological relations of conjugal pili', *Plasmid* **4**: 155–69.

Brantl, S., Behnke, D. and Alonso, J.C. (1990) 'Molecular analysis of the replication region of the conjugative *Streptococcus agalactiae* plasmid pIL501 in *Bacillus subtilis*. Comparison with plasmids pAMβ1 and pSM19035', *Nucleic Acids Research* **18**: 4783–90.

Bräutigam, M., Hertel, C. and Hammes, W.P. (1997) 'Evidence for natural transformation of *Bacillus subtilis* in foodstuffs', *FEMS Microbiology Letters* **155**, 93–98.

Brockman, E., Jacobsen, B.L., Hertel, C. *et al.* (1996) 'Monitoring of genetically modified *Lactococcus lactis* in gnotobiotic and conventional rats by using antibiotic resistance markers and specific probe or primer based methods', *Systems in Applied Microbiology* **19**: 203–12.

Castellanos, M.I., Chauvet, A., Deschamps, A. *et al.* (1996) 'PCR methods for identification and specific detection of probiotic lactic acid bacteria', *Current Microbiology* **33**: 100–103.

Claverys, J.-P., Dintilhac, A., Mortier-Barrière, I. *et al.* (1997) 'Regulation of competence for genetic transformation in *Streptococcus pneumoniae*', *Journal of Applied Microbiology Symposium Supplement* **83**: 32S–41S.

Clewell, D.B. (1990) 'Movable genetic elements and antibiotic resistance in enterococci', *European Journal of Clinical Microbiology and Infectious Diseases* **9**: 90–102.

Clewell, D.B. (ed.) (1993a) *Bacterial Conjugation*. New York: Plenum.

Clewell, D.B. (1993b) 'Bacterial sex pheromone induced plasmid transfer', *Cell* **73**: 9–12.

Clewell, D.B. and Gawron-Burke, C. (1985) 'Conjugative transposons and the dissemination of antibiotic resistance in streptococci', *Annual Reviews in Microbiology* **40**: 635–59.

Clewell, D.B., Tomich, P.K., Gawron-Burke, M.C. *et al.* (1982) 'Mapping of Streptococcus faecalis plasmids pAD1 and pAD2 and studies relating to transposition of Tn917', *Journal of Bacteriology* **152**: 1220–30.

Conway, P.L. (1995a) 'The use of *in vitro* models to assess interactions between human microbiota and genetically modified microorganisms', *Microbial Ecology in Health and Disease* **8**(S1): S27–S32.

Conway, P.L. (1995b) 'Microbial ecology of the human large intestine', in G.R. Gibson and G.T. Macfarlane (eds) *Human Colonic Bacteria, Role in Nutrition, Physiology and Pathology*. Boca Raton: CRC Press.

Corpet, D.E. (1998) 'Antibiotic resistant bacteria in human food', *Revue de de Médecine et Véterinaire* **149**(8–9): 819–22.

Crisona, N.J., Novak, J.A., Nagaiashi, H. *et al.* (1980) 'Transposon-mediated conjugational transmission of non-conjugative plasmids', *Journal of Bacteriology* **142**: 701–13.

Dahlberg, C., Bergström, M. and Hermansson, M. (1998) '*In situ* detection of high levels of horizontal plasmid transfer in marine bacterial communities', *Applied Environmental Microbiology* **64**: 2670–75.

Day, M.J. and Fry, J.C. (1992) 'Gene transfer in the environment: conjugation', in J.C. Fry and M.J. Day (eds) *Release of Genetically Engineered and Other Micro-organisms*. Cambridge: Cambridge University Press, pp. 40–53.

De Freire Bastos, M. Do-C., Tanimoto, K. and Clewell, D.B. (1997) 'Regulation of transfer of the *Enterococcus faecalis* pheromone-responding plasmid pAD1: temperature-sensitive transfer mutants and identification of a new regulatory determinant, *tra*D', *Journal of Bacteriology* **179**: 3250–59.

De Vries, J. and Wackernagel, W. (1998) 'Detection of *nptII* (kanamycin resistance) genes in genomes of transgenic plants by marker-rescue transformation', *Molecular and General Genetics* **257**: 606–13.

Doig, P., Yao, R.J., Burr, D.H. *et al.* (1996) 'An environmentally regulated pilus-like appendage in *Campylobacter pathogenesis*', *Molecular Microbiology* **20**: 885–94.

Drasar, B.S. (1988) 'The bacterial flora of the intestine', in I.R. Rowland (ed.) *Role of the Gut Microflora in Toxicology and Cancer*. London: Academic Press, p. 24.

Duval-Iflah, Y., Raibaud, P., Tancrede, C. *et al.* (1980) 'R-plasmid transfer from *Serratia liquefaciens* to *Escherichia coli in vitro* and *in vivo* in the digestive tract of gnotobiotic mice associated with human faecal flora', *Infection and Immunity* **28**: 981–90.

Duval-Iflah, Y., Gainche, I., Ouriet, M.-F. *et al.* (1994) 'Recombinant DNA transfer to *Escherichia coli* of human faecal origin *in vitro* and in digestive tract of gnotobiotic mice', *FEMS Microbiology and Ecology* **15**: 79–88.

Duval-Iflah, Y., Maisonneuve, S. and Ouriet, M.-F. (1998) 'Effect of fermented milk intake on plasmid transfer and on the persistence of transconjugants in the digestive tract of gnotobiotic mice', *Antonie von Leeuwenhoek* **73**: 95–102.

Estruch, J.J., Chilton, M-D., Lotstein, R. *et al.* (1997) 'Safety of transgenic corn', *Nature* **385**: 109.

EU Scientific Committee for Food (1997) 'Opinion on the potential for adverse health effects from the consumption of genetically modified maize (*Zea mays* L.)'. Report of the Scientific Committee for Food, 41st series (1997). Catalogue no. GT 07 97660-EN-DE-FR. Luxembourg: Office for Official Publications of the European Community.

Farrar, W.E., Eidson, M., Guerry, P. *et al.* (1972) 'Interbacterial transfer of R-factor in the human intestine: *in vivo* acquisition of R-factor-mediated kanamycin resistance by a multi-resistant strain of *Shigella sonnei*', *Journal of Infectious Diseases* **126**: 27–33.

Fernandez-Astorga, A., Muela, A., Cisterna, R. et al. (1992) 'Biotic and abiotic factors affecting plasmid transfer in Escherichia coli strains', Applied Environmental Microbiology 58: 392–98.

Freter, R., Freter, R.R. and Brickner, H. (1983) 'Experimental and mathematical models of Escherichia coli plasmid transfer in vitro and in vivo', Infection and Immunity 39: 60–84.

Fuller, R. (1997) Probiotics 2: Applications and Practical Aspects. London: Chapman & Hall.

Fuller, R. and Gibson, G.R. (1997) 'Modification of the intestinal microflora using probiotics and prebiotics', Scandinavian Journal of Gastroenterology 32(222): 28–31.

Gasson, M.J. (1996) 'Lytic systems in lactic acid bacteria and their bacteriophages', Antonie van Leeuwenhoek 70: 147–59.

Gasson, M.J. and Davies, F.L. (1980) 'High-frequency conjugation associated with Streptococcus lactis donor cell aggregation', Journal of Bacteriology 143: 1260–64.

Gabin-Gauthier, K., Gratadoux, J.J. and Richard, J. (1991) 'Conjugal plasmid transfer between lactococci on solid surface matings and during cheese making', FEMS Microbiology Letters 85: 133–40.

Gerhard, F. and Smalla, K. (1998) 'Transformation of Acinetobacter sp. Strain BD413 by transgenic sugar beet DNA', Applied Environmental Microbiology 64: 1550–54.

Gibson, G.R., Cummings, J.H. and Macfarlane, G.T. (1988) 'Use of a three-stage continuous culture system to study the effect of mucin on dissimilatory sulfate reduction and methanogenesis by mixed populations of human gut bacteria', Applied Environmental Microbiology 54: 2750–55.

Goodgal, S.H. (1982) 'DNA uptake in Haemophilus transformation', Annual Reviews in Genetics 16: 169–92.

Goodman, A.E., Marshall, K.C. and Hermansson, M. (1994) 'Gene transfer among bacteria under conditions of nutrient depletion in simulated and natural environments', FEMS Microbiology and Ecology 15: 55–69.

Gruzza, M., Duval-Iflah, Y. and Ducluzeau, R. (1990) 'In vivo establishment of genetically engineered Lactococci in gnotobiotic mice; plasmid transfer to Enterococcus faecalis', Microecology and Therapy 20: 465–68.

Gruzza, M., Duval-Iflah, Y. and Ducluzeau, R. (1992) 'Colonization of the digestive tract of germ-free mice by genetically engineered strains of Lactococcus lactis: study of recombinant DNA stability', Microbial Releases 1: 165–71.

Gruzza, M., Fons, M., Ouriet, M.F. et al. (1994) 'Study of gene transfer in vitro and in the digestive tract of gnotobiotic mice from Lactococcus lactis strains to various strains belonging to human intestinal flora', Microbial Releases 2: 183–89.

Guiney, D.G. and Lanka, E. (1989) 'Conjugative transfer of IncP plasmids', in C.M. Thomas (ed.) Promiscuous Plasmids of Gram-negative Bacteria. New York: Academic Press, pp. 27–56.

Harmsen, H.J.M., Gibson, G.R., Elfferich, P. et al. (1998) 'Comparison of viable cell counts and fluorescent in situ hybridization using specific rRNA-based probes for the quantification of human faecal bacteria', FEMS Microbiology Ecology 183: 125–29.

Heaton, M.P., Discotto, L.F., Pucci, M.J. et al. (1996) 'Mobilization of vancomycin resistance by transposon-mediated fusion of a VanA plasmid with an Enterococcus faecium sex pheromone-response plasmid', Gene 171: 9–17.

Heller, K.J., Geis, A. and Neve, H. (1995) 'Behaviour of genetically modified microorganisms in yoghurt', Systems in Applied Microbiology 18: 504–509.

Henschke, R.B. and Schmidt, F.R.J. (1990) 'Plasmid mobilization from genetically engineered bacteria to members of the indigenous soil microflora in situ', Current Microbiology 20: 105–10.

Hentges, D.J. (1992) 'Gut flora and disease resistance', in R. Fuller (ed.) Probiotics: the Scientific Basis. London: Chapman & Hall, pp. 87–110.

Hentges, D.J., Petschow, B.W., Dougherty, S.H. *et al.* (1995) 'Animal models to assess the pathogenicity of genetically modified microorganisms for humans', *Microbial Ecology Health and Disease* 8: S23–S26.

Hertel, C., Probst, A.J., Cavadini, C. *et al.* (1995) 'Safety evaluation of genetically modified micoorganisms applied in meat fermentation', *Systems in Applied Microbiology* 18: 469–76.

Hirsch, P.R. (1990) 'Factors limiting gene transfer in bacteria', in J.C. Fry and M.J. Day (eds) *Bacterial Genetics in Natural Environments*. London: Chapman & Hall, pp. 31–40.

Hodson, R.E., Dustman, W.A., Garg, R.P. *et al.* (1995) '*In situ* PCR for visualisation of microscale distribution of specific genes and gene products in Prokaryotic communities', *Applied Environmental Microbiology* 61: 4074–82.

Hoffmann, A., Thimm, T., Droge, M. *et al.* (1998) 'Intergeneric transfer of conjugative and mobilizable plasmids harbored by *Escherichia coli* in the gut of the soil microarthropod *Folsomia candida* (Collembola)', *Applied Environmental Microbiology* 64: 2652–59.

Kado, C.I. (1998) 'Origin and evolution of plasmids', *Antonie van Leeuwenhoek* 73: 117–26.

Kahn, M.E. and Smith, H.O. (1984) 'Transformation in haemophilus: a problem in membrane biology', *Journal of Membrane Biology* 81: 89–103.

Klaenhammer, T.R. and Fitzgerald, G.F. (1992) 'Bacteriophages and bacteriophage resistance', in M.J. Gasson and W.M. de Vos (eds) *Genetics and Biotechnology of Lactic Acid Bacteria*. London: Blackie Academic & Professional, pp.106–63.

Klecner, N. (1981) 'Transposable elements in prokaryotes', *Annual Reviews in Genetics* 15: 341–404.

Klijn, N., Weerkamp, A.H. and De Vos, W.M. (1995) 'Biosafety assessment of the application of genetically modified *Lactococcus lactis* spp. in the production of fermented milk products', *Systems in Applied Microbiology* 18: 286–492.

Klingmuller, W. (1994) 'Metabolic deprivation: a lead to containment in bacterial releases', *Microbial Releases* 2: 289–92.

Kokjohn, T.A. (1989) 'Transduction: mechanisms and potential for gene transfer in the environment', in S.B. Levy and R.V. Miller (eds) *Gene Transfer in the Environment*. New York: McGraw-Hill, pp. 73–98.

Kokjohn, T.A. and Miller, R.V. (1992) 'Gene transfer in the environment: transduction', in J.C. Fry and M.J. Day (eds) *Release of Genetically Engineered and Other Micro-organisms*. Cambridge: Cambridge University Press, pp. 54–81.

Kokjohn, T.A., Sayler, G.S. and Miller, R.V. (1991) 'Attachment and replication of *Pseudomonas aeruginosa* bacteriophages under conditions simulating aquatic environments', *Journal of General Microbiology* 137: 661–66.

Langendijk, P.S., Schut, F., Jansen, G.J. *et al.* (1995) 'Quantitative fluorescence *in situ* hybridization of *Bifidobacterium* spp. with genus-specific 16S rRNA-targeted probes and its application in faecal samples', *Applied Environmental Microbiology* 61: 3069–75.

Leisinger, T. and Meile, L. (1990) 'Approaches to gene transfer in methanogenic bacteria', in J.-P. Bélaich, M. Bruschi and J.-L. Garcia (eds) *Microbiology and Biochemistry of Strict Anaerobes Involved in Interspecies Hydrogen Transfer*. New York: Plenum Press, p. 11.

Levine, M.M., Kaper, J.B., Lockman, H. *et al.* (1983) 'Recombinant DNA risk assessment studies in humans: efficacy of poorly mobilizable plasmids in biologic containment', *Journal of Infectious Diseases* 148: 699–709.

Lorenz, M.G. and Wackernagel, W. (1994) 'Bacterial gene transfer by natural genetic transformation in the environment', *Microbiological Reviews* 58: 563–602.

Marshall, B.M., Schluederberg, S., Tachibana, C. *et al.* (1981) 'Survival and transfer in the human gut of poorly mobilizable (pBR322) and of transferable plasmids from the same carrier *E. coli*', *Gene* 14: 145–54.

Mazodier, P. and Davies, J. (1991) 'Gene transfer between distantly related bacteria', *Annual Reviews in Genetics* 25: 147–71.

MacDonald, J.A., Smets, B.F. and Rittmann, B.E. (1992) 'The effects of energy availability on the conjugative-transfer kinetics of plasmid RP4', *Water Research* **26**: 461–68.

Macfarlane, G.T., Macfarlane, S. and Gibson, G.R. (1998) 'Validation of a three-stage compound continuous culture system for investigating the effect of retention time on the ecology and metabolism of bacteria in the human colon', *Microbial Ecology* **35**: 180–87.

McKay, L.L. and Baldwin, K.A. (1990) 'Applications for biotechnology: present and future improvements in lactis acid bacteria', *FEMS Microbiological Reviews* **87**: 3–14.

Mercer, D.K., Scott, K.P., Bruce-Johnson, W.A. *et al.* (1999) 'Fate of free DNA and transformation of the oral bacterium *Streptococcus gordonii* DL1 by plasmid DNA in human saliva', *Applied Environmental Microbiology* **65**: 6–10.

Mergeay, M., Lejeune, P., Sadouk, A. *et al.* (1987) 'Shuttle transfer (or retrotransfer) of chromosomal markers mediated by plasmid pULB113', *Molecular and General Genetics* **209**: 61–70.

Minekus, M., Marteau, P., Havenaar, R. *et al.* (1995) 'A multicompartmental dynamic computer-controlled model simulating the stomach and small intestine', *ATLA* **23**: 197–209.

Molin, S., Boe, L., Jensen, L.B. *et al.* (1993) 'Suicidal genetic elements and their use in biological containments of bacteria', *Annual Review of Microbiology* **47**: 139–66.

Morelli, L., Sarra, P.G. and Bottazzi, V. (1988) '*In vivo* transfer of pAMβ1 from *Lactobacillus reuteri* to *Enterococcus faecalis*', *Journal of Applied Bacteriology* **65**: 371–75.

Neilson, K.M., Gebhard, F., Smalla, K. *et al.* (1997) 'Evaluation of possible horizontal gene transfer from transgenic plants to the soil bacterium, *Acinetobacter calcoaceticus* BD413', *Theoretical and Applied Genetics* **95**: 815–21.

Neu, H.C., Huber, P.J. and Winshell, E.B. (1973) 'Interbacterial transfer of R-factor in humans', *Antimicrobial Agents and Chemotherapy* **3**: 542–44.

Ogunseitan, O.A., Salyer, G.S. and Miller, R.V. (1990) 'Dynamic interaction of *Pseudomonas aeruginosa* and bacteriophages in lake water', *Microbial Ecology* **19**: 171–85.

Perkins, C.D., Davidson, A.M., Day, M.J. *et al.* (1994) 'Retrotransfer kinetics of R300B by pQKH6, a conjugative plasmid from river epilithon', *FEMS Microbiology and Ecology* **15**: 33–44.

Petrocheilou, V., Grinsted, J. and Richmond, M.H. (1976) 'R-plasmid transfer *in vivo* in the absence of antibiotic selection pressure', *Antimicrobial Agents and Chemotherapy* **10**: 753–61.

Prosser, J.I. (1994) 'Molecular marker systems for detection of genetically engineered microorganisms in the environment', *Microbiology* **140**: 5–17.

Rapp, B.J. and Wall, J. (1987) 'Genetic transfer in *Desulfovibrio desulfuricans*', *Proceedings of the National Academy of Sciences of the USA* **84**: 9128.

Raibaud, P., Ducluzeau, R., Dubos, F. *et al.* (1980) 'Implantation of bacteria from the digestive tract of man and various animals into gnotobiotic mice', *American Journal of Clinical Nutrition* **33**: 2440–47.

Rang, C.U., Kennan, R.M., Midtvedt, T. *et al.* (1996) 'Transfer of the plasmid RP1 *in vivo* in germ free mice and *in vitro* in gut extracts and laboratory media', *FEMS Microbiology and Ecology* **19**: 133–40.

Raya, R.R., Kleenman, E.G., Luchansky, E.G. *et al.* (1989) 'Characterization of the temperate bacteriophage fadh and plasmid transduction in *Lactobacillus acidophilus* ADH', *Applied Environmental Microbiology* **55**: 2206.

Rosenshine, I., Tchelet, R. and Mevarech, M. (1989) 'The mechanism of DNA transfer in the mating system of an archaebacterium', *Science* **245**: 1387–89.

Rowland, I.R. (1988) 'Factors affecting metabolic activity of the intestinal microflora', *Drug Metabolism Reviews* **19**: 243–61.

Rumney, C.J. and Rowland, I.R. (1992) '*In vivo* and *in vitro* models of the human colonic flora', *Critical Reviews in Food Science and Nutrition* **31**: 299–331.

Salyers, A.A. (1993) 'Gene transfer in the mammalian intestinal tract', *Current Opinion in Biotechnology* 4: 294–98.

Salyers, A.A. and Shoemaker, N.B. (1994) 'Broad host range gene transfer: plasmids and conjugative transposons', *FEMS Microbiology and Ecology* 15: 15–22.

Salyers, A.A. and Shoemaker, N.B. (1995) 'Conjugative transposons: the force behind the spread of antibiotic resistance genes among *Bacteroides* clinical isolates', *Anaerobe* 1: 143–50.

Salyers, A.A., Shoemaker, N.B., Stevens, A.M. *et al.* (1995) 'Conjugative transposons: an unusual and diverse set of integrated gene transfer elements', *Microbiological Reviews* 59: 579–90.

Sandt, C.H. and Herson, D.S. (1991) 'Mobilisation of the genetically engineered plasmid pHSV106 from Escherichia coli HB101(pHSV106) to *Enterobacter cloacae* in drinking water', *Applied and Environmental Microbiology* 57: 194–200.

Saunders, J.R. and Saunders, V.A. (1988) 'Bacterial transformation with plasmid DNA', *Methods in Microbiology* 21: 79–128.

Saunders, J.R., Morgan, J.A.W., Winstanley, C. *et al.* (1990) 'Genetic approaches to the study of gene transfer in microbial communities', in J.C. Fry and M.J. Day (eds) *Bacterial Genetics in Natural Environments*. London: Chapman and Hall, pp. 3–21.

Saye, D.J., Ogunseitan, O.A., Salyer, G.S. *et al.* (1990) 'Transduction of linked chromosomal genes between *Pseudomonas aeruginosa* strains during incubation in situ in a fresh-water habitat', *Applied Environmental Microbiology* 56: 140–45.

Schäfer, A., Kalinowski, J. and Pühler, A. (1994) 'Increased fertility of *Corynebacterium glutamicum* recipients in intergeneric matings with *Escherichia coli* after stress exposure', *Applied Environmental Microbiology* 60: 756–59.

Schlütter, K., Fütterer, J. and Potrykus, I. (1995) ' "Horizontal" gene transfer from a transgenic potato line to a bacterial pathogen (*Erwinia chrysanthemi*) occurs -if at all- at an extremely low frequency', *BioTechnology* 13: 1094–98.

Schlundt, J., Saadbye, P., Lohmann, B. *et al.* (1994) 'Conjugal transfer of plasmid DNA between *Lactococcus lactis* strains and distribution of transconjugants in the digestive tract of gnotobiotic rats', *Microbial Ecology Health and Disease* 7: 59–69.

Schubbert, R., Renz, D., Schmitz, B. *et al.* (1997) 'Foreign (M13) DNA ingested by mice reaches peripheral leukocytes, spleen and liver via the intestinal wall mucosa and can be covalently linked to mouse DNA', *Proceedings of the National Academy of Sciences of the USA* 94: 961–66.

Scott, J.R. (1992) 'Sex and the single circle: conjugative transposition', *Journal of Bacteriology* 174: 6005–10.

Scott, K.P., Mercer, D.K., Glover, L.A. *et al.* (1998) 'The green fluorescent protein as a visible marker for lactic acid bacteria in complex ecosystems', *FEMS Microbiology and Ecology* 26: 219–30.

Simonsen, L. (1991) 'The existence conditions for bacterial plasmids: theory and reality', *Microbial Ecology* 22: 187–205.

Smith, H.W. (1969) 'Transfer of antibiotic resistance from animal and human strains of E. coli to resistant E. coli in the alimentary tract of man', *Lancet* 2: 1174–76.

Steer, T., Carpenter, H., Tuohy, K. *et al.* (2000) 'Perspectives on the role of the human gut microbiota and its modulation by pro- and prebiotics', *Nutrition Research Reviews* 13: 229–54.

Stewart, G.J. (1992) 'Transformation in natural environments', in E.M.H. Wellington and J.D. van Elsas (eds) *Genetic Interactions among Microorganisms in the Natural Environment*. Oxford: Pergamon, pp. 216–34.

Stewart, G.J. and Carlson, C.A. (1986) 'The biology of natural transformation', *Annual Reviews in Microbiology* 40: 211–35.

Sun, L., Bazin, M.J. and Lynch, J.M. (1993) 'Plasmid dynamics in a model soil column', *Molecular Ecology* **2**: 2–15.

Tani, K., Kurokawa, K. and Nasu, M. (1998) 'Development of direct *in situ* PCR method for detection of specific bacteria in natural environments', *Applied Environmental Microbiology* **64**: 1536–40.

Thomas, C.M. (1989) *Promiscuous Plasmids of Gram-negative Bacteria*. New York: Academic Press.

Tohyama, K., Sakurai, T. and Arai, H. (1971) 'Transduction by temperate phage PLS-1 in *Lactobacillus salivarius*', *Japanese Journal of Bacteriology* **26**: 482.

Trevors, J.T., Barkay, T. and Bourquin, A.W. (1986) 'Gene transfer among bacteria in soil and aquatic environments: a review', *Canadian Journal of Microbiology* **33**: 191–98.

Tsuda, M., Karita, M. and Nakazawa, T. (1993) 'Genetic transformation in *Helicobacter pylori*', *Microbiology and Immunology* **37**: 85–89.

Tuohy, K. (2000) 'An investigation into the dissemination of mobile genetic elements in vivo in HFA rats: implications for genetic modification in the food industry', PhD thesis, University of Surrey, Guildford, Surrey, UK.

Tuohy, K., Davies, M., Rumsby, P. *et al.* (1997) 'The transfer of plasmid DNA from *Lactococcus lactis* strains to members of the human gut microflora in Human Flora Associated (HFA) rats', paper presented at Lactic '97 (Lactic Acid Bacteria: Which strains? For which products?), Congress Centre Caen, Normandy, France 10th, 11th and 12th September.

Valdivia, E., Martín-Sánchez, I., Quirantes, R. *et al.* (1996) 'Incidence of antibiotic resistance and sex pheromone response among enterococci isolated from clinical human samples and from municipal waste water', *Journal of Applied Bacteriology* **81**: 538–44.

Van der Vossen, J.M., Havekes, W.A., Koster, D.S. *et al.* (1998) 'Development and application of in vitro intestinal tract model for safety evaluation of genetically modified foods as a basis for market introduction', Ministry of Economic Affairs, The Netherlands.

Veal, D.A., Stokes, H.W. and Daggard, G. (1992) 'Genetic exchange in natural microbial communities', *Advances in Microbial Ecology* **12**: 383–430.

Verrips, C.T. and van den Berg, D.J.C. (1996) 'Barriers to application of genetically modified lactis acid bacteria', *Antonie Van Leeuwenhoek* **70**: 299–316.

Viljanen, P. and Boratynski, J. (1991) 'The susceptibility of conjugative resistance transfer in Gram-negative bacteria to physicochemical and biochemical agents', *FEMS Microbiological Reviews* **88**: 43–54.

Walsh, P.M. and McKay, L.L. (1981) 'Recombinant plasmid associated cell aggregation and high-frequency conjugation of *Streptococcus lactis* ML3', *Journal of Bacteriology* **146**: 937–44.

Wells, J.M. and Allison, C. (1995) 'Molecular genetics of intestinal anaerobes', in G.R. Gibson and G.T. Macfarlane (eds) *Human Colonic Bacteria, Role in Nutrition, Physiology and Pathology*. Boca Raton: CRC Press, pp. 25–61.

Wells, J.M., Robinson, K., Chamberlain, L.M. *et al.* (1996) 'Lactic acid bacteria as vaccine delivery vehicles', *Antonie Van Leeuwenhoek* **70**: 317–30.

Wilkins, B.M. (1988) 'Organization and plasticity of enterobacterial genomes', *Journal of Applied Bacteriology Symposium Supplement* 51S–69S.

Wilkins, B.M. (1992) 'Factors influencing the dissemination of DNA by bacterial conjugation', in J.C. Fry and M.J. Day (eds) *Bacterial Genetics in Natural Environments*. London: Chapman and Hall, pp. 22–29.

Williams, P.H. (1977) 'Plasmid transfer in the human alimentary tract', *FEMS Microbiology Letters* **2**: 91–95.

Wilson, K.H. and Blitchington, R.B. (1996) 'Human colonic biota studied by ribosomal DNA sequence analysis', *Applied and Environmental Microbiology* **62**: 2273–78.

Yin, X. and Stotzky, G. (1997) 'Gene transfer among bacteria in natural environments', *Advances in Applied Microbiology* **45**: 153–212.

Zatyka, M. and Thomas, C.M. (1998) 'Control of genes for conjugative transfer of plasmids and other mobile elements', *FEMS Microbiology Reviews* **21**: 291–319.

Zenz, K.I., Neve, H., Geis, A. *et al.* (1998) '*Bacillus subtilis* develops competence for uptake of plasmid DNA when growing in milk products', *Systems in Applied Microbiology* **21**: 28–32.

Zoetendal, E.G., Akkermans, A.D.L. and De Vos, W.M. (1998) 'Temperature gradient gel electrophoresis analysis of 16S rRNA from human faecal samples reveals stable and host-specific communities of active bacteria', *Applied and Environmental Microbiology* **64**: 3854–59.

Chapter 7

Case study: canola tolerant to Roundup® herbicide

An assessment of its substantial equivalence compared to non-modified canola

Thomas E. Nickson and Bruce G. Hammond

Introduction

Glyphosate (*N*-phosphonomethylglycine), the active ingredient in Roundup® herbicide, is a broad-spectrum, post-emergent herbicide that is highly regarded for its efficacy and environmental safety (Malik *et al.*, 1989). Until recently, Roundup has not been available for in-crop, selective control of the weeds which affect the yield and quality of canola crops. Developments in biotechnology have made it possible to insert two genes into canola which provide tolerance to glyphosate and thus offer Roundup Ready™ Canola (RRC) to Canadian farmers. Before RRC could be used by farmers, its food, feed and environmental safety had to be established and approved by the Canadian regulatory authorities. The focus of the safety assessment was to establish that RRC was 'substantially equivalent' to conventional canola, allowing for the tolerance to glyphosate. This paper describes the results of compositional studies that were conducted with RRC canola grown in 1992 and 1993 in Canada. The compositional studies measured the levels of important nutrients in canola seed, processed canola meal and in canola oil. The levels of anti-nutrients in canola seed and in processed meal were also determined. Results of a four-week rat feeding study with RRC canola processed meal that confirm 'substantial equivalence' are also presented.

The original line (transformation event) upon which the safety tests were performed is termed GT73. It is a selection from *Brassica napus* L. cv. Westar (Klassen *et al.*, 1987) transformed using *Agrobacterium tumefaciens* (White, 1989; Howard *et al.*, 1990). Line GT73 expresses two proteins conferring tolerance to Roundup herbicide (Padgette *et al.*, 1996). The proteins are CP4 5-enolpyruvylshikimate-3-phosphate synthase (CP4 EPSPS) and glyphosate oxidoreductase (GOX) and they confer tolerance through two different mechanisms (Figure 7.1). The first tolerance mechanism involves a decreased sensitivity of the site of herbicidal action to glyphosate accomplished through the discovery of CP4 EPSPS, an EPSPS with a significantly higher K_i for glyphosate (Barry *et al.*, 1992). The presence of CP4 EPSPS protein in RRC enables the plant to make aromatic amino acids despite application of glyphosate, which inhibits endogenous EPSPS (Padgette *et al.*, 1996). The second mechanism entails the breakdown of glyphosate to aminomethylphosphonic acid (AMPA) and glyoxylic acid by GOX (Padgette *et al.*, 1996). Line GT73 has demonstrated commercial levels

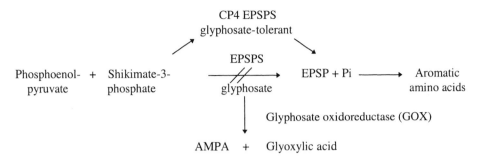

Figure 7.1 Mechanisms of glyphosate tolerance in Roundup Ready Canola (RRC).

of tolerance to Roundup herbicide in numerous field trials. Roundup herbicide has a long history of safe use in agriculture based on the low toxicity of its active ingredient, glyphosate, to a wide variety of vertebrate and invertebrate species (US EPA, 1993). AMPA has a safety profile similar to that of glyphosate (JMPR, 1997).

The safety assessment of the CP4 EPSPS enzyme introduced into canola has been addressed elsewhere (Harrison *et al.*, 1996). CP4 EPSPS protein was shown to be: (1) readily degraded in simulated digestive fluids; (2) non-toxic when administered orally to mice at an acute dose thousands of times higher than potential human exposure to CP4 EPSPS in foods; and (3) not structurally or functionally related to known protein allergens or toxins based on amino acid sequence homology searches of protein toxin and allergen databases (SwissProt, PIR and Genpept).

GOX enzyme has been subjected to the same safety evaluation and has also been shown to be safe for consumption in food or feed (data not presented).

Since the use of biotechnology to develop new crop varieties is relatively new, nations such as Canada have developed guidelines to assess the safety of genetically modified plants. These guidelines are science-based, applied case-by-case, and are designed to assess the feed and environmental safety of the plant (regulated by the Canadian Food Inspection Agency) as well as the safety of human food derived from the plant (regulated by Health Canada). Like most national guidelines, the Canadian guidelines are founded on the concept of 'substantial equivalence'. The premise of 'substantial equivalence', as defined by the Organisation for Economic Co-operation and Development (OECD, 1992) and the World Health Organization (WHO, 1995) and used in Canada to regulate genetically modified plants, is that existing plant varieties can be used as a basis of comparison to establish safety. It is presumed that if the agronomic performance and nutrient/anti-nutrient composition of GT73 is similar to that of *B. napus* canola from which it is derived, GT73 would be considered substantially equivalent and 'as safe as' other canola varieties. This assumes that the safety of proteins introduced into GT73 have already been established. Furthermore, new *B. napus* varieties of RRC that have been developed using traditional breeding and have not been otherwise modified will be regulated as any other canola variety.

We assessed the substantial equivalence of GT73 in field tests evaluating its agronomic performance (data not presented) and extensive compositional analyses of key nutritional and anti-nutritional components known to exist in canola. Agronomic performance was assessed in Western Canadian Canola/Rapeseed Recommending

Committee Co-Op Tests conducted in 1993 at 24 test locations; test plots were replicated four times.

This paper focuses upon summarizing the compositional analyses in addition to confirmatory feeding studies conducted with rats. Substantial equivalence was concluded when results obtained for GT73 were within commercially acceptable ranges for *B. napus* varieties of canola. The principal comparison was to the limits that have been established for canola and to the ranges published for *B. napus*. Data for Westar from Monsanto-conducted field trials were used to assess substantial equivalence. Additional data for Westar from the 1992 and 1993 Western Canadian Canola/Rapeseed Recommending Committee Co-Operative Test (Co-Op Test) was also used extensively for comparisons. Results of a four-week rat feeding study with an RRC line (RU3) derived from GT73 canola are also presented. In this study, results from a feeding of RU3 to rats are compared to those from feeding the parental (non-transformed line) and commercial canola/oilseed rape varieties grown in Canada, Europe and Chile. This study was undertaken, in part, to resolve organ weight findings from two previous four-week rat feeding studies. The first study found no differences in liver and kidney weights between the rats fed glyphosate-tolerant canola and those fed control, non-transgenic canola, but this study had to be repeated due to technical problems. In the repeat study, rats fed glyphosate-tolerant canola exhibited a slight (16%) increase in liver/body weight ratio when compared with rats fed control, non-transgenic canola meal.

Specifically described in this chapter are compositional data on % fat, % crude protein, % crude fiber, % ash, sinapines, fatty acid, amino acid and glucosinolate composition in the seed. In addition to the data obtained from harvested seed, representative samples of GT73 and Westar were processed into meal and refined oil and analyzed for comparison to accepted values for these components. Analyses of processing fractions and the four-week rat feeding study are considered confirmatory studies since they confirm the primary assessment of agronomic performance and food/feed composition used to establish substantial equivalence of GT73 canola.

Historical background of canola/oilseed rape

Oilseed rape varieties have been cultivated in north-western Europe for centuries (Scarisbrick *et al.*, 1986), primarily as a source of fuel. While the oil has been used for centuries for cooking in Asia and Mediterranean regions, widespread use of cultivated oilseed rape varieties as a source of human food and animal feed in western countries has occurred more recently. In 1989, rapeseed ranked third in the world production of oilseed crops, surpassed only by soybean and cottonseed (Shahididi and Naczk, 1990). It is grown primarily for the production of oil, which represents more than 40% of the seed weight (Downey and Bell, 1990). The oil is separated from the seed by crushing and processing and is used predominately in cooking oils, margarines and fat. The protein-rich meal remaining behind is employed as a feedstuff for livestock and poultry.

The seeds also contain varying levels of the following anti-nutrients: a fatty acid (C22:1) called 'erucic acid' and glucosinolates; a mixture of thioglycosides containing either an aliphatic, aromatic or heteroaromatic side-chain. Over 100 structural forms of glucosinolates have been identified. Glucosinolates are hydrolyzed to thiocyanates, isothiocyanates, cyclic sulfur compounds and nitriles by myrosinase, an enzyme found in *Brassica* seed cells (Fenwick *et al.*, 1989). Glucosinolates can also be hydrolyzed or otherwise broken down by microflora in the gastrointestinal tract.

Oil from early varieties of oilseed rape were quite high in erucic acids. Oil containing high erucic acid levels (up to 50% of total fatty acids – designated as HEAR oil) has produced myocardial lesions when fed to animals (Sauer and Kramer, 1983).

Oil from more recently developed low erucic acid oilseed rape varieties does not produce myocardial lesions in animals except when fed at high dietary levels to male rats of the Sprague–Dawley strain. Feeding high dietary levels of other vegetable oils to male Sprague–Dawley rats also produces myocardial lesions (Kramer and Sauer, 1983). These pathological changes are attributed to the sex- and species-specific sensitivity of this strain of rat to consumption of high levels of vegetable oils. Myocardial lesions have not been observed in other animal species fed oil from low erucic acid oilseed rape varieties; thus the rat is considered an unsuitable model for assessing the safety of rapeseed oil for humans (US FDA, 1985).

The glucosinolates have been traditionally considered to be anti-nutrients since they are metabolized to products such as isothiocyanates which can act as goitrogens. When swine, cattle, poultry and rats were fed varieties of oilseed rape containing high glucosinolate levels, various deleterious effects (e.g. reduced palatability and growth performance, goitrogenicity, liver hypertrophy, etc.) were observed (Paik *et al.*, 1980; Fenwick *et al.*, 1982, 1989). Commercial processing of oilseed rape reduces the levels of glucosinolates and their metabolites, due in part to inactivation of myrosinase. However, some glucosinolates remain in the meal and metabolites may be formed by microbial flora in the digestive tract. Nutritionists have encouraged breeders to develop oilseed rape varieties lower in glucosinolates to avoid anti-nutritional effects.

Significant progress has been made in recent years to reduce the levels of erucic acid and glucosinolates in oilseed rape through classical breeding approaches. Canadian varieties of oilseed rape which contain low levels of erucic acid (less than 2% of the total fatty acids present in the oil) and low levels of alkyl glucosinolates (less than 30 µmol/g in defatted meal) may be sold under the trademark designation 'Canola', which is owned by the Canola Council of Canada. Oil derived from the low erucic acid oilseed rape varieties was determined by FDA to be 'generally recognized as safe' (GRAS) for use as a human food (US FDA, 1985). Processed canola meal has been widely used in Canada as an animal feed for beef and dairy cattle, swine and poultry. The reduced levels of glucosinolates and erucic acid in canola varieties have made it possible for farmers to take advantage of the high protein content of processed canola meal for use in animal feed. The canola varieties are nutritionally superior to older varieties of oilseed rape that had higher glucosinolate and erucic acid content (Clandinin, 1990). New varieties of oilseed rape referred to as 'double-zero' have been developed recently in Europe. These are also low in erucic acid and glucosinolate levels compared with canola varieties (Baudet *et al.*, 1988; NIAB, 1993).

While glucosinolates have been considered to be anti-nutrients, recent evidence suggest that certain glucosinolates (e.g. 3-methylsulfinylpropylglucosinolate) and their metabolites may have beneficial health effects. There is epidemiological evidence that consumption of cruciferous vegetables (e.g. broccoli), which contain a variety of glucosinolates, can reduce the incidence of certain cancers in humans (Kore *et al.*, 1993; Wallig *et al.*, 1998). Considerable research is under way to investigate the effects of glucosinolates on cancer development in animal models and the mechanisms by which glucosinolates act as anti-carcinogens. These studies may help scientists elucidate whether there are human health benefits from consumption of glucosinolates.

Materials and methods

Preparation of canola seed for compositional analysis

All plots were planted and managed under normal agronomic practices for canola using seed provided by Monsanto Company and were maintained in strict compliance with all permit requirements of the Canadian Food Inspection Agency. At each location, GT73 of generation R_2 or later was planted with a plot of non-modified Westar of the parental variety. The plot sizes ranged from 7.5 m² to 100 m², and their locations are given below. All GT73 samples that underwent compositional analysis were obtained from 100-m² plots which were completely harvested and subsequently sampled. These plots were not replicated because it was assumed that site-to-site variation in compositional data would be greater than intra-site variation. Additional data for Westar variety canola were obtained in 1992 (seven locations) and in 1993 (four locations) from the Western Canadian Canola/Rapeseed Recommending Committee Co-Op Field Tests. Certified Westar canola was used as a check (non-genetically modified) variety in these trials. In 1985, Westar was used to plant over 80% of all *Brassica napus* canola acres in Canada and has been used as a standard in the Western Canadian Cooperative Rapeseed Tests (Co-Op Test) until 1994.

Compositional analyses

All analyses were conducted on seed obtained from either the seven field locations in 1992 (Saskatoon, Scott, Melfort, Watrous, Saskatchewan; Minto and Portage la Prairie, Manitoba and Guelph, Ontario) or the four locations in 1993 (Saskatoon, Melfort, Saskatchewan; Minto and LaSalle, Manitoba). Crude protein in defatted meal, % fat (whole seed), glucosinolate and fatty acid composition and sinapine analyses were conducted at Agriculture and Agri-Food Canada's Research Station in Saskatoon (Ag Canada). In addition, proximate analyses (crude protein, fat, fiber, moisture and ash) of seed from the same field trials were conducted under the US EPA Good Laboratory Practices (GLP) Regulations at Ralston Analytical Labs (RAL) in St. Louis, Missouri. Analyses of meal and refined oil were conducted at RAL and POS (Protein, Oil and Starch Pilot Plant Corporation in Saskatoon, Saskatchewan). The methods used were either validated or based on published methods. References for the specific analytical methods used to analyze seed and meal are listed in Table 7.1. Defatted meal samples were prepared for analysis at Ag Canada using a Raney grinder (Raney *et al.*, 1987).

Processing study

Processing was conducted on 80-kg samples of GT73 and Westar at POS. Both samples were composites of seed grown at four locations in Canada in 1993 (see above). The meal and refined oil samples were produced using procedures developed by POS and equipment suitable for the scale of production. All samples were representative of commercially produced material.

Table 7.1 References to analytical methods used to analyze seed and meal

Analysis	Literature method
Ag Canada Laboratory in Saskatoon.	
Fat	Conway and Earle, 1963
Protein	Edeling, 1968; Sweeney and Rexroad, 1987; Sweeney, 1989
Fatty acid composition	Thies, 1971; Hougen and Bodo, 1973; Freedman *et al.*, 1986; Bannon *et al.*, 1982; Conacher and Chadha, 1974
Glucosinolates	Thies, 1980; Heaney and Fenwick, 1980
Sinapines	Bjerg *et al.*, 1984
Ralston Analytical Laboratories in St. Louis.	
Fat	AOAC, 1990a; Foster and Gonzales, 1992; Bhatly, 1985
Protein	AOAC, 1990c
Crude fiber	AOAC, 1984b
Ash	AOAC, 1984a
Moisture	AOAC, 1984c; AACC, 1987
Amino acid composition	Moore and Stein, 1954, 1963; Blackburn, 1978
Tryptophan	Hiroyuki *et al.*, 1984; Jones *et al.*, 1981
Cysteine/methionine	AOAC, 1990d; Schram *et al.*, 1954; Moore, 1963
Nitrogen solubility	AOCS, 1989
Mineral screen	AOAC, 1989; Winge *et al.*, 1985
Phytic acid	AOAC, 1990b; Harland and Oberleas, 1977; Ellis and Morris, 1983

Statistical analyses

To determine if any differences in fat, protein, crude fiber, ash, phenylalanine, tryptophan and tyrosine levels between GT73 and Westar were statistically significant, a randomized complete block design model was used. Each variable in the model $Y_{ij} = \mu_j + L_i + E_{ij}$, where Y_{ij} is the value obtained for group j at location I, μ_j is the true mean of group j over all locations, L_i is the random effect of location I and E_{ij} is the random effect for group j within location i, was fitted to the data using the GLM procedure in SAS. SAS ESTIMATE statements in GLM procedure were used to specify the overall means, the mean difference from the Westar mean and the *t*-test to determine if this difference is statistically significant. For alkyl glucosinolates and erucic acid in GT73, the overall mean, standard error and the 95% confidence interval for the mean were computed using the MEANS procedure in SAS. To determine if there were any differences in alkyl glucosinolates, indolyl glucosinolates and sinapine values, a paired *t*-test was performed using Microsoft Excel version 5.0 software. The Westar control used in all statistical analyses was planted and grown in an adjacent plot to GT73 at each field location. Compositional data for Westar from the Co-Op Tests conducted in 1992 and 1993 were used for comparative purposes in graphs and tables but were not used in statistical analyses.

Four-week rat feeding study

Processed canola meal was added to rodent diets at a level of 10% (w/w) and fed to rats for four weeks. Ten varieties of canola were tested as follows: (1) RU3 (RRC line derived from GT73) and the parental variety (designated as Alliance) grown side-by-side with RU3 in Chile early 1996; (2) canola seed from five commercial varieties grown in different geographical locations across Canada in 1995 (designated Can 5, 6, 7, 8, 11) and (3) three commercial oilseed rape varieties grown in Europe (United Kingdom, France, Germany) in 1995 (designated OSR 1, 2, 3). The Roundup Ready trait from GT73, the original transformation event, was backcrossed into the Alberta Wheat Pool variety known as Alliance. After three backcrosses, the new variety was selfed several times to create what was identified in the Western Canadian Cooperative Rapeseed Recommending Trials (Co-Op Test) as RU3.

There were a total of 15 low erucic acid canola/oilseed rape varieties originally collected for the rat feeding study. Glucosinolate and compositional data from the lines were reviewed by Dr Sylvie Rabot of INRA Centre de Recherches de Jouy-en-Josas, France. Based on the data, Dr Rabot selected 10 varieties for testing that represent the normal variability in glucosinolate levels that are considered acceptable for commercial use. Seed from the 10 canola/oilseed rape lines was sent to POS for processing. Each lot of seed was divided into duplicate batches of approximately 75 kg of seed so that a total of 20 batches of seed were processed. Random numbers were assigned to each batch so that processing of batches was randomized. The canola seed was flaked, cooked (to inactivate enzymes such as myrosinase), pressed (to remove oil) and the press cake solvent was extracted (hexane) to remove the remaining oil. The canola meal was further desolventized (toasted) to remove residual hexane. Processing conditions were continually monitored to maintain uniformity across all batches.

The toasted canola/oilseed rape meal samples were subsequently analyzed at POS using AOAC methods: (1) crude protein; (2) crude fiber; (3) ash; (4) moisture; (5) acid detergent; (6) neutral detergent fiber; (7) nitrogen solubility; (8) myrosinase activity (non-AOAC method – Dr Lloyd Campbell, Department of Animal Science, University of Manitoba, Winnipeg, Manitoba) and (9) total and individual glucosinolates: 3-butenyl, 4-pentenyl, 2-hydroxy-3-butenyl, 2-hydroxy-4-pentenyl, 3-methylindol and 4-hydroxy-3-methylindol (HPLC AOAC method). The toasted canola/oilseed rape meal samples were sent to Monsanto's Environmental Health Laboratory (St. Louis, MO) for conduct of the rat feeding study. The toasted canola/oilseed rape meal samples were added to commercial rodent diets (PMI Inc., St. Louis, MO) at a level of 10% w/w. Male and female Sprague–Dawley rats were approximately 5–7 weeks of age at study initiation. Ten (10) male and ten (10) female rats were randomly assigned to each group. There were a total of 21 groups in the study. Twenty (20) groups were fed duplicate batches of either RU3, Alliance or one of the five Canadian or three European varieties of toasted canola/oilseed rape meal. One group was fed only rodent diet (diet control).

Animals were observed twice daily for signs of toxicity. Body weight and food consumption were recorded weekly for each animal. At the end of four weeks, all test animals were sacrificed by carbon dioxide asphyxiation, and the livers and kidneys were removed and weighed. These organs were weighed since it had been reported in the literature that feeding rats canola meal caused increased liver and kidney weights (Vermorel et al., 1987, 1988; Smith and Bray, 1992). The in-life phase of the rat feeding study was conducted in conformance with Good Laboratory Practice Standards of the FDA/EPA (US EPA, 1988; FDA, 1989) and Good Laboratory Principles of OECD (1981).

The following statistical procedure was used to detect statistically significant differences between the groups. A mixed linear model (Crowder and Hand, 1993) was used to compare group mean body weights, cumulative weight gain, food consumption, liver and kidneys weights and liver or kidney/body weight ratios. The SAS statistical program (SAS, 1990, 1992) was used to process both the in-life and organ weight data files to extract and compute the data necessary for analysis.

Results

Compositional analyses

Crude protein and fat

Measurement of protein and fat levels is a requirement of variety registration for any new canola variety in Canada. Crude protein in the defatted meal and the % fat in the whole seed were determined on seed obtained from Monsanto field trials in 1992 and 1993. Measurement of the % protein and fat in canola seed was conducted at RAL. Other nutrient parameters that were measured included ash, moisture, fiber and carbohydrate (calculated). There were no statistical differences ($p \leq 0.05$) between GT73 and Westar in seed protein or fat content for 1992 and 1993. There were also no differences in the other measured parameters. Since the results were very similar for both years, proximate determinations for 1993 are presented in Figure 7.2. The proximate values are typical of the published values for *B. napus* canola (Canola Council, 1991).

Protein and fat were also measured at Ag Canada Research Station Labs in Saskatoon using the same methods as those used to evaluate new canola varieties in the Co-Op Field Tests. The results of the Ag Canada analysis for canola seed obtained from the 1992 and 1993 field trials and the Co-Op tests are summarized in Figure 7.3. In 1992, the protein value (% of defatted meal) in GT73 was slightly, but statistically significantly ($p \leq 0.05$) higher compared to Westar; while in 1993, the fat

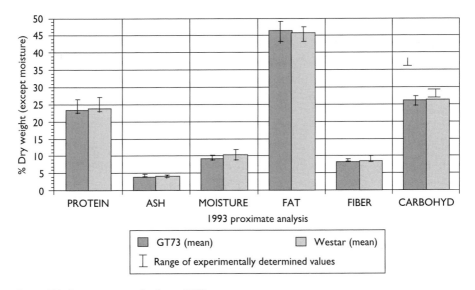

Figure 7.2 Proximate results from 1993.

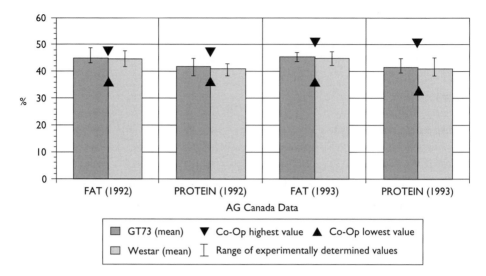

Figure 7.3 Crude protein and percentage fat values for RRC from 1992 and 1993.
Note: Fat is given as percentage of whole seed; protein is given as percentage of the defatted meal.

level (% of seed) in GT73 was slightly, but statistically significantly ($p \leq 0.05$) higher compared to Westar. The results are attributed to normal biological variation since increased protein (% of toasted meal) and increased fat (% of seed) were not consistently observed in both 1992 and 1993. Moreover, the increase in fat (% of seed) observed in 1993 (Ag Canada analysis) was not apparent in the aforementioned RAL analysis of fat from canola seed collected from the same 1992 and 1993 Monsanto field trials (compare Figures 7.2 and 7.3). In Figure 7.3, a range of values for protein and fat for Westar from the Co-Op Field Tests is also presented. In general, the range of protein and fat values for Westar from the Co-Op Tests ($n = 52$ in 1992 and $n = 87$ in 1993) encompass the values for GT73 and Westar from the 1992 and 1993 Monsanto field tests.

Amino acid composition

Samples of seed from three of seven Monsanto field trial sites in 1992 and the four Monsanto field trial sites in 1993 were analyzed for amino acid composition. Results of these analyses expressed on a per seed basis were very similar for both years. Results from 1993 are summarized in Figure 7.4.

Values obtained from the three sites in 1992 found comparable levels of amino acids for GT73 and Westar (data not shown). The aromatic amino acids were evaluated more closely because RRC expresses a glyphosate tolerant form of EPSPS which, in addition to endogenous plant EPSPS, is also involved in aromatic amino acid biosynthesis (Figure 7.1). As shown in Table 7.2, phenylalanine, tryptophan and tyrosine levels in RRC were not different when calculated either on a whole seed or on a per unit protein basis.

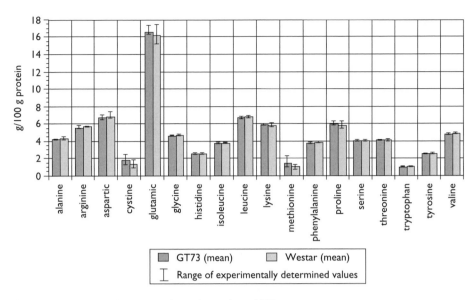

Figure 7.4 Amino acid analysis of canola seeds in 1993.

Table 7.2 Values of the aromatic amino acids in GT73 and Westar on a per seed and per protein basis from 1992 and 1993

	GT73		Westar	
	Mean	Range	Mean	Range
1992				
Amino acid				
Per seed basis[a]				
Phenylalanine	0.97	0.91–1.03	0.93	0.79–1.01
Tryptophan	0.25	0.23–0.27	0.24	0.21–0.26
Tyrosine	0.65	0.60–0.69	0.62	0.52–0.69
Protein basis[b]				
Phenylalanine	3.88	3.79–4.01	3.82	3.71–3.94
Tryptophan	1.01	0.98–1.03	0.98	0.97–0.99
Tyrosine	2.57	2.47–2.70	2.55	2.43–2.63
1993				
Per seed basis[a]				
Phenylalanine	0.88	0.82–1.01	0.91	0.86–1.01
Tryptophan	0.24*	0.23–0.28	0.26	0.24–0.29
Tyrosine	0.59	0.56–0.66	0.62	0.59–0.67
Protein basis[b]				
Phenylalanine	3.77	3.69–3.83	3.84	3.76–3.95
Tryptophan	1.03*	1.02–1.06	1.09	1.04–1.12
Tyrosine	2.53	2.50–2.57	2.58	2.50–2.68

* Significantly different from the value for Westar ($p \leq 0.05$).
[a] Values are given as g/100 g of seed on a dry weight basis, $n = 3$ in 1992 and $n = 4$ in 1993.
[b] Values are reported as g/100 g protein calculated by dividing the g/100 g seed by the determined protein value.

Values obtained from four sites in 1993 for all amino acids in GT73 and Westar were similar to those found in 1992 (data not shown). Statistical analysis of the aromatic amino acids was again conducted and a small but statistically significant ($p \leq 0.05$) decrease in the level of tryptophan was detected as shown in Table 7.2. Since the differences in tryptophan levels were not consistent between both years, they were attributed to normal biological variation. Moreover, the levels of all amino acids including tryptophan are well within the normal variation expected from canola (Baidoo and Aherne, 1985; Shahididi and Naczk, 1990; Bell and Keith, 1991; Canola Council, 1991).

Fatty acid composition

Seed samples from all seven 1992 Monsanto field trial sites and all four 1993 Monsanto field trial sites were analyzed for fatty acid composition at Ag Canada Research Station in Saskatoon. Since the fatty acid levels for GT73 and Westar were similar for both years, results from 1993 are presented in Figure 7.5. Also included in each graph are the ranges of values for the individual fatty acids from the 1993 Co-Op Tests ($n = 9$ in 1993). Statistical analysis of the four major fatty acids in canola oil (C16:0, palmitic acid; C18:1, oleic acid; C18:2, linoleic acid and C18:3, linolenic acid) found no differences between GT73 and Westar, with the exception of linoleic acid (C18:2). Linoleic acid in oil from GT73 was statistically significantly ($p \leq 0.05$) lower than Westar in both 1992 and 1993. The magnitude of the difference was small: 0.56% in 1992 (19.23% for GT73 compared to 19.79% for Westar) and 1.01% in 1993 (18.73% for GT73 compared to 19.74% for Westar). The levels of linoleic acid for GT73, while slightly different from Westar, were still within the range of linoleic acid values (18.3–22.1% of total fat content) from the 1992 and 1993 Co-Op trials for

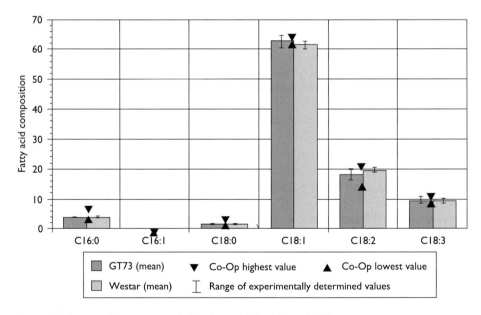

Figure 7.5 Fatty acid compositional data from 1993. C16 and C18.

Table 7.3 Values for erucic acid in 1992 and 1993 from GT73 and Westar

Year	Erucic acid mean		Erucic acid high		Erucic acid low	
	GT73	Westar	GT73	Westar	GT73	Westar
1992 (7 sites)	0.24	0.39	0.5	0.6	0.1	0.3
1993 (4 sites)	0.04	0.42	0.1	0.6	0.0	0.1

Westar. The levels for the other fatty acids were also generally within the normal ranges for the Westar variety as determined in the 1993 Co-Op tests.

An analysis of the level of erucic acid (C22:1) was also conducted for seed collected from the 1992 and 1993 Monsanto field test sites as illustrated in Table 7.3. Since erucic acid has been reported to produce myocardial lesions when fed to animals (Dupont *et al.*, 1989), the Canola Council of Canada (1991) recommended that the level of erucic acid in new canola varieties should not exceed 2% of total fatty acids. As shown in Table 7.3, the erucic acid levels were well below the 2% limit; and GT73 erucic acid levels were numerically lower than the Westar variety. Statistical analysis of erucic acid levels from the 1992 and 1993 field trials confirmed that at the 95% confidence limit, erucic acid levels for GT73 would not exceed the 2% recommended limit for canola oil.

Glucosinolates

Glucosinolate levels were analyzed in seed from GT73 and Westar in 1992 and 1993 Monsanto field trials. In Figure 7.6, total alkyl and indolyl glucosinolates (μmol/g

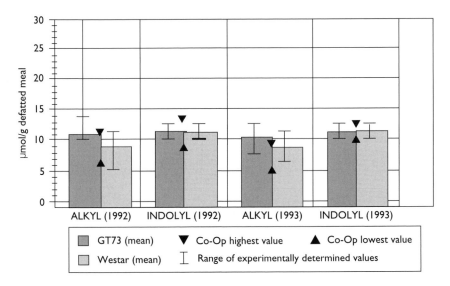

Figure 7.6 Alkyl and indolyl glucosinolate levels from 1992 and 1993.

Table 7.4 Glucosinolate profiles of GT73 and Westar in 1992 and 1993 (values are % of total glucosinolates)

Glucosinolate	GT73 (1992)	Westar (1992)	GT73 (1993)	Westar (1993)
Alkyls:				
3-Butenyl	14.4	13.0	14.0	12.7
4-Pentenyl	1.9	1.8	1.4	1.5
Hydroxybutenyl	31.8	27.6	31.8	28.8
Hydroxypentenyl	0.4	0.4	0.1	0.1
Thioalkyls:				
Methylthiobutenyl	0.9	1.0	0.8	0.9
Methylthiopentenyl	0.3	0.3	0.4	0.4
Indolyls:				
Methylindolyl	3.0	4.2	3.5	5.5
Hydroxyindolyl	47.3	51.8	48.0	50.2

defatted meal) are presented along with the range of glucosinolate values from Westar in the Co-Op Test. In both 1992 and 1993, the level of alkyl glucosinolates in GT73, while well below the canola limit of 30 μmol/g defatted meal, was statistically significantly higher than Westar. There were no statistically significant differences between Westar and GT73 in the total levels of indolyl glucosinolates. Statistical analysis of alkyl glucosinolate levels from the 1992 and 1993 field trials confirmed that at the 95% confidence limit, alkyl glucosinolate levels for GT73 would not exceed the 30 μmol/g defatted meal recommended limit for canola.

A detailed analysis of the levels of individual glucosinolates in GT73 and Westar was carried out to understand what was contributing to the differences in alkyl glucosinolate levels. The glucosinolate profiles are presented in Table 7.4. Most of the increase in alkyl glucosinolates for GT73 was accounted for by hydroxybutenyl glucosinolate, although the levels are well within the expected range for canola (Downey, personal communication) and well below the Canola Council recommended target level of 30 μmol/g defatted meal. This is further supported by analysis of glucosinolates in seed and defatted meal from canola/oilseed rape varieties used in the four-week rat feeding study.

Levels of glucosinolates in seed and processed (defatted) meal from canola/ oilseed rape varieties used in the four-week rat feeding study were also measured at POS. The levels of hydroxybutenyl glucosinolate in RU3 seed were lower than seed from some of the commercial lines of canola/oilseed rape used in the rat feeding study, e.g. 6.8 μmol/g seed for RU3 compared with 4.7 μmol/g seed for Alliance, 3.7–15.7 μmol/g seed for Canadian varieties and 9.6–20.0 μmol/g seed for European varieties.

The most compelling evidence that the aforementioned differences in alkyl glucosinolate levels between GT73 and Westar were not biologically meaningful are based on subsequent glucosinolate analysis of canola lines backcrossed with GT73. Alkyl glucosinolate levels in defatted meal from several backcrossed lines from five different canola seed companies were compared to the parental lines from which they were derived (Table 7.5). The levels of alkyl glucosinolates were generally comparable between the parental lines and their progeny resulting from backcrosses with GT73.

Table 7.5 Comparison of alkyl glucosinolates in canola lines from five different canola seed companies backcrossed with line GT73

Backcrossed line number	Alkenyl glucosinolates (μmol/g)[a]	
	Parental line	Backcrossed line
1–1 (company 1)	13.2	13.4
1–2	15.2	12.6
1–3	14.3	13.8
1–4	13.5	12.7
2–1 (company 2)	15.4	15.7
2–2	14.0	13.4
3–1 (company 3)	11.4	12.8
3–2	12.5	13.3
4–1 (company 4)	8.5	11.4
4–2	11.3	11.7
4–3	10.2	11.5
5–1 (company 5)	13.5	13.3
5–2	11.4	9.5
5–3	12.4	13.4
5–4	11.1	11.5
5–5	16.6	13.1

[a] Alkenyl glucosinolates are presented on a μmol/g defatted meal basis. They are the total of 3-butenyl, hydroxybutenyl, 4-pentenyl and hydroxypentenyl glucosinolates.
[b] The parent listed in the table refers to the female parent variety from the respective seed company, not the Westar parental variety for glyphosate-tolerant line GT73.

Sinapines

Sinapines are choline esters of naturally occurring plant phenolics derived from cinnamic acid derivatives. Furthermore, these compounds are downstream products of aromatic amino acid synthesis in canola. They are typically not monitored in new varieties of canola. Their presence in canola meal can impact poultry feed since they are known to render an off-odor in chicken eggs (Fenwick *et al.*, 1984). The results of our analyses in 1992 and 1993 are presented in Figure 7.7. Based on statistical analysis, no differences between GT73 and Westar in levels of sinapines were detected.

Processing study

Toasted meal

Toasted meal samples derived from GT73 and Westar seed were prepared at POS and extensively analyzed at RAL as illustrated in Table 7.6. In addition to the proximate and nitrogen solubility analyses, an extensive analysis of minerals and phytic acid was conducted at RAL. The results of these analyses and literature ranges for comparison are given in Table 7.7.

The mineral values from GT73-derived meal, as shown in Table 7.7, were all within the range of literature values and were in agreement with those values obtained from the Westar-derived sample. In addition, the proximate values for GT73-derived meal

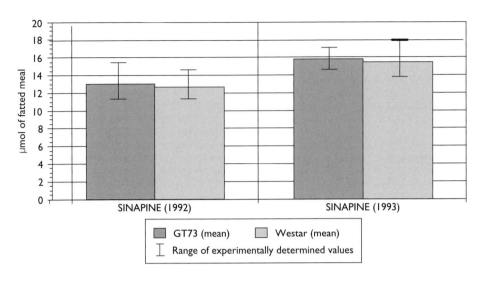

Figure 7.7 Sinapine levels from 1992 and 1993.

Table 7.6 Proximate analysis of toasted meal fractions

Analysis[a]	Line designation	
	GT73	Westar
Protein, % DW	41.3	41.4
Ash, % DW	7.24	7.33
Moisture, %	10.8	10.0
Fat, % DW	4.5	3.4
Crude fiber, % DW	14.5	14.6
Carbohydrate, % DW[b]	47.0	47.8
Calories, kcal/100 g[c]	332	330
Nitrogen solubility[d]	28.3	29.1

[a] Processing was conducted on a composite sample from four sites.
    All samples were dried prior to analysis.
[b] Carbohydrates were calculated by subtraction of protein, fat and ash.
[c] Calories are calculated from the protein, fat and carbohydrate values.
[d] Nitrogen solubility was determined from water extraction at neutral pH.

Table 7.7 Mineral and phytic acid analysis of toasted meal fractions

Analysis[a]	Line designation		
	GT73	Westar	Literature ranges[b]
Calcium, % DW	0.82	0.76	0.57–0.82
Copper, µg/g	7.56	7.06	4.9–8.0
Iron, µg/g	160	194	116–204
Magnesium, % DW	0.64	0.57	0.49–0.64
Manganese, µg/g	53.2	48.5	30.0–62.9
Phosphorus, % DW	1.23	1.19	1.08–1.33
Potassium, % DW	1.42	1.38	1.20–1.46
Zinc, µg/g	64.7	55.0	59.0–80.9
Phytic Acid, % DW	3.33	3.09	2.0–5.0

[a] Processing was conducted on a composite sample from four sites. All samples were dried prior to analysis.
[b] Literature values obtained from Bell and Keith (1987, 1991) and Bell *et al.* (1984).

(Table 7.6) were representative of canola and comparable to the values from the Westar-derived sample. These results confirm that meal derived from GT73 is substantially equivalent to meal derived from Westar canola.

Refined oil

Refined oils derived from GT73 and Westar seed were analyzed at RAL and compared with standards published in the Food Chemical Codex (1981) (Table 7.8). With the exception of fatty acids C22:0, C24:0 and C24:1, the fatty acid levels in GT73 and Westar oil were within the definition of canola oil.

Table 7.8 Analysis of refined oil fractions

Analysis[a]	Line designation		
	GT73	Westar	Codex specifications
Fatty acid analysis			
14:0	0.05	0.05	<0.2
16:0	3.99	4.01	<6.0
16:1	0.20	0.18	<1.0
18:1	2.10	2.05	<2.5
18:1	64.0	63.3	>50.0
18:2	19.0	19.8	<40.0
18:3	6.74	6.27	<14.0
20:0	1.06	1.02	<1.0
20:1	1.67	1.82	<2.0
22:0	0.52	0.51	<0.5
22:1	0.19	0.39	<2.0
24:0	0.23	0.24	<0.2
24:1	0.31	0.30	<0.2
Other tests			
Acid value	0.03	0.03	≤6
Arsenic[b]	<0.2	<0.2	≤0.5 mg/kg
Cold test	Pass	Pass	Pass
Lovibond color	0.2 red; 2 yellow	0.2 red; 2 yellow	≤1.5 red; ≤15 yellow
Erucic acid	0.19	0.39	≤2.0%
Free fatty acids[c]	0.02	0.02	≤0.05%
Heavy metals[d]	3: <5; 1: >5	<5	≤5 mg/kg
Iodine value	110	111	110–126
Lead	<0.1	<0.1	≤0.1 mg/kg
Linolenic acid	6.74	6.27	≤15.0%
Peroxide value	0.7	0.7	≤10 meq/kg
Refractive index	1.4657	1.4658	1.465–1.467
Saponifiable value	190.1	189.3	178–193
Stability	17	17	≥7 h
Sulfur	2.12	4.04	≤10 mg/kg
Unsaponifiable matter	0.460	0.400	≤1.5%
Moisture	<0.1	<0.1	≤0.1%

[a] All analyses were performed by Ralston Analytical Laboratories.
[b] This method measures both inorganic and organic arsenicals.
[c] Free fatty acids were calculated as oleic acid.
[d] Values were calculated as lead sulfide.

Table 7.9 Comparison of replicates, dry basis, oil-free results for toasted meal

Variety:	RU3		Alliance		Can 05		Can 06		Can 07		Can 08		Can 11		OSR 1		OSR 2		OSR 3	
Sample no:	7	18	8	20	9	11	6	16	13	21	12	19	3	15	1	17	2	5	10	14
Protein %	46.2	46.2	47.7	47.1	42.3	41.9	40.8	41.3	49.3	47.7	43.7	43.6	42.9	43.4	40.7	41.2	38.4	39.1	40.1	41.4
Ash %	8.5	8.5	9.3	9.2	8.5	8.4	8.5	8.5	7.3	7.3	8.0	7.8	8.1	8.2	7.7	7.6	7.4	7.5	7.5	7.4
Crude fiber %	13.2	13.0	11.2	11.4	12.7	11.9	14.2	14.7	13.1	13.8	13.9	13.5	13.8	13.5	17.1	16.2	17.0	16.5	17.1	15.8
NSI[a]	22.2	22.3	21.1	30.3	25.7	26.9	27.0	22.2	33.2	30.7	28.9	27.6	29.3	27.4	27.9	25.4	19.5	18.5	21.0	18.3
Glucosin, µmol/g																				
2-OH-3-Butenyl	6.3	5.8	4.5	4.4	12.2	11.9	13.2	13.6	8.9	7.9	8.6	8.1	3.7	3.9	6.9	8.3	16.7	15.1	11.1	11.7
4-OH-3-Methyindol	4.6	4.4	6.2	7.3	8.0	8.1	5.0	5.0	9.4	5.5	8.4	6.4	11.3	9.3	5.5	4.3	7.5	5.1	5.2	4.1
Total[b] glucosin	14.3	13.6	14.1	15.1	27.6	27.5	24.0	24.8	24.0	18.7	21.7	19.2	18.3	16.8	16.0	17.0	32.1	27.3	27.6	27.9

[a] Nitrogen solubility index.
[b] Total glucosinolates, 2-OH-3-butenyl, 4-OH-3-methyindol. (Other glucosinolates were also measured and included in the total. Their values were low and are not summarized here. They include 3-butenyl, 4-pentenyl, 2-OH-4-pentenyl, 3-methylindol glucosinolates.)

Although the levels of these three fatty acids were slightly higher than the Codex standards for both the GT73- and Westar-derived oil samples, they are minor components of the oil and are not considered essential oils for human nutrition. Thus, no impact on human health would be expected because of slightly higher values than Codex specifications. The slight variation in their levels relative to the Codex Specifications may be a result of breeding lower levels of erucic acid in the Westar variety from which GT73 was derived. The results of the analysis of refined oil derived from GT73 seed confirm that GT73 is substantially equivalent to canola that is currently being marketed. Furthermore, the safety of refined oil from RRC is equivalent to oil refined from non-modified canola.

Four-week rat feeding study

Results of compositional analysis for duplicate samples of toasted canola meal/oilseed rape are found in Table 7.9. While there were occasional numerical differences between different canola/oilseed rape varieties for some measured parameters, there was generally good agreement between duplicates of the same variety. Processing of canola/oilseed rape to toasted meal produced, on average, a 30% reduction in total glucosinolate levels. Minimal or no detectable myrosinase activity was detected in the toasted meal samples, indicating that processing conditions were sufficient to inactivate myrosinase and reduce glucosinolate levels.

During the course of the four-week feeding study, no test article-related adverse clinical signs were observed. Consistent with the literature, body weights and weight gains were slightly reduced for nearly all of the groups that were fed canola/oilseed rape meal (Vermorel *et al.*, 1988; Smith and Bray, 1992) when compared with the diet control. Food consumption was comparable for males in all groups compared with the diet control. For females, food consumption was slightly lower for some groups fed canola/oilseed rape compared with the diet control (data not presented).

Liver/body and kidney/body weights were slightly increased for most groups fed canola/oilseed rape meal when compared with the diet control pictured in Figure 7.8 and Figure 7.9. These increases were due, in part, to the slight reduction in body weights since absolute liver and kidney weights were generally similar to those of diet controls. There were no statistically significant differences ($p \leq 0.05$) in body weight, body weight gain, food consumption, absolute liver and kidney weight, liver and kidney/body weights for male and female rats fed RU3 canola meal when compared with the Alliance variety and the population mean for the commercial canola/oilseed rape varieties that were fed to rats. Data summarizing comparisons of terminal body weight and liver and kidney/body weight between RU3 and the population mean for commercial varieties or the Alliance variety are found in Table 7.10.

Discussion

The basis of our safety assessment of RRC and the widely accepted basis of assessing the safety of any transgenic food/feed crop is establishing that the initial transgenic line is substantially equivalent to its parental counterpart (WHO, 1995). The key parameters that should be assessed to confirm substantial equivalence are the agronomic properties of the crop, the compositional analysis of critical nutrients and, if

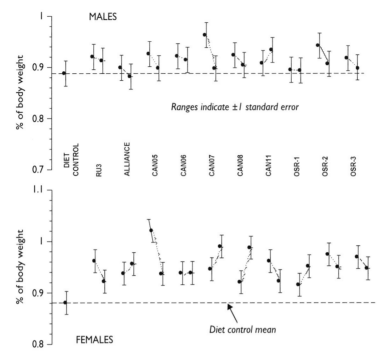

Figure 7.8 Paired kidney/body weight means (± standard error) for each experimental group. Standard errors are based on common within-group variance. Replicate batch means for the same variety are joined by a dotted line with the earlier processed batch first.

present, important anti-nutrients. The nutritional or anti-nutritional components that are selected for analysis should be those that would have a meaningful impact on human or animal health should levels be significantly altered in the plant. Regulators in the United States and Europe are collaborating to develop a harmonized list of important nutrients and anti-nutrients/toxicants for each major food/feed crop that could be analyzed to confirm substantial equivalence for a biotech crop.

With regard to evaluation of agronomic properties, a number of parameters have been assessed over consecutive growing seasons, including germination; time of flowering; dormancy; over-wintering capacity; pod shattering, pollen fertility time to maturity, seed production; disease and insect susceptibility. GT73 is typical of *B. napus* canola in all these measured properties (data not presented).

The focus of the food and feed safety assessment of RRC is compositional analysis. Data presented demonstrate that GT73 is substantially equivalent to non-modified *B. napus* canola. Only two components showed consistent and statistically significant differences between GT73 and Westar: the levels of linolenic acid and alkyl gluco-sinolates. In both cases, the differences were of small magnitude and are attributed to the fact that genetic transformation to produce line GT73 represents the selection of a cell from a single seed from the Westar population. Since seed from Westar canola exhibit a range in concentration of nutrients and anti-nutrients, the seed selected for transformation, while within this range, might be expected to exhibit some differences

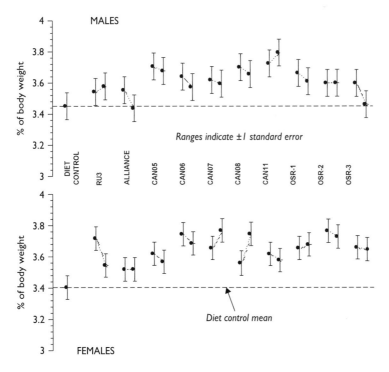

Figure 7.9 Liver/body weight means (± standard error) for each experimental group. Standard errors are based on common within-group variance. Replicate batch means for the same variety are joined by a dotted line with the earlier processed batch first.

Table 7.10 Comparison of terminal body weight (g) and liver and kidney weights (g or % body wt) of RU3-fed rats compared with the alliance variety and the population mean for commercial canola/oilseed rape varieties

	RU3	C/OSR	Difference p-value[a]	Alliance	Difference p-value
Males	Mean	Mean		Mean	
Body weight, g	369	368	0.881	373	0.651
Kidney weight, g	3.38	3.35	0.640	3.31	0.440
Kidney weight, % body weight	0.917	0.913	0.824	0.890	0.293
Liver weight, g	13.1	13.3	0.730	13.0	0.806
Liver weight, % body weight	3.56	3.62	0.468	3.50	0.446
Females					
Body weight, g	219	213	0.392	222	0.697
Kidney weight, g	2.06	2.03	0.467	2.10	0.535
Kidney weight, % body weight	0.94	0.95	0.560	0.95	0.874
Liver weight, g	7.96	7.77	0.438	7.80	0.528
Liver weight, % body weight	3.63	3.65	0.790	3.52	0.139

[a] The observed significance level for a test that the RU3 and C/OSR or Alliance means are equal. A *p*-value less than 0.05, for example, would indicate that the difference is significant at the 5% level.

from the Westar population mean for nutrients and anti-nutrients. The nutrient and anti-nutrient composition data were reviewed in detail by food safety experts in Health Canada and the United Kingdom Ministry of Agriculture, Fisheries and Food who concluded that these differences were not biologically meaningful and approved the oil for human consumption.

The alkyl glucosinolate results were also reviewed by plant breeders at the Canadian Food Inspection Agency who developed the Westar variety. They concluded: 'In our opinion, this is a very minor deviation and one which would be expected when single plants are selected from the heterozygous plant population that constitutes the cultivar Westar' (Downey, personal communication, 1994). Subsequent to this testing, the levels of the alkyl glucosinolates in backcrossing events with GT73 and other canola varieties have been monitored by canola seed companies and found to be comparable to the glucosinolate levels in the parental varieties. To date, a number of varieties of RRC have been approved for cultivation in Canada. In all cases, the fatty acid and glucosinolate composition have been reviewed by experts responsible for canola variety registration in Canada (Western Canada Canola Rape Seed Recommending Committee) and judged to be acceptable for canola.

Additional compositional analysis was completed on processing fractions derived from GT73 and Westar seed samples grown at the same locations in 1993. Samples representative of commercially prepared toasted meal and refined oil were obtained and extensively analyzed. These confirmatory results demonstrated that processed fractions derived from GT73 and Westar are substantially equivalent.

The results of the four-week rat feeding study also confirmed the equivalence of RU3 (RRC line derived from GT73) to other commercial canola/oilseed rape varieties. The slight reductions in body weights and associated increases in liver and kidney/body weights observed in all canola/oilseed rape diets have been reported in other rat feeding studies with oilseed rape (Vermorel et al., 1987, 1988; Smith and Bray, 1992). There does not always appear to be a dose relationship between the levels of glucosinolates and the changes in relative liver and kidney weights. Smith and Bray (1992) found no dose-related increase in liver/body weight as the levels of canola meal in the rat diets were increased incrementally from 10% to 40% (w/w). Vermorel et al. (1987) found no dose-related change in liver/body weights for rats fed rapeseed meal with glucosinolate concentrations in meal ranging from 4 to 36 μmol/g.

Duplicate samples of toasted canola/oilseed rape were randomly processed in order to test the variation that might occur during processing. The processing of the batches was carefully controlled to maintain uniformity across all batches as nitrogen solubility indices, an indicator of the degree of processing was comparable across batches.

It was also of interest to assess the responses of animals to feeding duplicate batches of toasted canola/oilseed rape meal. As shown in Figures 7.8 and 7.9, liver and kidney weights were generally within 1 standard error for most duplicates. However, for 5 out of 40 duplicate measurements, means for liver or kidney/body weights were greater than 1 standard error between duplicate samples. This points out the random variation that can occur in feeding studies that can sometimes confound the safety evaluation.

The absence of any statistical differences in measured parameters between rats fed RU3 (GT73 derived) toasted canola meal at 10% (w/w) in the diet and rats fed processed meal from nine commercial varieties provides confirmatory evidence of the substantial equivalence of RRC to conventional oilseed rape/canola varieties.

Conclusion

This paper outlines a detailed experimental strategy to assess the safety of canola that has been genetically modified to be tolerant to the herbicide glyphosate. The basis for assessing safety is focused on extensive compositional analysis, the results of which lead to the scientific conclusion of substantial equivalence. Stated another way, RRC is 'as-safe-as' traditional canola. This conclusion is based on the results from over 1000 individual assays conducted over two years in addition to the detailed safety assessment of the introduced proteins and molecular analysis. In addition to establishing the substantial equivalence through field studies (measuring agronomic performance), compositional analyses confirmatory studies, such as the analysis of processing fractions and the four-week rat feeding study, further established that RRC is as acceptable for production of food and feed as non-modified canola.

Acknowledgements

This paper is dedicated to the late Dr J. Milton Bell,
a consumate scientist and gentleman.

We acknowledge help received from Aaron Mitchell (Monsanto Canada Ltd., Saskatoon, Saskatchewan), Roy Fuchs and Mary Taylor (Monsanto Company, St. Louis), Dr J.M. Bell (University of Saskatchewan), Dr J. Daun (Canadian Grains Laboratory), Drs R.K. Downey, D.I. MacGregor and J.P. Raney (Agriculture and AgriFood Canada Research Station, Saskatoon, Saskatchewan) and Mr R. Kruger (POS Pilot Plant Corporation, Saskatoon, Saskatchewan). In addition, special thanks go to Drs Steve Padgette and Ganesh Kishore of Monsanto Company, St. Louis, and Rob Neyedley, Clint Jurke and Mayson Maerz of Monsanto Canada Inc., Saskatoon, Saskatchewan for their support and invaluable contributions.

The statistical analysis of the compositional data and the data from the rat feeding study was completed by Larry Holden, Monsanto, St. Louis.

Mark W. Naylor was the study director for the four-week rat feeding study conducted at Monsanto's Environmental Health Laboratory.

References

AACC (1987) 'Moisture-air-oven methods, method 44-15A', in *Approved Methods of the AACC*. St. Paul: American Association of Cereal Chemists.
AOAC (1984a) 'Ash of flour, direct method, final action, method 14.006; Ash, method 10.178; Ash, method 14.063; Ash of bread, method 14.098; Baked products, ash, method 14.117; Ash, method 14.130; Ash, method 16.216', in *AOAC Official Methods of Analysis*. Washington, DC: AOAC International.
AOAC (1984b) 'Fiber (crude) in animal feed, ceramic fiber filter method, method 7.066–7.070', in *AOAC Official Methods of Analysis*. Washington, DC: AOAC International.
AOAC (1984c) 'Plants, preparation of sample, final action, method 3.002; Sampling of animal feed, drying at 135°, method 7.007; Soybean flour, final action, method 14.081', in *AOAC Official Methods of Analysis*. Washington, DC: AOAC International.
AOAC (1989) 'Methods 965.09, 968.08 and 984.27', in *AOAC Official Methods of Analysis*. Washington, DC: AOAC International.

AOAC (1990a) 'Fat (crude) or ether extract in animal feed, method 920.39C (modified)', in *AOAC Official Methods of Analysis*. Washington, DC: AOAC International.

AOAC (1990b) 'Method 986.11', in *AOAC Official Methods of Analysis*. Washington, DC: AOAC International.

AOAC (1990c) 'Protein (crude) in animal feed, $CuSO_4/TiO_2$ mixed catalyst Kjeldahl method, method 988.05', in *AOAC Official Methods of Analysis*. Washington, DC: AOAC International.

AOAC (1990d) 'Sulfur amino acids in food and feed ingredients, method 985.28', in *AOAC Official Methods of Analysis*. Washington, DC: AOAC International.

AOCS (1989) 'Method Ba 11-65', in *Official and Tentative Methods of the AOCS*. Champaign, IL: American Oil Chemists Society.

Baidoo, S.K. and Aherne, R.X. (1985) 'Canola meal for livestock and poultry', *Agricultural and Forestry Bulletin* 8: 21–26.

Bannon, C.D., Breen, G.J., Craske, J.D. *et al.* (1982) 'Analysis of fatty acid methyl esters with high accuracy and reliability III, Literature review of and investigations into the development of rapid procedures for the methoxide-catalysed methanolysis of fats and oils', *Journal of Chromatography* 247: 71–89.

Barry, G., Kishore, G., Padgette, S. *et al.* (1992) 'Inhibitors of amino acid biosynthesis: strategies for imparting glyphosate tolerance to crops plants', in B.K. Singh, H.E. Flores and J.C. Shannon (eds) *Biosynthesis and Molecular Regulation of Amino Acids in Plants*. Rochrille, MD: American Society of Plant Physiologists, pp. 39–145.

Baudet, J.J., Burghart, P. and Evrard, J. (1988) *Colza Cahier Technique. Tourteau: a basse tenuer en glucosinolates*. Paris: CETIOM.

Bell, J.M. and Keith, M.O. (1987) 'Feeding value for pigs of canola meal derived from Westar and Triazine-tolerant cultivars', *Canadian Journal of Animal Sciences* 67: 811–19.

Bell, J.M. and Keith, M.O. (1991) 'A survey of variation in the chemical composition of commercial canola meal produced in western canadian crushing plants', *Canadian Journal of Animal Sciences* 71: 469–80.

Bell, J.M., Keith, M.O., Blake, J.A. *et al.* (1984) 'Nutritional evaluation of ammoniated mustard meal for use in swine feeds', *Canadian Journal of Animal Sciences* 64: 1023–33.

Bhatly, R.S. (1985) 'Comparison of the soxtec and goldfisch systems for determination of oil in grain species', *Canadian Institute of Food Science and Technology Journal* 18: 181–84.

Bjerg, B., Olsen, O., Rasmussen, K.W. *et al.* (1984) 'New principles of ion-exchange techniques suitable to sample preparation and group separation of natural products prior to liquid chromatography', *Journal of Liquid Chromatography* 7: 691–707.

Blackburn, S. (1978) *Amino Acid Determination – Methods and Techniques*. New York: Marcel Dekker.

Canola Council of Canada (1991) *Canada's Canola*. Winnipeg, Manitoba: Canola Council of Canada.

Clandinin, D.R. (ed.) (1990) *Canola Meal for Livestock and Poultry*. Winnipeg, Manitoba: Canola Council of Canada, pp. 1–24.

Conacher, H.B.S. and Chadha, R.K. (1974) 'Determination of docosenoic acids in fats and oils by gas-liquid chromatography', *Journal of AOAC* 57: 1161–64.

Conway, T.F. and Earle, F.R. (1963) 'Nuclear magnetic resonance for determining oil content of seeds', *Journal of the American Oil Chemists' Society* 40: 265–68.

Crowder, M.J. and Hand, D.J. (1993) '*Analysis of Repeated Measures*'. New York: Chapman and Hall.

Downey, J.K. and Bell, J.M. (1990) 'New developments in canola research', in F. Shahidi (ed.) *Canola and Rapeseed, Production, Chemistry, Nutrition, and Processing Technology*. New York: Van Nostrand Reinhold, pp. 37–46.

Dupont, J., White, P.J., Johnston, K.M. *et al.* (1989) 'Food safety and health effects of canola oil', *Journal of the American College of Nutrition* 8: 360–75.

Edeling, M.E. (1968) 'The Dumas Method for nitrogen in feeds', *Journal of the Association of Official Analytical Chemists* 51: 766–70.

Ellis, R. and Morris, E.R. (1983) 'Improved ion-exchange phytate method', *Cereal Chemistry* 60: 121–24.

Fenwick, G.R., Heaney, R.K. and Mullin, W.J. (1982) 'Glucosinolates and their breakdown products in food and food plants', *CRC Critical Reviews in Food Science and Nutrition* 18: 123–201.

Fenwick, G.R., Curl, C.L., Pearson, A.W. *et al.* (1984) 'The treatment of rapeseed meal and its effect on the chemical composition and egg tainting potential', *Journal of the Science of Food and Agriculture* 35: 757–61.

Fenwick, G.R., Heaney, R.K. and Mawson, R. (1989) 'Glucosinolates', in P.R. Cheeke (ed.) *Toxicants of Plant Origin, Vol. II: Glycosides.* Boca Raton, FL: CRC Press.

Food Chemical Codex (1981) *Third Supplement III. Food Chemical Codex,* pp. 104–106.

Foster, M.L. and Gonzales, S.E. (1992) 'Soxtec fat analyzer for the determination of total fat in meat, Collaborative Study', *Journal of AOAC International* 75: 288–92.

Freedman, B., Kwolek, W.F. and Pryde, E.H. (1986) 'Quantitation in the analysis of trans-esterified soybean oil by capillary gas chromatography', *Journal of the American Oil Chemists' Society* 63: 1370–74.

Harland, B.F. and Oberleas, D. (1977) 'A modified method for phytate analysis using an ion-exchange procedure: application to textured vegetable proteins', *Cereal Chemistry* 54: 827–32.

Harrison, L.A., Bailey, M.R., Naylor, M.W. *et al.* (1996) 'The expressed protein in glyphosate-tolerant soybean, 5-enolpruvylshikimate-3-phosphate synthase from *Agrobacterium* sp. strain CP4, is rapidly digested in vitro and is not toxic to acutely gavaged mice', *Journal of Nutrition* 126: 728–40.

Heaney, R.K. and Fenwick, G.R. (1980) 'The analysis of glucosinolates in *Brassica* species using gas chromatography. Direct determination of the thiocyanate ion precursors, glucobrassicin and neoglucobrassicin', *Journal of the Science of Food and Agriculture* 31: 593–99.

Hiroyuki, S., Seino, T., Kobayashi, T. *et al.* (1984) 'Determination of the tryptophan content of feed and feedstuffs by ion exchange liquid chromatography', *Agricultural Biology and Chemistry* 48: 2961–69.

Hougen, F.W. and Bodo, V. (1973) 'Extraction and methanolysis of oil from whole crushed rapeseed for fatty acid analysis', *Journal of the American Oil Chemists' Society* 50: 230–34.

Howard, E., Citovsky, B. and Zambryski, P. (1990) 'Transformation: the T-complex of *Agrobacterium tumefaciens*', in C.J. Lamb and R.N. Ceachy (eds) *Plant Gene Transfer.* New York: Alan R. Liss, pp. 1–12.

JMPR (1998) 'Joint Meeting on Pesticide Residues (JMPR), Lyon France, September 22–October 1st (1997)', *FAO Plant Production and Protection Paper* 145: 125–34.

Jones, A.D., Hitchcock C.H.S. and Jones, G.H. (1981) 'Determination of tryptophan in feeds and feed ingredients by high-performance liquid chromatography', *Analyst, London* 106: 968–73.

Klassen, A.J., Downey, R.K. and Capcara, J.J. (1987) 'Westar summer rape', *Canadian Journal of Plant Science* 67: 491–93.

Kore, A.M., Spencer, G.F. and Wallig, M.A. (1993) 'Purification of the ω – (methylsulfinyl)alkyl glucosinolate hydrolysis products: 1-isothiocyanoto–3-(methylsulfinyl)propane, 1-isothiocyanoto-4-(methylsulfinyl)butane, 4-(methylsulfinyl)butanenitrile, and 5-(methylsulfinyl)pentanenitrile from broccoli and *Lesquerella fendler*', *Journal of Agriculture and Food Chemistry* 41: 89–95.

Kramer, J.K.G. and Sauer, F.D. (1983) 'Results obtained with feeding low erucic acid rapeseed oils and other vegetable oils to rats and other species', in J.K.G. Kramer, F.D. Sauer and W.P. Pigden (eds) *High and Low Erucic Acid Rapeseed Oils. Production, Usage, Chemistry, and Toxicological Evaluation*. New York: Academic Press, pp. 413–74.

Malik, J., Barry, B. and Kishore, B. (1989) 'The herbicide glyphosate', *BioFactors* 2: 17–25.

Moore, S. (1963) 'On the determination of cystine as cysteic acid', *Journal of Biological Chemistry* 238: 235–37.

Moore, S. and Stein, W.H. (1954) 'Procedures for the chromatographic determination of amino acids on four per cent cross-linked sulfonated polystyrene resins', *Journal of Biological Chemistry* 211: 893–906.

Moore, S. and Stein, W.H. (1963) 'Chromatographic determination of amino acids by the use of automatic recording equipment', *Methods in Enzymology* VI: 819–31.

NIAB (1993) *Varieties of Oilseed Crops, Farmers Leaflet No. 9*. Cambridge, UK: National Institute of Agricultural Botany.

OECD (1981) *Principles of Good Laboratory Practice*, Annex 2, (81)30.

OECD (1992) *OECD Safety Evaluation of Foods Derived by Modern Biotechnology: Concepts and Principles*. Paris: Organisation for Economic Co-operation and Development.

Padgette, S.R., Re, D.B., Barry, G.F. *et al.* (1996) 'New weed control opportunities: Development of soybeans with a *Roundup Ready™* gene', in S.O. Duke, (ed.) *Herbicide Resistant Crops*. Boca Raton: CRC Press, pp. 53–84.

Paik, I.K., Roblee, A.R. and Clandinin, D.R. (1980) 'Products of the hydrolysis of rapeseed glucosinolates', *Canadian Journal of Animal Sciences* 60: 481–93.

Raney, J.P., Love, H.K., Rakow, G.F.W. *et al.* (1987) 'An apparatus for rapid preparation of oil and oil-free meal from Brassica seed', *Fat Science and Technology* 235–37.

SAS (1990) *SAS Language: Reference*, Version 6, 1st edition. Cary, NC: SAS Institute, Inc.

SAS (1992) 'The MIXED procedure', chapter 16 in *SAS Technical Report P-229. SAS/STAT Software: Changes and Enhancements, Release 6.07*. Cary, NC: SAS Institute, Inc.

Sauer, F.D. and Kramer, J.K.G. (1983) 'The problems associated with the feeding of high erucic acid rapeseed oils and some fish oils to experimental animals', in J.K.G. Kramer, F.D. Sauer and W.P. Pigden (eds) *High and Low Erucic Acid Rapeseed Oils. Production, Usage, Chemistry, and Toxicological Evaluation*. New York: Academic Press, pp. 254–88.

Scarisbrick, D.H. and Daniels, R.W. (eds) (1986) 'Oilseed rape', *Collins Professional and Technical Books*. Chatham, Kent: MacKays.

Schram, E., Moore, S. and Bigwood, E.J. (1954) 'Chromatographic determination of cystine as cysteic acid', *Biochemistry Journal* 57: 33–37.

Shahididi, F. and Naczk, M. (1990) 'North American production of canola', in F. Shahididi (ed.) *Canola and Rapeseed, Production, Chemistry, Nutrition, and Processing Technology*. New York: Van Nostrand Rheinhold.

Smith, T.K. and Bray, T.M. (1992) 'Effect of dietary cysteine supplements on canola meal toxicity and altered hepatic glutathione metabolism in the rat', *Journal of Animal Sciences* 70: 2510–15.

Sorenson, H. (1990) 'Glucosinolates: structure-properties-function', in F. Shahididi (ed.) *Canola and Rapeseed, Production, Chemistry, Nutrition, and Processing Technology*. New York: Van Nostrand Rheinhold.

Sweeney, R.A. (1989) 'Generic combustion method for determination of crude protein in feeds, Collaborative Study', *Journal of AOAC* 72: 770–74.

Sweeney, R.A. and Rexroad, P.R. (1987) 'Comparison of LECO FP-228 "Nitrogen Determinator" with AOAC copper catalyst kjeldahl method for crude protein', *Journal of AOAC* 70: 1028–30.

Thies, W. (1971) 'Rapid and simple analyses of the fatty-acid composition in individual rape cotyledons', *I. Zeitschrift für Pflanzenzuchtg* 65: 181–202.

Thies, W. (1980) 'Analysis of glucosinolates via "on column" desulfation', in *Analytical Chemistry of Rapeseed and Its Products*. Winnipeg: Canola Council of Canada, pp. 66–71.

US EPA (1988) *Good Laboratory Practice Standards*, 40 Code of Federal Regulations, part 160.

US EPA (1993) *Reregistration Eligibility Document (RED): Glyphosate*. Washington, DC: Office of Prevention, Pesticides and Toxic Substances.

US FDA (1989) *Good Laboratory Practice for Nonclinical Laboratory Standards*, 21 Code of Federal Regulations, part 58.

US FDA (1985) 'Direct food substances affirmed as generally recognized as safe; low erucic acid rapeseed oil. US Food and Drug Administration Bureau of Foods', *Federal Register* 50(18): 3745–44.

Vermorel, M., Davicco, M.J. and Evrad, J. (1987) 'Valorization of rapeseed meal. 3. Effects of glucosinolate content on food intake, weight gain, liver weight and plasma thyroid hormone levels in growing rats', *Reproduction, Nutrition, Development* 27(1A): 57–66.

Vermorel, M., Heaney, R.K. and Fenwick, G.R. (1988) 'Antinutritional effects of the rapeseed Mrals, Darmor and Jet Neuf, and progoitrin together with myrosinase, in the growing rat', *Journal of the Science of Food and Agriculture* 44: 321–34.

Wallig, M.A., Kingston, S., Staack, R. *et al.* (1998) 'Induction of rat pancreatic glutathione S-transferase and quinone reductase activities by a mixture of glucosinolate breakdown derivatives found in Brussel sprouts', *Food and Chemical Toxicology* 36(5): 365–73.

White, F. F. (1989) 'Vectors for gene transfer in higher plants', in S. Kung and C.J. Arntzen (eds) *Plant Biotechnology*. Boston, MA: Butterworth, pp. 3–34.

WHO (1995) *Application of the Principles of Substantial Equivalence to the Safety Evaluation of Foods or Food Components from Plants Derived from Modern Biotechnology. A Report of a WHO Workshop*. WHO/FNU/FOS/95.1. Geneva: World Health Organization, Food Safety Unit.

Winge, R.K., Fassel, V.A., Peterson, V.J. *et al.* (1985) *Inductively Coupled Plasma Atomic Emission Spectroscopy, an Atlas of Spectral Information*. New York: Elsevier.

Chapter 8

Case study: Bt crops

A novel mode of insect control

Brian A. Federici

Introduction

The bacterium *Bacillus thuringiensis* (Bt) is distinguished from other bacilli by the production of large crystals during sporulation composed of one or more insecticidal proteins. These proteins have served as the principal active ingredient of numerous commercial bacterial insecticides used worldwide for more than 40 years to protect crops against caterpillar pests, and more recently to control coleopterous insects and mosquito and blackfly vectors of human and animal diseases (Federici, 1999). The first genes encoding insecticidal Bt proteins were cloned early in the 1980s. This led quickly to their use to construct recombinant bacterial insecticides containing novel combinations of these proteins and insect-resistant Bt crops (Estruch *et al.*, 1996). The first Bt crops became commercially available in the United States in the mid-1990s, and have been widely adopted by farmers despite higher seed costs in comparison with conventional crops. During the 1999 growing season, farmers in the United States planted over 20 million hectares of insect-resistant Bt transgenic crops, including over 10 million hectares of Bt maize (corn), 1.5 million hectares of Bt cotton, and about 30 000 hectares of Bt potatoes (Thayer, 1999; Shelton *et al.*, 2002). This acreage is expected to grow to about 15 million hectares of corn and 3 million hectares of cotton within five years, representing, respectively, about a third of the corn and half of the cotton acreage in the United States. Bt crops are also being grown in China and Argentina, and their potential deployment is being assessed in several other countries. In addition, more than 30 other Bt crops, including rice, sorghum, most major vegetable crops as well as many tree crops used for fruit, nut and fiber production are under development (Schuler *et al.*, 1998).

For farmers, Bt crops offer advantages over conventional crops in that the insecticidal proteins are produced directly by the plants, and continuously during most of the growing season, thereby reducing the material and application costs of using synthetic chemical insecticides. Growers have reported improved profit margins over conventional varieties averaging $60 to $100 per hectare during the first few years of Bt cotton plantings in the southeastern United States. Consumers and the environment benefit from reductions in the use of synthetic chemical insecticides. For example, it is estimated that the use of Bt cotton in the United States reduced chemical insecticide applications in cotton for lepidopterous pests by 2.7 million pounds in 1999 (Shelton *et al.*, 2002). This represented a reduction of 15 million insecticide applications (22%) in comparison to the number used in 1995 prior to the use of Bt cotton. These reductions preserve populations of insect natural enemies such as spiders and predatory and parasitic insects, and eliminate exposure of non-target vertebrates, including humans, to these chemicals. The agronomic and environmental advantages of Bt crops are widely recognized in the agricultural community. In fact, the United States Environmental Protection Agency has declared that Bt crops are an "environmental asset". However, concerns about the long-term impact of their use have been raised by environmentalists and scientists in regard to the development of resistance in insect populations, and the safety of insecticidal proteins produced by Bt crops to humans and other non-target organisms. The issues of the development of resistance and resistance management have been written about widely, and interested readers should consult one of the recent reviews to access the primary literature and earlier reviews on this important topic (Gould, 1998; Frutos *et al.*, 1999; Shelton *et al.*, 2002). In the present chapter, I discuss the subject of Bt safety to humans and other non-target organisms, especially the safety of the insecticidal Bt proteins as they occur in bacterial insecticides and transgenic crops.

Important questions have been raised about the safety of Bt crops, but most of the specific test protocols and resultant data relevant to Bt proteins produced by these crops are scarce and are only now being published in refereed scientific journals. This is because these data were developed primarily to fulfill governmental registration processes required in most countries to sell products, processes which do not require the data to be published in journals. Nevertheless, an extensive database exists in the literature on Bt strains, proteins and products and this information is reviewed here along with data available on Bt crop safety.

To address the issue of safety, it is important to understand the role that Bt's insecticidal proteins play in its general biology and the mechanisms by which these proteins exert their toxic effects on insects. Bt and its insecticidal proteins have a high degree of specificity that accounts for their safety to most organisms, especially in comparison with synthetic chemical insecticides. An understanding of Bt's mode of action, specificity and toxicology, only rarely treated in discussions of safety, is important in assessing any risks. Thus, to place the safety of Bt insecticidal proteins and Bt crops to non-target vertebrates and invertebrates in perspective, I will first summarize our knowledge of the most critical aspects of Bt relevant to understanding this issue, and then deal with the data on the safety of Bt crops.

The general biology of *Bacillus thuringiensis*

Bacillus thuringiensis (Bt) is a common Gram-positive, spore-forming bacterium that can be readily isolated on simple media such as nutrient agar from a variety of habitats including soil, water, plants, grain dust, dead insects and insect feces. Its life cycle is simple. When nutrients are sufficient for growth, the spore germinates, producing a vegetative cell that grows and reproduces by binary fission. The bacterium continues to multiply until one or more nutrients, such as sugars, certain amino acids and oxygen, become insufficient for continued vegetative growth. Under these conditions, the bacterium sporulates to produce a spore and parasporal body, the latter composed primarily of insecticidal proteins (Figure 8.1). These are commonly referred to in the literature as protein toxins or endotoxins.

Although Bt can be isolated from many environmental sources, and is often referred to as a 'soil bacterium', several features indicate that its principal ecological niche is insects (Federici, 1993, 1999; Meadows, 1993). The original isolations of Bt,

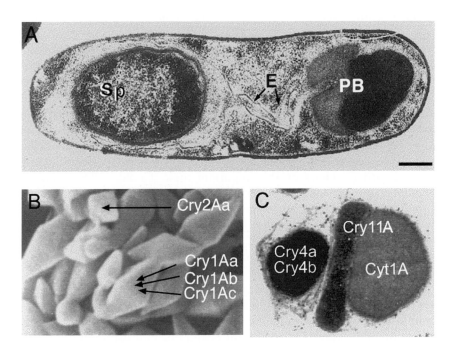

Figure 8.1 Sporulating cells of *Bacillus thuringiensis* and representative endotoxin-containing parasporal bodies. (A) A sporulating cell of *B. thuringiensis* subsp. *israelensis* showing the developing spore (Sp) exosporium membrane (E) and parasporal body. (B) Bipyramidal and cuboidal crystals characteristic of the HD1 isolate of *B. thuringiensis kurstaki* (H 3a3b) used to control lepidopterous pests. The bipyramidal crystal is assembled by the co-crystallization of three Cry proteins, Cry1Aa, Cry1Ab, and Cry1Ac. The latter two proteins serve as the basis for, respectively, Bt maize and Bt cotton. Cry2Aa is assembled into a cuboidal crystal, which is often embedded in the bipyramidal crystal. (C) Parasporal body of *Bacillus thuringiensis* subsp. *israelensis* exhibiting mosquitocidal Cry and Cyt protein inclusions. The Cyt protein masks resistance to mosquitocidal Cry proteins. Bar in A represents 250nm.

for example, were made from diseased caterpillars, and these remain a good source of isolates. More importantly, Bt produces a range of protein toxins and toxin synergists that are very effective at killing certain species of insects, especially larvae of lepidopterous insects, providing a rich substrate for Bt's reproduction. The principal toxins are protein δ-endotoxins, the 'δ' designating a particular class of toxins, and endotoxin referring to their localization within the bacterial cell after production. Many isolates of Bt also produce the β-exotoxin (a competitive inhibitor of messenger RNA polymerase), insecticidal proteins produced during vegetative growth (Vips), the synergist zwittermicin A, and enzymes such as phospholipases that enhance the activity of the δ-endotoxins. In addition, the spore itself can synergize the activity of Bt proteins in some insects, especially those not very sensitive to δ-endotoxins. In many insect species, especially grain-feeding lepidopterans, Bt reproduces to very high levels after insect death, with millions of spores being produced per insect. Thus, Bt's ecology and reproductive biology suggest that its toxins and toxin synergists evolved to debilitate or kill directly a range of insect species, thereby providing a substrate for reproduction of this bacterium.

The diversity of Bacillus thuringiensis subspecies

While commonly referred to in the singular as 'Bt', *B. thuringiensis* is actually a large group of subspecies, all characterized by the production of one or a few parasporal bodies per cell during sporulation. Each parasporal body contains one or more protein endotoxins, typically as crystalline inclusions, and each of these is typically toxic to a limited range of insects (Figure 8.1). The endotoxins occur in the parasporal body as protoxins, which after ingestion, dissolve and are converted to active toxins through cleavage by proteolytic enzymes in the insect stomach (midgut). The activated toxins bind to receptors on the midgut microvillar membrane in sensitive insects, lysing the cells and destroying much of the midgut epithelium, as described in more detail below, causing insect death. In essence, Bt endotoxins are stomach poisons selective for insects and certain other invertebrates. At present, there are more than 70 subspecies of Bt, distinguished from one another by immunological differences in flagellar (H antigen) serotype (de Barjac and Frachon, 1990; Thiery and Frachon, 1997). Each subspecific name corresponds with a specific H antigen number (Table 8.1). For example, *B. thuringiensis* subspecies *kurstaki* is H3a3b, whereas *B. thuringiensis* subspecies *morrisoni* is H8a8b. In the literature, the term 'variety' is also used in place of subspecies, as is occasionally the term 'strain'. Because the H antigen serotype-subspecific name often does not correlate with insecticidal properties, acronyms and numbers are often used to designate specific isolates, especially those with important insecticidal properties. For example, HD1 (isolate number 1 from Howard Dulmage), is used to designate a specific isolate of *B. thuringiensis* subsp. *kurstaki* (H3a3b) that produces four major endotoxin proteins and has a broad spectrum of activity against lepidopteran pests. Another isolate of *B. thuringiensis* subsp. *kurstaki* (H3a3b) is HD73. This produces only a single endotoxin protein (Cry1Ac) and as a result has a much narrower spectrum of activity against insects than HD1. Historically, HD1 was the first Bt isolate developed commercially for the control of lepidopterous pests, and remains the most widely used in commercial products today. This isolate has also been the source of much of our knowledge

Table 8.1 Important subspecies of *Bacillus thuringiensis* used in bacterial insecticides

Subspecies/serovariety[a]	H-antigen	Major endotoxin proteins (mass in kDa)	Insect spectrum[b] (target group)
kurstaki	3a3b	Cry1Aa (133), Cry1Ab (131)* Cry1Ac (133)*, Cry2Aa (72)[d]	Lepidoptera
aizawai	7	Cry1Aa (133), Cry1Ab (131) Cry1Ca (135), Cry1D (133)	Lepidoptera
morrisoni[c]	8a8b	Cry3Aa (73)*	Coleoptera
israelensis	14	Cry4Aa (134), Cry4Ab (128) Cry11Aa (72), Cyt1Aa (27)	Diptera[e]

[a] From de Barjac and Frachon (1990).
[b] See Table 8.3 for examples of commercials products based on these subspecies.
[c] Strain tenebrionis, commonly referred to as *B. t.* subsp. *tenebrionis* or *san diego.*
[d] Also toxic to larvae of nematoceran dipterans (e.g. mosquitoes and blackflies).
[e] Only toxic to species of the dipteran suborder Nematocera (e.g. mosquitoes and blackflies).
* Used to construct insect-resistant transgenic crops.

of Bt genetics and molecular biology as well as endotoxin genes used in transgenic bacteria and plants.

Despite the large number of Bt subspecies that have been described, the taxonomic validity of these as well as maintaining *B. thuringiensis* as a species separate from *B. cereus* has been in question for many years. At the species level, the primary problem is that the only phenotypic character that clearly differentiates *B. cereus* from *B. thuringiensis* is the parasporal body synthesized by the latter species during sporulation (Baumann *et al.*, 1984). In most Bt subspecies, the information for endotoxin production is encoded on large transmissible plasmids (see Hofte and Whiteley, 1989, for review). When these plasmids are lost naturally or cured from Bt subspecies by growing cells at 42°C, no endotoxins are produced. Cured Bt strains that hosted these plasmids cannot be reliably distinguished from *B. cereus*. At the subspecies level, phenotypic biochemical differences among many are minor, indicating that many subspecies may not be valid using accepted standards of differentiating bacterial subspecies. In other cases, the differences in biochemical properties and insecticidal activity are so significant that some subspecies could be viewed as distinct species.

From the standpoint of Bt's use as an insecticide, taxonomic studies, especially flagellar serotyping, have aided isolate classification but have proven unreliable as accurate predictors of insecticidal activity. Perhaps the best example of this is found in *B. thuringiensis* subsp. *morrisoni* (H8a8b), which includes isolates active against lepidopterous (isolate HD12), coleopterous (isolate DSM2803 and others) or dipterous (isolate PG14) insects. Another example of this is found in the occurrence of the 128-kb plasmid that encodes mosquitocidal Cry and Cyt proteins among numerous subspecies of Bt, including *israelensis*, *morrisoni*, *entomocidus*, *kenyae* and *thompsoni* (Delecluse *et al.*, 2000).

To summarize the current status of Bt systematics, most molecular evidence is in agreement with more classical biochemical and physiological studies which indicate that *B. thuringiensis* and *B. cereus* are the same species, and that the latter becomes the former when it acquires one or more plasmids that express genes for insecticidal proteins. Nevertheless, maintaining Bt as a separate species has practical value because of its insecticidal properties, as does dividing Bt isolates into subspecies based on flagellar antigens. The latter is useful in cataloguing the more than 20 000 isolates of Bt collected to date. However, to understand the insecticidal properties of an isolate, regardless of its subspecies/serotype or other designation, knowledge of the insecticidal protein genes encoded and expressed is required.

Insecticidal proteins of B. thuringiensis

Because the β-exotoxin is not permitted in commercial bacterial insecticides in the United States, the two principal active components in commercial Bt preparations are the spore and parasporal body. The endotoxins in the parasporal body account for most of the formulation's insecticidal activity, including initial paralysis followed by death. The lethal effects of these proteins have been known since their discovery in the 1950s (see Luthy and Ebersold, 1981). In the early 1980s, shortly after the development of recombinant DNA techniques, it was discovered that Bt δ-endotoxins were encoded by genes carried on plasmids. This discovery led to a major research

Table 8.2 Toxicity of Bt Cry proteins to first instars of lepidopteran and dipteran pest species[a]

Cry Protein	LC_{50} (ng/cm^2 of diet or water)[b]				
	Tobacco hornworm	Tobacco budworm	Cotton leafworm	Yellow fever mosquito	Colorado potato beetle
CryIAa	5.2	90	>1350	>5000	>5000
CryIAb	8.6	10	>1350	>5000	>5000
CryIAc	5.3	1.6	>1350	>5000	>5000
CryIC	>128	>256	104	>5000	>5000
CryIIA	>5000	>5000	>5000	60	>5000
Cry3A	>5000	>5000	>5000	>5000	<200

Modified from Hofte and Whiteley (1989).

[a] Tobacco hornworm (*Manduca sexta*), tobacco budworm (*Heliothis virescens*), cotton leafworm (*Spodoptera littoralis*), yellow fever mosquito (*Aedes aegypti*), colorado potato beetle (*Leptinotarsa decimlineata*).

[b] Values of >5000 indicate a lack of toxicity at high doses, that is doses equivalent to field applications rates that would not be economical. Lack of toxicity at these rates, however, also illustrates the high degree of insect specificity demonstrated by Cry endotoxins.

effort in many laboratories around the world aimed at understanding the genetic and molecular biology of these toxins. This effort resulted in the cloning and sequencing of numerous Bt genes, over 100 to date, and characterization of the toxicity of the proteins each encodes. Pertinent information derived from these studies through 1997 is summarized in a comprehensive review by Schnepf *et al.* (1998), but an earlier review by Hofte and Whiteley (1989) remains worthy reading (Table 8.2).

At the time Hofte and Whiteley wrote their review, a wide variety of confusing names and acronyms were being used to refer to Bt's endotoxin genes and proteins. Computer analyses showed that the nucleotide sequences of most of these genes were quite similar. To standardize the terminology, Hofte and Whiteley (1989) proposed a simplified nomenclature for naming all insecticidal Bt genes and proteins. In this nomenclature, the proteins are referred to as Cry (for crystal) and Cyt (for cytolytic) proteins. Though modified recently (Crickmore *et al.*, 1998), this system is still in use today, and is described briefly here along with its modifications and additions.

The Cry and Cyt nomenclature developed by Hofte and Whiteley (1989) was originally based on the spectrum of activity of the proteins as well as their size and apparent relatedness as deduced from nucleotide and amino acid sequence data, and protein gel analyses. At that time, with the exception of a 27-kDa cytolytic protein from *B. thuringiensis* subsp. *israelensis*, all proteins appeared to be related, and probably derived from the same ancestral protein. Therefore Hofte and Whiteley termed the encoding genes '*cry*' genes and the proteins they encode 'Cry' proteins. This designation was followed by a Roman numeral which indicates pathotype (I and II for toxicity to lepidopterans, III for coleopterans and IV for dipterans), followed by an upper-case letter indicating the chronological order in which genes with significant differences in nucleotide sequences were described. The I and II for lepidopteran-toxic proteins also indicate size differences, with the I referring to proteins with a mass of about 130 kDa and the II designating those of 65–70 kDa. Some epithets also include a lower-case letter in parentheses which indicates minor differences in the nucleotide sequence within a gene/protein type. Thus, CryIA referred to a 130 kDa protein toxic to lepidopterous insects for which the first gene (*cryIA*) was sequenced, whereas CryIVD referred to a 72-kDa protein with mosquitocidal activity for which the encoding gene was the fourth from this pathotype sequenced.

The 27-kDa Cyt1A protein first isolated from *B. thuringiensis* subsp. *israelensis* differs from other Bt proteins not only in its smaller size, but also in that it is highly cytolytic to a wide range of cell types *in vitro*, including those of vertebrates (see Federici *et al.*, 1990 and Chilcott *et al.*, 1990 for reviews). In addition, it shares no apparent relatedness to Cry proteins. Owing to these differences and its broad cytolytic activity, Hofte and Whiteley (1989) referred to this as the CytA protein encoded by the *cytA* gene.

In their revision of Bt gene nomenclature, Hofte and Whiteley (1989) listed 38 published gene sequences that encoded 13 different Cry proteins and the single CytA protein. Since their publication, the number of Bt Cry endotoxins has more than tripled, and several new *cyt* genes have also been described (Crickmore *et al.*, 1998). As more and more *cry* genes were sequenced and analyzed, it was decided to name genes based on their relatedness as determined primarily from the degree of their deduced amino acid identity. As a result, the nomenclature developed by Hofte and Whiteley (1989) was modified in the following manner. The *cry* and *cyt* descriptors have been maintained, but the Roman numerals have been replaced with Arabic numbers to indicate major relationships (90% identity), with higher degrees of identity being indicated by following upper-case letters (95% identity), and minor variations of these alleles being designated by lower-case letters, with the parentheses around the latter eliminated. For example, what was CryIA(c) is now Cry1Ac (and the corresponding gene, *cry1Ac*), a relatively minor change. However, for some genes/proteins the changes are greater. For example, CryIVD does not cluster with the earlier CryIV (now Cry4) proteins, and is now the taxon Cry11Aa. The complete list of currently recognized Bt genes/proteins and references to the literature describing these can be obtained from http://www.biols.susx.ac.uk/Home/Neil_Crickmore/Bt/index.html.

Although the new designations carry no specific information regarding insecticidal spectrum, primary activity is still inferred because the numbers have been maintained for many of the genes, and there remains a high degree of correlation between relatedness and the spectrum of activity. For example, Cry1 still refers largely to toxicity to lepidopteran insects; Cry2 to lepidopteran toxicity and in some, dipteran activity; Cry3 to coleopteran toxicity, and Cry4 to dipteran toxicity.

Mode of action of Cry proteins

Aside from allergenicity, knowledge of the mode of action and structure of Bt's insecticidal proteins are the two interrelated topics that have most bearing on understanding the safety and risks of insecticides or transgenic crops based on these proteins. Although the spore can play an important role in the pathogenicity of *B. thuringiensis* to certain insect species, the Cry and Cyt proteins are responsible for the paralysis and death of most target insects (Huber and Luthy, 1981; Knowles and Ellar, 1987; Knowles and Dow, 1994; Schnepf *et al.*, 1998). The three-dimentional structure of three of these is known from X-ray crystallographic studies. Although the mode of action remains to be resolved at a detailed molecular level, these structures along with studies on the effects of the toxins *in vivo* and *in vitro* show how Bt proteins kill insects while being safe for most non-target organisms, especially vertebrates. Below I first provide an overview of Bt's mode of action, and then discuss the insights the three-dimensional structure provides into this mode of action.

In the typical Bt, *B. thuringiensis* subsp. *kurstaki*, for example, the endotoxin crystals dissolve after ingestion upon encountering the alkaline (pH 8–10) juices of the midgut. Dissolution requires the reduction of disulfide bridges that stabilize the Cry molecules in the parasporal crystal (Aronson, 1993). Most Cry toxins are actually protoxins of about 130–140 kDa (e.g. Cry1 and Cry4 proteins) from which an active toxin 'core' in the range of 60–70 kDa is released into the midgut by proteolytic cleavage. Proper activation results in cleavage of 26–29 amino acids from the N-terminus and about 600 amino acids from the C-terminus (Schnepf *et al.*, 1998). Activated toxin molecules pass through the peritrophic membrane and bind to specific receptors on the apical microvillar brush border membrane of midgut epithelial cells, which lies just beneath the peritrophic membrane. Binding is an essential step in the intoxication process, and in susceptible insects the toxicity of a particular Bt protein is correlated with the number of specific binding sites (i.e. receptors) on microvilli as well as the affinity of the Bt molecules for these sites (see Schnepf *et al.*, 1998 for review). However, binding by itself, even high-affinity binding, does not always lead to toxicity, indicating that insertion and probably post-insertional processing in the midgut membrane is required to obtain toxicity. For example, in three different studies it has been shown that the Cry1Ac toxin can bind with high affinity to microvillar membrane vesicles from *Lymantria dispar* (Wolfersberger, 1990), *Spodoptera frugiperda* (Garczynski *et al.*, 1991) and *Heliothis virescens* (Gould *et al.*, 1992), but with little or no subsequent toxicity. This indicates that insertion and likely post-insertional processing of Cry proteins is essential to intoxication.

In highly sensitive insect species, the microvilli lose their characteristic structure within minutes of toxin insertion, and the cells become vacuolated and begin to swell (Luthy and Ebersold, 1981). This swelling continues until the cells lyse and slough from the midgut basement membrane. As more and more cells slough, the alkaline gut juices leak into the hemocoel where, as a result, the hemolymph pH rises by a half unit or more. This causes the paralysis and eventual death of the insect (Heimpel and Angus, 1963).

Though this general picture of the mode of action has been known for some time, the details of the insecticidal process at the molecular level remain unresolved, especially the series of events that occur after the toxin binds to a receptor. There is good evidence for an immediate influx of potassium and calcium, in response to which the cell takes in water in an effort to balance these cations. To explain this cationic influx, it has been proposed that Bt molecules insert into the microvillar membrane forming transmembrane cation pores. For pores to form, it is postulated that during membrane insertion, the molecules undergo conformational changes that permit insertion. After insertion the toxin molecules associate to form a pore composed of six toxin molecules. As more and more Cry molecules enter the membrane additional pores form, leading to the influx of cations and water followed by cell hypertrophy and lysis. For a more in depth discussion of the studies dealing with the mode of action of Bt, the interested reader is referred to the reviews by Gill *et al.* (1992), Knowles and Dow (1993) and Schnepf *et al.* (1998).

Structure of Cry proteins

Analysis of *cry* gene sequences showed that the active portion of the Cry toxin molecule (i.e. essentially amino acids 30–630) contains five blocks of conserved amino acids distributed along the molecule, and a highly variable region within the C-terminal half (Hofte and Whiteley, 1989). The variable region was thought to be responsible for the insect spectrum of activity and experimental evidence from recombinant DNA studies was obtained in support of this hypothesis. For example, by swapping the highly variable regions of Cry1Aa and Cry1Ac, the insect spectrum of these molecules could be reversed (Ge *et al.*, 1989). In addition, binding studies showed that in most cases the degree of sensitivity of an insect to a particular Cry molecule was directly correlated with the number of high-affinity binding sites on the midgut microvillar membrane (Hofmann *et al.*, 1988; Van Rie *et al.*, 1989). Studies over the past decade have improved our knowledge of Cry protein molecular biology through determination of the structure of Cry3A, Cry1Aa and Cry2A, identification of regions on the Cry molecule involved in midgut binding and specificity, and identification of several glycoprotein receptors on the insect microvillar membrane.

The structure of the Cry3A molecule was the first to be solved, and similarities in conserved amino acid blocks among Cry toxins and conservation in hydrophobicity indicate that this structure is a good general model for Cry toxins. The information discussed below, therefore, combines interpretations from the crystal structure of the Cry3A molecule with recombinant DNA experiments on Cry1 molecules.

The solution of the crystal structure of the Cry3A molecule (Li *et al.*, 1991) showed that this protein is basically wedge-shaped and consists of three domains (Figure 8.2). Domain I is composed of amino acids 1–290 and contains a hydrophobic seven-helix amphipathic bundle, with six helices surrounding a central helix. This domain contains all of the first conserved block and a major portion of the second conserved block of amino acids noted above. Theoretical computer models of the helix bundle of this domain show that after insertion and rearrangement, aggregations of six of these domains could form a pore through the microvillar membrane (Li *et al.*, 1991; Gill *et al.*, 1992). Domain II extends through amino acids 291–500 and contains three antiparallel β-sheets around a hydrophobic core. This domain contains most of the hypervariable region and most of conserved blocks 3 and 4. The crystal structure of the molecule together with recombinant DNA experiments and binding studies indicate that the three extended loop structures in the β-sheets are responsible for initial recognition and binding of the toxin to binding sites on the microvillar membrane (Lee *et al.*, 1992). Domain III comprises amino acids 501–644 and consists of two antiparallel β-sheets, within which are found the remainder of conserved block number 3 along with blocks 4 and 5. The structure resolved by Li *et al.* (1991) indicates that this domain provides structural integrity to the molecule. Later site-directed mutagenesis studies of conserved amino acid block 5 in the Cry1Aa molecule suggest that this domain may also play a role in pore formation (Chen *et al.*, 1993). Though the structure of the molecule, experiments and computer modeling indicate that binding is attributed primarily to the hypervariable region of this domain, studies by Wu and Aronson (1992) also show that positively charged amino acids in domain I are important to binding and toxicity. Moreover, domain III has also been shown to play a role in receptor binding (DeMaagd *et al.*, 1998).

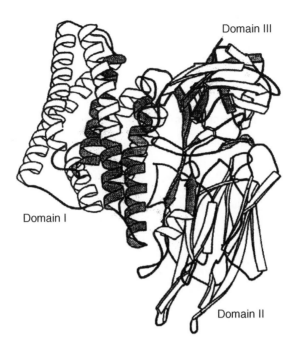

Domain III

Domain I

Domain II

Figure 8.2 Three-dimensional structure of Cry molecules as represented by the structure of Cry3Aa. Domain I is the pore-forming domain, whereas domain II is responsible for the recognition and binding of the molecule to receptors on the microvilli of the insect midgut epithelium. Domain II therefore accounts for much of the high selectivity and safety of Bt-Cry proteins because mammals and other vertebrates lack receptors for this domain. Domain III assists in stabilizing the structure of the molecule, and also can play a role in binding. From Knowles and Dow (1993).

An important aspect of understanding Bt's molecular biology and determining its mode of action at the molecular level is identification of the proteins to which the toxin molecules bind on the microvillar membrane. Sangadala *et al.* (1994) contributed to this area with the identification of the first two proteins in the midgut of *Manduca sexta* to which the Cry1Ac toxin binds. They identified an amino peptidase of 120 kDa as a major binding protein. More recently, cadherin-like proteins and glycoconjugates have also been identified as receptors for Bt endotoxins. These findings indicate that the initial recognition signals on the midgut of a sensitive insect are not ion channels, but one or more proteins or glycoproteins within or which extend from the microvillar membrane. Thus, if these observations hold for other Cry proteins, the midgut receptors are then essentially docking proteins for the toxins, indicating that the toxins do not directly affect ion permeability by binding to ion channels.

These studies of the structure of Bt toxins and their binding properties have identified key regions of the molecule responsible for binding and, putatively, the first binding proteins on the insect midgut epithelium. While further studies are needed to more clearly define the specificity of binding and binding sites, the most enigmatic area of Bt's molecular biology with respect to its mode of action is how the toxin causes toxicity after binding to the membrane. An understanding of this could provide information that might enable the development of insecticidal Cry proteins for use against insects against which we currently have none, such as cockroaches and grasshoppers.

Mode of action and structure of Cyt proteins

Several Cyt proteins are known, including Cyt1A, Cyt2A, Cyt1B and Cyt2B (Crickmore *et al.*, 1998), though most of our knowledge of these proteins is based on studies of Cyt1A and Cyt2A, and the crystal structure of the latter toxin (Li *et al.*, 1996). These proteins are highly hydrophobic and all have a mass in the range of 24–28 kDa. They share no significant amino acid sequence to identify with Cry proteins, and are thus unrelated. The first Cyt protein, Cyt1A, was identified as a component of the parasporal body of *B. thuringiensis* subsp. *israelensis*. Initially, it was thought that Cyt1A played little role in the toxicity of this subspecies to mosquito and black-fly larvae. However, there is now general agreement that it is not only toxic to these and related flies belonging to the dipteran suborder Nematocera, but that it is important as a synergist of Cry4 and Cry11 toxins, and in avoiding the development of insect resistance. While rare in comparison to Cry proteins, since the early work on Cyt1A in *B. thuringiensis* subsp. *israelensis*, Cyt proteins have been reported from mosquitocidal isolates of *B. thuringiensis* subsp. *kyushuensis* (H11a11c), *morrisoni* (H8a8b), *medellin* (H30) and *jegatheson* (H28a28c).

As in the case of Cry proteins, Cyt proteins are synthesized during sporulation as protoxins and assembled into crystalline inclusions that make up a portion of the parasporal body. They are always associated with Cry proteins, but do not co-crystallize with these, instead forming a separate inclusion. The shape of inclusion formed by Cyt proteins varies from hemispherical to angular because they assemble in a spherical parasporal body simultaneously along with inclusions formed by the other Cry proteins with which they occur. However, when produced alone in a recombinant *B. thuringiensis*, the Cyt1A protein forms a bipyramidal crystal with 12 faces (Wu and Federici, 1993).

Based on studies of Cyt1A, after ingestion the 27.3-kDa protoxin molecules dissolve from the crystal under the alkaline conditions present in the midgut of susceptible insects such as mosquito larvae. Proteolytic enzymes then activate the protoxin by cleaving amino acids at both the C- and N-terminus, releasing an active toxin of 24 kDa. This molecule passes through the peritrophic membrane where it then inserts into the microvillar brush border membrane of midgut epithelial cells. Unlike Cry molecules, it appears that Cyt proteins do not require a glycoprotein receptor for binding. Instead, Cyt proteins have a high affinity for the lipid portion of the membrane, specifically for unsaturated phospholipids such as cholesterol, phospholipidcholine and sphigomyelin (Thomas and Ellar, 1983). After binding to the microvillar membrane, the cells hypertrophy and subsequently lyse.

The specific mode of action of Cyt toxins at the molecular level is not known. The leading hypothesis is that Cyt molecules insert into the microvillar membrane and then assemble into clusters of as many as 18 molecules to form pores that act much like those proposed for Cry toxins (Gill *et al.*, 1992; Koni and Ellar, 1993). Lysis then results from the influx of cations and water. However, evidence has also been provided that Cyt molecules do not form pores, but rather act like detergents, binding to and perturbing the lipid bilayer and disrupting the structural arrangement of membrane proteins (Butko *et al.*, 1996, 1997), which would also cause lysis.

Unlike Cry proteins, Cyt proteins exhibit cytolytic activity *in vitro* against a wide range of invertebrate and vertebrate cell types, though at higher concentrations. However, toxicity is not observed when ingested *in vivo* by most insects tested, including larvae of lepidopterous species, non-nematoceran flies such as houseflies and fruitflies, and vertebrates such as mice. The mechanism of this *in vivo* specificity is not known, but may involve specific combinations of unsaturated phospholipids. Though Cyt proteins have been thought to be only toxic to nematocerous dipterans, it has recently been shown that at least one beetle species, *Chrysomela scripta*, is sensitive to the Cyt1A protein (Federici and Bauer, 1998).

As noted above, an important property of Cyt proteins, based primarily on studies of Cyt1A, is their ability to synergize the toxicity of the Cry proteins with which they occur. For example, combination of Cyt1A with the Cry4A, Cry4B or Cry11A proteins with which it occurs in the parasporal body of *B. thuringiensis* subsp. *israelensis* results in toxicities three- to fivefold higher than the specific toxicity of Cyt1A or any of the Cry proteins alone (Wu *et al.*, 1994; Crickmore *et al.*, 1995; Poncet *et al.*, 1995). The mechanism of this synergism is not known but probably involves cooperativity of the Cyt and Cry proteins in binding to and/or inserting into the microvillar membrane.

Bacterial insecticides and transgenic crops based on *Bacillus thuringiensis*

More than 30 Bt formulations are on the market in the United States, Europe, Asia and South America. Most of these are complex mixtures of sporulated cells produced by large-scale fermentation and formulation ingredients (stabilizing agents, spreaders, stickers, fillers). The fermentation products contain spores and an array of toxins found in either *B. thuringiensis* subsp. *kurstaki*, which is used to control a wide range of lepidopteran pests, or *B. thuringiensis* subsp. *israelensis*, used to control the larvae of numerous mosquito and blackfly species. Both owe their success to their broad spectrum of activity and the absence of any widespread insect resistance after many years of use. These characteristics are due to the complex composition of the parasporal body, which in each case contains four major proteins (Tables 8.1 and 8.2), and to the moderate frequency of usage. A list of the key bacterial insecticides based on Bt subspecies is shown in Table 8.3.

With respect to Bt transgenic crops, those on the market are based on full-length or truncated Cry proteins. These crops were constructed by transforming the plants with Bt *cry* genes modified to optimize their expression in plants by changing the codons to reflect plant codon usage. All Bt crops currently on the market produce only a single Cry protein, Cry1Ab, Cry1Ac or Cry3A. However, crops which produce two Cry proteins, such as Bt cotton producing Cry1Ac and Cry2Ab, are being tested in the field. In addition, a variety of insect-resistant crops which produce Bt non-Cry or non-Bt insecticidal proteins are under development, and submissions for regulatory approval have been made in the United States and Australia.

Table 8.3 Examples of commercially available microbial insecticides based on *Bacillus thuringiensis*

Target group	Bt Subspecies	Crop/habitat	Product	Producer
Lepidoptera (Caterpillars)	*B. t. kurstaki* (H 3a3b)	Vegetables, field crops and ornamentals	Able	Thermo-Trilogy
			BioBit	Valent BioSciences
			Condor	Ecogen
			Cutlass	Ecogen
			Dipel	Valent BioSciences
			Javelin	Thermo-Trilogy
			MVP	Dow AgroSciences
			Thuricide	Thermo-Trilogy
			Toarow CT	Toagosei
		Forestry	Dipel	Valent BioSciences
			Foray	Valent BioSciences
			Thuricide	Thermo-Trilogy
	B. t. aizawai (H 7)	Vegetables	XenTari	Valent BioSciences
		Bee hives	Certan	Thermo-Trilogy
Coleoptera (Beetles)	*B. t. morrisoni* (H 8a8b)	Vegetables	Foil	Ecogen
			M-Trak	Dow AgroSciences
			Novodor	Valent BioSciences
			Trident	Thermo-Trilogy
Diptera (Mosquitoes and blackflies)	*B. t. israelensis* (H 14)	Breeding waters	Teknar	Thermo-Trilogy
			Vectobac	Valent BioSciences
			Acrobe	Becker Microbials

Safety of *Bacillus thuringiensis* – General considerations

Knowledge of Bt's biology reviewed above, especially the information on the mode of action and structure of its δ-endotoxin proteins, forms a sound foundation for understanding and evaluating Bt safety. From these studies, it can be seen that these proteins have a series of 'built-in' levels of specificity that should provide Bt bacterial insecticides and Bt crops with a high degree of safety for vertebrates and most non-target invertebrates. For Bt insecticides based on Cry proteins, these specificities include the following.

A   Endotoxin crystals and spores must be ingested to have an effect; there is no 'contact' activity, as occurs with chemical insecticides. This is the reason sucking insects and other invertebrates such as spiders and mites are not sensitive to Bt.

B   After ingestion, the endotoxin crystals of Bt subspecies used to control lepidopterous pests and mosquitoes and blackflies require alkaline conditions, typically a pH in the range of 8–11 in the midgut lumen, to be solubilized in a form conducive to activation by midgut proteases. Under the highly acidic conditions that exist in the stomachs of many vertebrates, including humans, the endotoxin crystals may dissolve, but most of the solubilized protein is rapidly degraded, often within minutes, by gastric juices to non-toxic peptides.

C   Once in solution in the insect midgut lumen, the toxin must be properly activated through cleavage by midgut proteases. Improper cleavage results in inactive toxin.

D   Once activated, the toxin must bind to 'receptors' on midgut microvilli. Most chewing insects that ingest toxin crystals, even if they have alkaline midguts, including many lepidopterans, do not have the appropriate receptors, and thus are not sensitive to the toxins. Even insects sensitive to one class of Bt proteins, such as lepidopterans sensitive to Cry1 proteins, are not sensitive to the Cry3A proteins active against coleopterans, one reason being they lack the receptors for these.

E   After binding to a midgut receptor, the toxin must enter the membrane, change conformation in the process, and oligomerize to form pores that will be toxic.

With respect to level E, at present, the specific conformational changes that must take place to exert toxicity are not known. It is known, however, that high-affinity irreversible binding can occur in some insects, yet not lead to toxicity. This implies that a specific type of processing (i.e. another level of specificity) is required for toxicity that occurs as or after the toxin inserts in the membrane.

In Bt crops, only a portion of the second level (i.e. B) of the first five levels of specificity has been circumvented. When synthesized in plants, full-length and truncated Cry proteins do not form crystals, and even if quasi-crystalline inclusions do form, toxin remains in solution within the plant cells. Nevertheless, whether full-length or truncated protoxins, Cry proteins produced by Bt crops must still be properly activated after ingestion, and must successfully meet the criteria for binding and membrane insertion defined above by levels D and E to be toxic. Furthermore, with the one exception of the Cry9C protein, which was engineered to resist rapid proteolytic cleavage, Bt proteins produced in Bt crops are rapidly degraded under conditions that mimic the mammalian digestive system. Therefore, most of the inherent levels of specificity that account for the safety of Cry proteins used in commercial bacterial insecticides apply to these same proteins when used to make Bt crops resistant to insects.

For Cyt proteins, which are not used in any Bt crops at present, the specificity *in vivo* has the same levels or stages as described above in A–C. However, membrane insertion does not appear to require a protein receptor, but rather the presence of specific unsaturated fatty acids in the lipid portion of the microvillar membrane (Thomas and Ellar, 1983).

Lastly, an important concept of evaluating safety is to consider the route by which an organism is likely to encounter a toxin. Even though pulmonary (inhalation) and intraperitoneal injection studies are done with microbial Bt insecticides and proteins, their normal route of entry by target and non-target organisms is by ingestion. This is even more so for Bt proteins produced in Bt crops.

Safety testing for bacterial insecticides

In addition to their insecticidal efficacy, a major impetus for using Cry proteins in Bt crops was their long history of safety to non-target organisms, especially to vertebrates. The most important levels of Bt endotoxin specificity described above, i.e. activation, binding and membrane insertion, apply equally to evaluating the safety of

Cry proteins whether used in Bt crops or bacterial insecticides. Therefore, the tests and data that support a very high degree of safety for bacterial insecticides containing Cry proteins are relevant to assessing the safety of Bt crops. Extensive testing has been and remains required to meet rigorous safety requirements established by governmental agencies such as the US Environmental Protection Agency. Use of data from these tests is valid as a major approach to evaluating Bt crop safety, especially considering that many hundreds of safety tests have been conducted over several decades to register numerous bacterial insecticides based on different subspecies of Bt. The principal Bt subspecies evaluated in these tests have been *B. thuringiensis* subsp. *kurstaki*, *B. thuringiensis* subsp. *israelensis*, *B. thuringiensis* subsp. *aizawai* and *B. thuringiensis* subsp. *morrisoni* (strain tenebrionis). The materials evaluated have been the active ingredients, i.e. sporulated cultures containing spores and crystals of Cry and Cyt proteins, as well as formulated products. Among the materials tested are all of the Cry proteins used in commercial Bt crops currently on the market.

In determining what types of tests should be done to evaluate the safety of bacterial insecticides, early tests were based primarily on those used to evaluate chemical insecticides. However, the tests have evolved over the decades and are now designed to evaluate the risks of Bt, specifically the infectivity of the bacteria and toxicological properties of proteins used as active ingredients. The tests are grouped into three tiers, I–III (Betz *et al.*, 1990; Siegel, 1997). Tier I consists of a series of tests aimed primarily at determining whether an isolate of a Bt subspecies, as the unformulated material, poses a risk if used at high levels, typically at least 100 times the amount recommended for field use, to different classes of non-target organisms (Table 8.4). The principal tests include acute oral, acute pulmonary (inhalation) and acute intraperitoneal evaluations of the material against different vertebrate species, with durations from a week to more than a month, the length depending on the organism. In the most critical tests, the mammals are fed, injected with, and forced to inhale millions of Bt cells in a vegetative or sporulated form. Against invertebrates, the tests are primarily feeding and contact studies. Representative non-target vertebrates and invertebrates include mice, rats, rabbits, guinea-pigs, various bird species, fish, predatory and parasitic insects, beneficial insects such as the honeybee, aquatic and marine invertebrates, and plants. If infectivity or toxicity clearly results in any of these

Table 8.4 Tier I safety tests required for the registration of bacterial insecticides based on *Bacillus thuringiensis* in the United States and Canada[a]

Toxicology	Non-target organisms/environmental fate
Acute oral exposure	Avian oral exposure
Acute dermal exposure	Avian inhalation
Acute pulmonary exposure	Wild mammal
Acute intravenous exposure	Freshwater fish
Primary eye irritation	Freshwater aquatic invertebrates
Hypersensitivity	Estaurine and marine animal
	Non-target plants
	Non-target insects including honeybees

[a] From Betz *et al.* (1990).

tests, then the candidate bacterium would be rejected (Siegel, 1997). If uncertainty exists, then tier II tests must be conducted. These tests are similar to those of tier I, but require multiple consecutive exposures, especially to organisms where there was evidence of toxicity or infectivity in the tier I tests, as well as tests to determine if and when the bacterium was cleared from non-target tissues. If infectivity, toxicity, mutagenicity or teratogenicity is detected, then tier III tests must be undertaken. These consist of tests such as two-year feeding studies and additional testing of teratogenicity and mutagenicity. The tests can be tailored to further evaluate the hazard based on the organisms in which hazards were detected in the tier I and II tests.

To date, *none* of the registered bacterial insecticides based on Bt have had to undergo tier II testing. In other words, no moderate or significant hazards or risks have been detected with any Bt subspecies against any of the non-target organisms studied so far (McClintock *et al.*, 1995; EPA, 1998). As a result, all Bt insecticides are exempted from a tolerance requirement, i.e. a specific level of insecticide residue allowed on a crop just prior to harvest. Moreover, no washing or other requirements to reduce levels consumed by humans are required. In fact, Bt insecticides can be applied to crops such as lettuce, cabbage and tomatoes just prior to harvest. It is important to realize that such a statement cannot be made for just about any synthetic chemical insecticide. This does not mean that registered bacterial insecticides do not have any negative impacts on non-target populations or do not harm non-target organisms, but rather that these materials do not pose a significant risk to populations of these organisms or the environment. This concept is discussed in more detail in the next section.

Safety of Bt to non-target invertebrates

The concept of a non-target organism is a relative one and therefore requires some clarification. The term non-target organism generally refers to organisms outside the main target group. For example, with most organophosphate, carbamate and pyrethroid insecticides, because they often are capable of killing many different types of insects as well as other invertebrates such as spiders and crustacea, non-target organism usually refers to non-insect species. With Bt insecticides, owing to their high selectivity, the definition of a non-target organism typically is much broader and includes all insects outside the taxonomic order to which the primary target insects belong. Bt insecticides are so specific, even against insects, that their spectrum of activity is typically identified in a very narrow manner, such as 'lepidopteran-active', 'dipteran-active' or 'coleopteran-active'. Even then, Bt insecticides are so specific that a Bt subspecies generally characterized as 'lepidopteran-active' may be highly toxic to some lepidopteran species, but have only low or no toxicity to others. This point can be illustrated with the HD1 isolate of *B. thuringiensis* subsp. *kurstaki* (Btk), the isolate used widely in commercial formulations to control lepidopteran pests. Btk is highly toxic and very effective against larvae of the cabbage looper, *Trichoplusia ni*, a common pest of vegetable crops, but typically exhibits poor activity against the beet armyworm, *Spodoptera exigua*, another important caterpillar pest that belongs to the same taxonomic family. This is because none of the toxins produced by Btk (Cry1Aa, Cry1Ab, Cry1Ac and Cry2A) are very toxic to *Spodoptera* species (Table 8.2). For this reason, the product XenTari (Valent BioSciences) based on *B.*

thuringiensis susbp. *aizawai*, which produces a protein (Cry1Ca) of moderate toxicity to *Spodoptera* species, was developed for control of *Spodoptera* species. With respect to the 'dipteran-active' *B. thuringiensis* subsp. *israelensis*, this subspecies is only highly toxic to species of the suborder Nematocera ('long-horned flies' meaning flies with long antennae), a subdivision of the fly group, which, in addition to mosquitoes and blackflies, contains the mushroom flies, craneflies and chironomid midges. Flies such as houseflies, horseflies, stable flies, and the many types of fruitflies are not sensitive to toxins or spores of this subspecies.

As noted above, a high degree of specificity and hence safety is attributed to each Bt insecticide, meaning that a Bt subspecies that serves as the active ingredient is limited to being toxic primarily to the insect species of only one taxonomic order. Nevertheless, this would still mean that many non-target species of this order would be sensitive to the Bt endotoxins by the normal route of entry (i.e. ingestion). Thus what we consider a pest is an arbitrary concept as opposed to one based on taxonomy. This has led to considerable misunderstanding about the effects of 'lepidopteran-active' Bt subspecies used as insecticides, or the proteins derived from these that are used in Bt crops. An isolate like HD1 of *B. thuringiensis* subsp. *kurstaki* has a broad host range against lepidopteran species, due primarily to the four insecticidal proteins it produces (Tables 8.1 and 8.2). Therefore, when used in the field it will be capable of killing larvae of target as well as certain non-target lepidopterans. Among the targets are larvae of many moth species, especially those of the family Noctuidae (e.g. the corn earworm, the cotton budworm and bollworm, and the cabbage looper). Among the non-targets in certain geographical areas are the larvae of non-pest lepidopterans including larvae of the monarch butterfly, and many other species of moths and butterflies, some of which are endangered species. This can pose a dilemma for farmers as well as the governmental agencies, both regulatory agencies and local governments, in making decisions about the effects of Bt insecticides, and now Bt crops, on non-target organisms.

The choices in the case of endangered species are particularly difficult, and can be illustrated with control of the gypsy moth, *Lymantria dispar*. Most communities in the eastern and mid-western regions of the United States have banned the use of synthetic chemical insecticides to control this pest in forests and residential areas. Bt is used despite the protests of many environmentalists. The reason is that Bt represents the most environmentally compatible choice, even for endangered lepidopteran species. If synthetic chemical insecticides were used, their impact on all non-target invertebrates would be much more devastating than the use of Bt. If Bt is used, populations of the gypsy moth are controlled, at least temporarily, but there are also losses of sensitive non-target lepidopterans in the treated habitats. However, the option of not treating with Bt is worse in that gypsy moth larvae periodically defoliate most plants, thereby eliminating the food sources of all leaf-eating insects. Alternatively, at least some of the larvae of the non-target species will survive the Bt treatment, whereas few, if any would survive the use of a chemical insecticide.

With respect to specific evaluations of Bt insecticides against non-target invertebrates, there have been numerous studies in the laboratory as well as in field situations under operational pest and vector control conditions. Literally thousands of tons of Bt insecticides have been applied in the environment over the past four decades, and the overall record, especially considering the amounts applied, is one of remarkable safety. The key results of these studies are summarized below. For more specific details and references to the extensive primary literature on non-target studies, the interested reader is referred to the recent comprehensive text by Glare and O'Callaghan (2000), as well as review articles by Couch and Foss (1990), Dejoux and Elouard (1990), Lacey and Mulla (1990), Meadows (1993), Melin and Cozzi (1990), Mulla (1990) and Vinson (1990).

Bacterial insecticides based on different subspecies of Bt have been tested extensively in the laboratory against non-target invertebrates to meet registration requirements, and have also been evaluated in field situations to assess effects of formulated products under operational conditions. Both short-term (i.e. from a few days to several weeks) as well as long-term studies of more than a year have been conducted. In the laboratory studies, doses used to evaluate the effects on non-targets are typically as much as 1000-fold the amount that these invertebrates would encounter in the field, and in many cases the doses are much higher (Melin and Cozzi, 1990). Representative non-target invertebrates that have been studied include earthworms and microcrustacea such as daphnids and copepods that make up much of the zooplankton in treated areas. In addition, insects tested have included non-target Coleoptera (beetles), Diptera (flies), Neuroptera (lacewings), Odonata (dragonflies and damselflies), Trichoptera (caddisflies) and Hymenoptera (parasitic wasps), especially species that constitute the major predator and parasite groups that attack the insect pests or disease vectors that are the targets of the Bt applications. Larvae and adults of beneficial insects such as the honeybee, *Apis mellifera*, are also tested. In testing Bt products used against caterpillar pests, more emphasis has been placed on evaluating the effects on terrestrial non-target invertebrates. However, because these products can drift or be washed into streams and ponds, many aquatic invertebrates have been tested in laboratory studies and in natural habitats (Melin and Cozzi, 1990). In the case of *B. thuringiensis* subsp. *israelensis*, used to control mosquito and blackfly larvae, greater emphasis has been placed on evaluating the effects on aquatic non-target insects and other arthropods (Mulla, 1990). Examples of representative studies of the effects of these formulations on insect non-targets in rivers treated to control blackfly larvae are summarized in Table 8.5.

Summaries of these results and those of other studies carried out over the past 30 years show virtually no adverse direct or indirect effects, especially long-term effects, of Bt or formulated products of Bt on non-target populations (Lacey and Mulla, 1990; Melin and Cozzi, 1990; Vinson, 1990). The obvious exceptions are non-target species that are closely related to the target pests or vectors, or insects such as endoparasitic hymenopteran species that require the target lepidopteran pests as hosts. But even these are not affected in some cases (Vinson, 1990). Even in 'forced' feeding studies, the Bt subspecies did not have an effect on insects or non-target invertebrates, such as shrimp, that were outside the order of insects designated as the target group (Couch and Foss, 1990; Melin and Cozzi, 1990). In some of the earliest studies, effects were seen on earthworms and flies. But these early studies were conducted with strains that may have had the β-exotoxin, which has not been permitted in commercial formulations for decades (Melin and Cozzi, 1990; McClintock *et al.*, 1995).

In cases where the effects of Bt on non-target populations have been monitored under field conditions, the effects on non-target organisms were much less than those

Table 8.5 Effects of *Bacillus thuringiensis* subsp. *israelensis* on aquatic non-target organisms when used in river habitats for control of blackfly larvae[a]

Major groups studied[b]	Formulation	Sampling	Impact	Location
Mayflies, Caddisflies, Dragonflies, Damselflies, Molluscs	Powder	Drift	No adverse effects	Ivory Coast
Caddisflies, Stoneflies, Beetles, Mayflies, Dragonflies and Damselflies	Aqueous	Substrate analysis	No adverse effects	Newfoundland
Midges, Mayflies, Caddisflies	Aqueous	Drift	Increased drift, some midge reduction	Ivory Coast
Midges, Caddisflies, Mayflies, Stoneflies	Powder	Substrate analysis	No adverse effects	United States
Mayflies, Caddisflies, Stoneflies, Beetles, Midges	Aqueous	Substrate analysis	No adverse effects	New Zealand
Midges	Aqueous	Drift	No adverse effects	South Africa
Mayflies, Caddisflies Stoneflies, Midges, Gastropods	Aqueous	Drift and substrate analysis	Mayfly and midge mortality, some gastropod reduction	South Africa
Midges	Aqueous	Substrate analysis	Some reduction at 17 × recommended application rate	Germany
Midges, Stoneflies Mayflies, Caddisflies	Aqueous	Drift and substrate analysis	Increased drift of two midge types, no other effects	Canada
Mayflies, Stoneflies, Caddisflies, Midges, Beetles	Powder	Drift	No significant adverse effects	United States

[a] Modified from Lacey and Mulla (1990).
[b] Other than midges, most of these groups have been shown to not be sensitive to the toxins of *B. thuringiensis* subsp. *israelensis* based on laboratory studies. Therefore, the increase in drift after application has been attributed to the increase in formulation particulates in the water due to the application.

resulting from the use of chemical insecticides. The effects of Bt therefore must be viewed from the perspective of the consequences of using alternative control technologies. An appropriate example is the use of *B. thuringiensis* subsp. *israelensis* (Bti) in the Volta River Basin to control the larvae of *Simulium damnosum*, the blackfly vector of onchocerciasis, a blinding eye disease of humans. The onchocerciasis control program is sponsored by the World Health Organization and United Nations Development Program. After more than a decade of intensive use, it was concluded that Bti was of 'only the slightest of hazards' to any of the non-target organisms tested (Dejoux and Elouard, 1990). More specifically, when Bti formulations were applied to rivers, the 'drift' of invertebrates (i.e. the target and non-target invertebrates found floating in the rivers and presumably killed or disturbed by the application) increased two- to three-fold in comparison with untreated rivers. However, when chemical insecticides were applied under similar ecological conditions, the drift increased 20- to 40-fold (Dejoux and Elouard, 1990). In other words, the application of chemical insecticides was approximately ten times more detrimental to the non-target invertebrate populations than the use of Bti. In addition to the much greater impact of the chemical insecticides on non-target invertebrates in the rivers, the blackfly population began to develop resistance to these chemicals. Replacement of the latter with Bti-based insecticides, to which no resistance has developed, during the drier periods of the year ensured the success of this program, and allowed large fertile areas of the river valleys in West Africa to be returned to productive agriculture.

Safety of Bt to mammals

Commercial formulations of Bt insecticides are used to control caterpillar pests on many vegetable crops and in forests near residential areas. They are also used to control mosquito and blackfly larvae in rivers, reservoirs and other bodies of water used for human consumption. As a result, and because other bacilli such as *B. cereus* and *B. anthracis* can infect humans, the Bt subspecies that serve as the active ingredients as well as the formulations made from these have been tested extensively for their safety to vertebrates, especially mammals. These tests are designed primarily to assess whether the bacteria have the potential to infect mammals, and whether the endotoxins are toxic when administered through different routes of entry. In addition, eye and dermal irritability tests have also been conducted. In some cases, preparations have even been tested on human volunteers (Fisher and Rosner, 1959; McClintock *et al.*, 1995; Siegel, 1997, 2001).

As noted above, standard tests required by governmental agencies include feeding studies, pulmonary studies, and intraocular and intraperitoneal inoculation studies. Standard test animals include mice, rats and guinea-pigs, but non-mammals including chickens and other birds as well as fish have also undergone similar tests. A typical test involves a replicated series of trials in which mice and/or rats are injected intravenously or intraperitoneally with the candidate preparation (one million Bt cells per mouse, 10 million cells per rat). The test animals are then observed for several weeks for infection or toxic reactions. Attempts are made during and at the end of the study to isolate bacteria as a measure of the rate at which inoculated cells grow in, or are cleared from, the test animals. Successful isolation, however, is not an indication of infection because this can be due to the persistence of spores within tissues, as noted by Siegel and Shadduck (1990). Therefore, criteria for infection are isolation of vegetative cells and an increase in the Bt population. Tests have also been done where large numbers of cells have been injected into the mouse brain.

The results of numerous feeding, ocular, pulmonary and interperitoneal injection tests conducted with *B. thuringiensis* subsp. *kurstaki*, *B. thuringiensis* subsp. *israelensis* and *B. thuringiensis* subsp. *morrisoni* (strain tenebrionis) against mice and rats have shown no evidence of infection or toxicity to the test animals (Siegel and Shadduck, 1990a,b; Saik *et al.*, 1990; McClintock *et al.*, 1995; Siegel, 1997, 2001). Typical results from these tests are shown in Table 8.6. Even in immunodeficient mice, no infection was detected after interperitoneal injection of as many as 10 million colony forming units. In pulmonary tests, *B. thuringiensis* subsp. *israelensis* cells were cleared from the lungs within four days. However, in the intraperitoneal tests, viable Bt cells could be isolated from mice up to 10 weeks after inoculation. Nevertheless, there was

Table 8.6 Toxicity and infectivity of *Bacillus thuringiensis* to mammals based on studies submitted to the US Environmental Protection Agency[a]

Bacterial species	Animal/test	Dose per animal	Effect
B. t. susbp. kurstaki[b]	Rat/Acute oral	>10^{11} spores/kg	No toxicity or infectivity
	Rat/Acute dermal	>10^{11} spores/kg	No toxicity or infectivity
	Rat/Inhalation	>10^7 spore/L	No toxicity or infectivity
	Rat/2-year oral	8.4 g/kg per day	Weight loss, but no toxicity or infectivity
	Human/Acute oral	1 g/day for 3 days	No toxicity or infectivity
B. t. subsp. israelensis[c]	Rabbit/Acute oral	>10^9 spores	No infectivity
	Rabbit/Acute dermal	>6.3 g/kg	No toxicity or infectivity
	Rat/Acute oral	>10^{11} spores/kg	No pathogenicity or infectivity
	Rat/Acute dermal	>10^{11} spores/kg	No toxicity or infectivity
	Rat/Inhalation	8×10^7 spores	No infectivity

[a] Data from McClintock *et al.* (1995).
[b] Principal endotoxin proteins: Cry1Aa, Cry1Ab, Cry1Ac and Cry2Aa.
[c] Principal endotoxin proteins: Cry4Aa, Cry4Ab, Cry11Aa and Cyt1Aa.

no evidence of infection, as defined above, or toxicity after 10 weeks. There are limits to such aggressive testing procedures where large doses of microorganisms are administered. For example, Siegel and Shadduck (1990b) found that intracerebral injection of *B. thuringiensis* subsp. *kurstaki* and *B. thuringiensis* subsp. *israelensis* caused high mortality when 10 million cells were injected into the brains of weanling rats. But even in these tests, most rats injected with one million or fewer cells survived and showed no signs of infection. Intraperitoneal injection of mice with 100 million colony forming units did cause high mortality with some isolates of Btk and Bti, but the results were not consistent, and lower amounts of 10 million cells caused no illness or mortality (McClintock *et al.*, 1995). Similar tests with *B. subtilis*, which does not produce Cry toxins, produced comparable results.

In contrast to these results, there is one study in which an unidentified Bt Cry protein putatively caused abnormalities in the ileum of mice fed potato treated with Bt (Fares and El-Sayed, 1998). The micrographs presented in this paper showed clear evidence of cell damage. But the cause of this damage remains unclear. The strain was not characterized with respect to endotoxin content, and it is possible that the damage was due to β-exotoxin, which is present in many Bt isolates and known to damage vertebrate intestinal mucosa. This study is also suspect because the authors erroneously attribute cytolytic proteins and properties to *B. thuringiensis* subsp. *kurstaki* that are not characteristic of this subspecies, but rather are properties of *B. thuringiensis* subsp. *israelensis*. In addition, in other studies in which the ileum of mice was examined using immunocytochemical methods, no receptors for the Cry1Ab protein were found (Noteborn *et al.*, 1995).

Only a few tests have been conducted on human volunteers, and these were feeding studies in which each individual was fed 1 g of Bt (10 billion spores along with endotoxin crystals) per day for three consecutive days (McClintock *et al.*, 1995). This is at least 1000 times the dose a human would consume by eating fresh vegetables immediately after treatment with a Bt formulation at the recommended rate. Yet no infections or illness occurred. In addition, there have been no confirmed human infections or deaths during the 40 years that commercial preparations have been used for pest and vector control around the world. This statement cannot be made for chemical insecticides, where hundreds of cases of illness and 5–10 deaths occur yearly as a result of the misuse of methyl parathion alone (Levine, 1991). An overwhelming amount of evidence indicates that Bt products, owing to their highly specific mode of action, cannot be 'misused' in the sense that they could accidentally cause illness or death. There have been a few reports of humans 'infected' with Bt, or Bt causing infections in domesticated animals. However, aside from being extremely rare, these reports have not been substantiated, and the presence of Bt may have been due to persistence rather than infection (Siegel and Shadduck, 1990a). McClintock *et al.* (1995) concluded that Bt proteins were not involved in any of these rare reports. Considering the widespread use of Bt, and its simple nutritional requirements for growth, if it had even marginal potential as a pathogen for humans and domesticated animals it should have been isolated as the cause of infection much more frequently. It is appropriate to note that the consumption of fresh vegetables, especially organically grown lettuce, cabbage, broccoli and tomatoes, even if washed, results in routine human consumption of viable Bt spores and endotoxin crystals. In addition to being allowable as a spray on crops just prior to harvesting, Bt is common in the environment (Meadows, 1993; Glare and O'Callaghan, 2000).

Safety testing for Bt crops

The genetic engineering of crops to make them resistant to insects through the production of Bt Cry proteins has raised concerns about the safety of these crops to non-target invertebrates and vertebrates, especially their safety to humans. Proponents of Bt crops argue that the safety of Bt crops has been established by direct feeding studies as well as by the 40-year safety record established by Bt insecticides. Moreover, they argue that Bt crops are even better because they typically contain only one Cry toxin and thus lack the other components of Bt insecticides such as spores, fermentation residues and formulation additives. Alternatively, opponents of Bt crops counter that their safety concerns are valid because Bt crops produce Cry proteins in a form that is modified from what the bacteria produce. For example, Cry proteins do not form crystals in plants, but remain in solution, thereby eliminating the dissolution step from the different levels of specificity. In addition, they argue that Bt proteins are synthesized in a truncated form making them already active, and that it is possible that Cry proteins undergo processing in plants, such as glycosylation (which has not been demonstrated), that could make them toxic or allergenic to mammals. The opponents of Bt crops state therefore that the safety record of Bt insecticides cannot be relied upon to validate the safety of Bt crops.

The concerns raised by the opponents of Bt crops have been widely reported in the popular press, creating a public backlash against Bt crops, especially in several European countries. The backlash has been minimal in the United States and countries such as Canada, China and Argentina. However, the finding that sufficient quantities of Bt maize pollen can kill larvae of the popular monarch butterfly under certain laboratory conditions was reported in most newspapers in the United States in 1999.

Table 8.7 Cry proteins produced in insect-resistant Bt crops registered in the United States[a,b]

Crop	Cry protein[c]	Target insects
Cotton	Cry1Ac	Tobacco budworm, *Heliothis virescens*
	Cry2Ab2	Cotton bollworm, *Helicoverpa zea*
	Cry1F	Pink bollworm, *Pectinphora gossypiella*
Maize	Cry1Ab	European corn borer, *Ostrinia nubilalis*
		Southwestern corn borer, *Diatraea grandiosella*
		Corn earworm, *Helicoverpa zea*
Maize[d]	Cry1Ac	European corn borer, *Ostrinia nubilalis*
		Southwestern corn borer, *Diatraea grandiosella*
Maize	Cry9C	Tobacco budworm, *Heliothis virescens*
		European corn borer, *Ostinia nubilalis*
Potato	Cry3Aa	Colorado potato beetle, *Leptinotarsa decemlineata*

[a] Source: US Environmental Protection Agency. http://www.epa.gov/oppbppdl/biopesticides/factsheets/.
[b] See Schuler *et al.* (1998) for a list of other plants engineered to produce Bt Cry proteins.
[c] Most of these proteins are produced as full-length molecules similar in mass to those produced by the Bt subspecies from which they were derived. Some Bt maize varieties based on Cry1Ab or Cry9C produce a truncated version of this protein.
[d] Voluntarily withdrawn

This resulted in a higher level of concern by the public about Bt crops, at least about their potential effects on non-target organisms. These concerns have resulted in additional scrutiny of existing tests and data used by scientists and governmental agencies to determine the safety of Bt crops, as well as additional investigations under typical field conditions. Based on a recent series of studies conducted under field conditions, it was concluded that pollen from Bt maize would have only a negligible effect on monarch populations (Sears *et al.*, 2001).

Before reviewing the types of studies and scientific evidence used to access Bt crop safety, some additional background is provided here about the forms of Cry proteins produced by crops. Even if in solution, the toxins still must undergo activation and meet the other criteria for toxicity summarized earlier. Most crops currently on the market produce full-length Cry or partially truncated proteins, typically Cry1Ab or Cry1Ac, (Table 8.7). In crops that produce truncated Cry proteins, the truncated version lacks a major portion of the C-terminal half of the molecule. It has been assumed in some studies that this truncation results in the production of active toxin in plants. However, as noted above, activation of the toxin by midgut proteases also cleaves 26–29 amino acids from the N-terminus of the molecule, so even the truncated Cry proteins produced by plants are not fully activated, unless this fragment has been removed by plant proteases.

It should also be realized that even in bacterial insecticides, some activated toxin is present due to the action of bacterial proteases produced during sporulation. Most of the toxin in crystalline form is protected from activation, but toxin in solution at the time of lysis is subject to activation. Thus the safety tests carried out with the various Bt subspecies and formulated products based on these would have included activated Cry proteins. If these were toxic to non-target organisms, this toxicity should have been detected in the safety tests conducted over the past 3–4 decades, especially considering the high doses used in these tests (see Table 8.6).

The possibility remains that some plants may modify Cry proteins in a way that the bacteria from which they were derived do not. The developers of Bt crops have examined Bt proteins produced by various Bt crops for post-translational modification, including glycosylation, phosphorylation and acetylation. No indication has been found that these proteins have been modified post-translationally. Even if post-translational modification of Bt proteins is eventually found to occur in some Bt crops, there is no reason to assume that these modifications would make a Bt crop less safe to mammals or non-target invertebrates. Many plant proteins, for example, seed storage proteins that are widely consumed by humans, are glycosylated, yet cause no harmful effects. Furthermore, if post-translational modifications were to occur, owing to the complex mode of action of Cry proteins, they are just as likely to interfere with the toxicity of Cry proteins to target insects, possibly making the crops ineffective.

Present Bt crops represent an early phase of a new technology, and it is easy to exaggerate their potential benefits and shortcomings. Given the 40-year safety record of Bt insecticides along with the well-accepted empirical methods for testing the safety of chemical insecticides, bacterial insecticides and drugs, it is appropriate that a combination of prior studies and empirical methods be used to establish the safety or lack thereof of Bt crops. Over the past few years, studies of Bt crop safety using empirical methods have begun to appear in the scientific literature. These studies have examined the effects on non-target invertebrates and vertebrates including mammals in the laboratory and field. Under operational growing conditions, without exception, these studies show that Bt crops have no significant adverse consequences for non-target invertebrate populations, and if anything their use is beneficial because the amount of broad-spectrum chemical insecticides used is reduced. Replacement of chemical pesticides with Bt crops provides better protection of beneficial insect populations due to the much greater specificity of Bt proteins. The laboratory studies also provide a variety of evidence that Bt crops are safe for human consumption. The most critical of these studies are summarized below.

Safety of Bt crops to non-target invertebrates

Cry proteins produced by transgenic plants are not easily extractable in the amounts that would be required for studies designed to test the effects on non-target organisms. Thus, the effects are currently assessed by feeding test species Cry proteins produced in either *Escherichia coli*, a *Bacillus* species, or on various Bt crop tissues such as leaves or pollen. Tests of Cry proteins produced in *E. coli* are similar to those used to evaluate these proteins when produced by *B. thuringiensis*, except that in many cases an activated form of the toxin is used to produce what could be considered a 'worst case' risk assessment. To complement laboratory studies, several field studies have been conducted in which non-target insect populations were monitored on Bt crops, mainly Bt maize and Bt cotton, throughout the growing season.

Most of the laboratory studies have been performed in the United States, where a complex of non-target organisms serves as a standard group for which results are accepted by the US Environmental Protection Agency. The standard test invertebrates have included a range of terrestrial and freshwater aquatic organisms generally considered beneficial. These typically are larval and/or adults of one or more of the

Table 8.8 Toxicity of Cry1Ab produced in *Escherichia coli* or Bt maize to non-target invertebrates and non-mammalian vertebrates[a,b]

Non-target organism	No effects level[c]
Invertebrates	
Insects	
Honeybee, *Apis mellifera*	20 ppm
Ladybird beetles, *Hippodamia convergens*	20 ppm
Green lacewing, *Chrysoperla carnea*	16 ppm
Wasp parasite, *Brachymeria intermedia*	20 ppm
Springtail, *Folsomia candida*	50 µg/g leaf tissue
Earthworms	
Earthworm, *Eisenia fetida*	200 mg/kg soil
Freshwater crustacea	
Daphnid, *Daphnia magna*[d]	100 mg pollen/liter
Vertebrates	
Northern bobwhite quail[e]	100 000 ppm
Channel catfish[e]	>3 µg/g maize feed
Broiler chickens[e]	>3 µg/g maize feed

[a] From Sanders *et al.* (1998), Brake and Vlachos (1998), Yu *et al.* (1997), and 'Factsheets' produced by the US Environmental Protection Agency and available on the following Agency website: http://www.epa.gov/oppbppdl/biopesticides/factsheets/.
[b] Results of similar tests on other Cry proteins used in Bt crops are similar, and can be viewed on the above website.
[c] Tests are conducted using a single high level of toxin much higher than that it is estimated the test organisms would likely encounter under field conditions. This is referred to as the no-observed-effect-level (NOEL).
[d] Bt maize pollen.
[e] Fed Bt maize grain.

following organisms: the honeybee, parasitic wasps, predatory ladybird beetles and lacewings, 'soil-dwelling' springtails, earthworms, and as a representative of a freshwater aquatic crustacean, a daphnid (Table 8.8). The non-mammalian non-target vertebrates have typically been a bird and a fish, respectively, the bobwhite quail and the channel catfish. In these tests, the non-target organisms were typically exposed to or fed amounts of toxin that were in the range of at least a hundred to several thousand times the amount they would be exposed to or consume under natural conditions. In such tests, when no effects are observed at the highest dose or rate tested, this amount is referred to as the no-observed-effect-level (NOEL). For a crop like Bt maize, the amount of Cry protein in a maturing field is estimated to be about 500 g per hectare, and thus the test levels are adjusted to ensure a dose at least 1000 times this level. Most of the studies conducted so far have been short-term studies, lasting from several days to a few weeks. The results of these studies, summarized in Table 8.8, have shown no adverse effects of Bt Cry proteins on the organisms tested.

In addition to these studies aimed at testing high levels of exposure to Cry proteins under situations that resemble a natural situation, there have been tests where non-target invertebrates have been forced to feed on Bt crops. In Italy populations of the aphid *Rhodopalosiphum padi* were force-reared on Bt maize (Cry1Ab). Then its natural predator, the green lacewing, *Chrysoperla carnea*, was fed on these aphids (Lozzia *et al.*, 1998a,b). In comparison to control populations, no adverse effects were noted on either the aphid or lacewing populations with respect to survival and fecundity. In contrast to these results, other laboratory studies have shown that force-feeding can result in mortality to certain non-target insect species. For example, it was shown that immature *C. carnea* fed on prey that had been fed Bt maize (Cry1Ab) suffered greater mortality than control lacewings fed on non-Bt maize-fed prey (Hilbeck *et al.*, 1998a). Only 37% of the *C. carnea* fed Bt maize-fed larvae of the cotton leafroller, *Spodoptera littoralis*, or the European corn borer, *Ostrinia nubilalis*, survived, whereas 62% of the control group fed on non-Bt maize-fed caterpillars survived. In a subsequent study, using an artificial liquid diet it was determined that immature *C. carnea* were sensitive to the Cry1Ab toxin at a level of 100 µg/ml of diet (Hilbeck *et al.*, 1998b). However, the level of Cry1Ab in maize is about 4 µg/g fresh weight, which is considerably less than 100 µg/ml (Hilbeck *et al.*, 1998b). At present there appears to be no explanation for the discrepancy between the artifical diet and exposure via prey. The apparent sensitivity of *C. carnea* to the Cry1Ab is an interesting laboratory observation, and should be validated with histological and receptor-binding studies.

The most widely publicized study of the potential impact of Bt maize on a non-target insect is the finding that larvae of the monarch butterfly, *Danaus plexippus*, are sensitive to certain doses of Bt maize pollen (Losey *et al.*, 1999). In this laboratory study, milkweed leaves were covered with Bt-maize pollen and then fed to larvae. Control larvae were fed on milkweed leaves covered with non-Bt pollen or untreated milkweed. The key finding of the study was that the larvae fed milkweed leaves treated with unquantified amounts of Bt pollen had a lower survival rate (56%) than controls (100%). The authors stated that their results had 'potentially profound implications for the conservation of monarch butterflies' because the central corn belt where Bt maize adoption by farmers continues to grow is also an important habitat for monarchs that migrate to the US from Mexico each year.

In assessing the relevance of these findings on the green lacewing and monarch larvae, or other non-target organisms for that matter, it should be kept in mind that bacterial insecticides based on Bt should be just as toxic if not more so because they contain more Cry proteins than Bt crops. In other words, monarch larvae that feed under field conditions on milkweed leaves treated with a product that contains *B. thuringiensis* subsp. *kurstaki*, from which the Cry1Ab protein gene was derived, will be equally if not more sensitive to the bacterial insecticide. Similarly, *C. carnea* immatures that feed on caterpillars intoxicated as a result of feeding on a Bt insecticide will be equally sensitive to the activated toxins in these larvae. So the issue here is not so much one of Bt crops, but whether Cry proteins will impact beneficial insects regardless of the source. Despite the sensationalistic attention given to the monarch study by the popular and even the scientific press, the design of this study has been criticized by scientists familiar with monarch biology and maize production. The principal criticisms are that monarch larvae were not fed well-quantified amounts of Bt maize pollen, that larvae given a choice may find milkweed contaminated with Bt pollen unpalatable and thus should have been given a choice, that Bt pollen is only present for 7–10 days of the growing season, and that the most Bt pollen dispersed remains within several meters of the dispersal in the field. Thus, maximum monarch exposure would normally be limited to larvae developing within or at the edges of corn fields, but not in other habitats. As a result, only monarch larvae developing in very close to Bt corn fields during pollen-shed would typically encounter quantities of Bt pollen grains on the surface of milkweed leaves.

Extraordinary attention was given to the preliminary findings in the scientific and popular press on the potential negative effects of Bt pollen on monarch populations. A benefit of this attention was that it resulted in a series of collaborative studies in 1999 and 2000 devoted to a much more rigorous assessment of these potential negative effects under field conditions throughout the US corn belt and Canada, as well as in the laboratory (Hellmich *et al.*, 2001; Pleasants *et al.*, 2001; Sears *et al.*, 2001; Stanley-Horn *et al.*, 2001). The overall conclusion of these studies is that the effects of Bt maize on monarch populations will be negligible, especially in comparison with the effects of using chemical insecticides to control maize pests. The basis for this conclusion was that most of the registered Bt maize varieties produced only a low level of Bt protein, and even moderately high levels of consumption of this pollen by larvae did not affect their development or survival (Sears *et al.*, 2001). Other 'mitigating' factors that limited exposure of larvae to Bt proteins in pollen included limited overlap between larval feeding periods throughout the corn belt and the period of pollen shed, the relatively limited adoption of Bt maize (19%), and the widespread availability of milkweed outside areas where maize is grown. Significantly, larvae in maize plots treated with the chemical insecticide (λ-cyhalothrin) showed no weight gain because all were killed by the insecticide within 24 hours of treatment (Stanley-Horn *et al.*, 2001). The experimental design used to assess risk in these studies provides a good model for other studies in that the critical aspects of the assessment focused on experiments under field conditions, further supported by laboratory experiments. Moreover, the relative risks to a non-target organism of using Bt maize or a chemical insecticide were compared, thereby eliminating the 'double-standard' that is often applied to assessing risks of Bt crops, in other words, applying standards to Bt crops that are much higher than those applied to chemical insecticides.

Regardless of whether the results obtained against non-target organisms in laboratory studies show favorable, unfavorable or neutral effects, these must be followed by long-term studies under field conditions. The reason is that laboratory studies are designed to reveal any potential adverse effects by exposing non-target organisms to excessively high levels of Bt proteins, levels that would not be encountered under field conditions. Moreover, field studies should include comparisons to current agricultural practices, which often will include the use of chemical insecticides. At present, there have only been a few studies so far that have evaluated the effects of Bt crops on non-target organisms under field conditions over the length of the growing season. These studies are nevertheless important because non-target organisms and their prey were exposed to Bt Cry proteins in the form synthesized in the crop and over a continuous period at an operational level. The non-target organisms studied under field conditions have all been insects and the test crops either Bt maize or Bt cotton. The insects consisted of a plant bug and four beneficial insects, specifically two parasites and two predators, one of which was the lacewing, *C. carnea*. In these season-long studies, no adverse effects were observed on any of the non-targets under field conditions (Table 8.9), even on *C. carnea*, which as noted above is apparently sensitive to Cry1Ab.

Other non-target organisms that have recently received attention with respect to the environmental effects of Bt crops are soil microorganisms. In a series of studies,

Table 8.9 Effects of Bt crops on non-target invertebrates under field conditions

Non-target (Insect order)	Crop	Cry protein	Adverse effects	Reference
Lygus lineolaris (Heteroptera)	Cotton	Cry1Ac	None	Hardee and Bryan (1997)
Coleomegilla maculata (Coleoptera)	Maize	Cry1Ab	None	Pilcher et al. (1997)
Orius insidiosus (Heteroptera)	Maize	Cry1Ab	None	Pilcher et al. (1997)
Chrysoperla carnea (Neuroptera)	Maize	Cry1Ab	None	Pilcher et al. (1997)
Eriborus tenebrans (Hymenoptera)	Maize	Cry1Ab	None	Orr and Landis (1997)
Macrocentrus grandi (Hymenoptera)	Maize	Cry1Ab	None	Orr and Landis (1997)

some in the laboratory, others under quasi-field conditions, it has been shown that Bt proteins such as Cry1Ab released into soil can bind to various types of soil particles and remain active for periods of at least several months (Crecchio and Stotzky, 1998; Tapp and Stotzky, 1998). The consequences of this accumulation remain unknown, but recent studies by Stotzky and his colleagues indicate there were no significant effects on soil microorganisms or earthworms (Stotzky, 2001, personal communication). Evidence from other studies indicates that Bt Cry proteins are rapidly degraded after incorporation of Bt crop tissue into soil, although low levels can persist for at least several months (Palm et al., 1996). While these studies indicate that the temporary accumulation of Bt Cry proteins in soil have no adverse effects, longer term studies are needed to further evaluate these initial results.

Safety of Bt crops to non-mammalian vertebrates

With respect to non-mammalian vertebrates, tests of Bt maize have been carried out using bobwhite quail, channel catfish and broiler chickens (Table 8.8), the latter in a 38-day study in which the chickens were fed on Bt maize grain (Brake and Vlachos, 1998). In the chicken studies, chickens fed Bt maize grain were compared with chickens fed non-Bt grain. The variables examined were survival, weight gain and feed efficiency. No significant differences were found between the chickens fed Bt maize grain versus non-Bt maize grain. Similar results were obtained in other studies in which the effects of feeding Bt maize versus non-Bt maize to chickens were studied (Halle et al., 1998).

Safety of Bt crops to mammals

There have been very few studies published in the scientific literature in which mammals were fed on Bt crops and then monitored for effects. Most of the safety data used to register Bt crops, as noted above, is based on studies of Cry proteins tested as bacterial insecticides or produced in E. coli. But there has been at least one

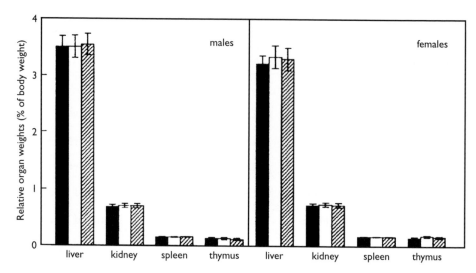

Figure 8.3 Comparison of organ weights from Wistar rats fed on transgenic Cry1Ab tomato for 91 days. No differences were found between rats fed on Bt tomatoes and the non-Bt tomato controls. From Noteborn et al. (1995).

comprehensive study in which the effect of feeding rats on Cry1Ab tomatoes for 91 days was examined, and amplified with other studies using Cry1Ab produced in *E. coli* (Noteborn *et al.*, 1995). In this study, weanling Wistar rats were fed for 91 days on a lyophilized powder of Cry1Ab tomatoes (40.6 ng Cry1Ab/mg protein) that made up 10% of the diet. Control rats were fed a similar diet consisting of a 10% non-Bt lyophilized tomato powder. In terms of human consumption, this amount was calculated to be the equivalent of a human eating 13 kg of tomatoes per day. No significant differences were observed between the treated and control rats in terms of survival, food intake, body and organ weights, or other gross macroscopic examinations of tissues (Figure 8.3). The lack of any apparent adverse effects resulting from feeding rats on a high dose of Cry1Ab tomato corresponds to results obtained in feeding rats bacterial toxins containing Cry1Ab (Table 8.6).

In related studies, these investigators used Cry1Ab produced in *E. coli* to study the effects and fate of this protein on various mammals and mammalian tissues (Noteborn *et al.*, 1995). Histological binding assays were conducted to determine whether the Cry1Ab could bind to the intestinal mucosa of mice, rats, rhesus monkeys and humans. Cry1Ab did not bind to microvilli in these mammals in any of the intestinal regions, including the esophagus, stomach, duodenum, jejunum, ileum and colon, though the toxin did bind to microvilli along the entire length of the larval midgut in the tobacco hormworm, *Manduca sexta*, the lepidopteran used as the insect control. In a separate study, gastrointestinal tissues were examined 7 hours after rats were fed an amount of Cry1Ab equivalent to a human consuming 2000 kg of tomatoes in a single day. No binding was detected in any of the above intestinal regions. These results indicate that none of the mammalian species tested had receptors for Cry1Ab on their intestines. In other studies, it was found that the active toxin form of Cry1Ab (66–68 kDa) was degraded to peptides of 15 kDa or less within 2 hours when incubated in simulated human gastric juices at pH 2. In studies using Cry1Ab consumed *ad libidum* in drinking water for 31 days by New Zealand white rabbits in an amount equivalent to humans consuming 60 kg of tomatoes per day, no differences were found between treated and control groups. Histological studies of the stomach and intestines showed no abnormalities, and no antibodies were induced against Cry1Ab in the treated rabbits.

In another study of a Bt crop, Bt potato producing an unidentified Cry1 protein was fed to one-month-old mice for two weeks, after which the mice were sacrificed and sections of the ileum were examined by light and electron microscopy (Fares and El-Sayed, 1998). The control mice were fed on non-Bt potato. No statistically significant differences were observed between the mice fed Bt potato and the control mice.

In another Bt potato study, raw potatoes containing the Cry3A protein (used to control the Colorado potato beetle) were fed to rats for 28 days (Larvik *et al.*, 1995). The control was the same potato variety that did not produce Cry3A. The rats were fed levels of Cry3A corresponding to a human consuming several kilograms of potato per day for 28 days. No adverse effects were observed. The treated and control rat groups were similar in weight gain and behavior.

In other studies on the effects of Bt grain on the health and perfomance of other animals, no significant differences were found between groups fed Bt versus non-Bt grains in dairy cows (Faust, 1998, 1999; Mayer and Rutzmoser, 1999), beef cows (Russell and Peterson, Russell *et al.*, 2000) or steers (Folmer *et al.*, 2000).

This combination of studies using a variety of methods and feeding studies that include Bt crops and candidate Cry proteins validate the use of these methods for assessing safety and risk, and support the use of data developed for the registration of bacterial insecticides to support the safety of Bt crops.

Most studies of Bt crop safety have focused on tests designed to detect potential harmful effects. There have been few studies of potential improvements in safety, though it is often noted that Bt crops should be safer because they reduce chemical insecticide usage. Yet one notable improvement in safety that has not received much attention is that Bt maize grain contains lower levels of certain fungal toxins known as fumonisins. These toxins can cause illness and death in horses and pigs, and have been implicated in certain forms of liver and esophageal cancer in humans. Fumonisins result from ear rot infections of maize by *Fusarium* species, which invade tissues damaged by the European corn borer, *O. nubilalis*. In Bt maize grain, the levels of fumonisins are reduced by as much as 93% in comparison with non-Bt maize grown under similar conditions (Munkvold *et al.*, 1997, 1999; Masoero *et al.*, 1999).

Summary and the future of Bt crop safety in perspective

Insecticides based on endotoxin proteins of *B. thuringiensis* have been in use for 40 years, and have a safety record for non-target invertebrates and vertebrates including mammals that far surpasses that of any synthetic chemical insecticide. This safety record, combined with the efficacy of certain Bt Cry proteins and the advent of recombinant DNA technology, led to the development of transgenic insect-protected Bt crops that are being adopted rapidly, especially Bt cotton and Bt maize, by farmers in the United States and a few other countries. The safety of these crops for non-target organisms is based primarily on data from numerous toxicological studies carried out on bacterial isolates used to register Bt insecticides, and to some extent on Cry proteins produced in *E. coli* or *Bacillus* species. There have been comparatively few studies so far that have directly evaluated Bt crops in feeding studies. Those that have been done provide strong evidence that Bt crops are safe for humans and other animals as well as for most non-target invertebrates, with the exception of insect species closely related to target insect pests, and a few other species. There is some evidence based on forced-feeding laboratory studies that non-target organisms such as larvae of the monarch butterfly and lacewings may be adversely affected or killed by Bt proteins produced in Bt crops. But the limited number of field trials conducted to date have found no evidence that Bt crops adversely affect generalist insect pre-dator populations or parasite populations. In fact, evidence is mounting that Bt crops may enhance biological control and integrated pest management programs by reducing the amounts of synthetic chemical insecticides used to control insect pests, thereby maintaining the populations of beneficial insects, which are typically decimated by the use of chemical insecticides. In addition, Bt crops such as Bt maize may actually be safer for consumption by mammals than non-Bt crops because the control of lepidopteran pests reduces the level of certain mycotoxins that result from fungal infections following insect damage.

Whereas existing evidence attests to a high degree of safety for Bt crops, these should be more extensively evaluated directly through detailed feeding and toxico-logical studies, and through more extensive and long-term ecological studies to deter-mine their effects on non-target invertebrate populations under operational field conditions. Given the public controversy that has arisen over genetically engineered crops, numerous such studies are underway, and will find their way into the refereed scientific literature in the coming years.

In designing safety studies for non-target mammals, more emphasis should be placed on feeding studies in which animals are fed Bt crops, rather than relying solely on surrogate data developed from feeding studies using Bt insecticides or Cry pro-teins produced in *E. coli*. This should at least be done for a range of Cry proteins used or destined for use in a variety of Bt crops to allay the concerns being raised by the public. The costs will be minor compared to the investments being made in Bt crops and the advantages that will accrue long term from having high public confid-ence in the safety of transgenic crops. Though there are technical hurdles that must be overcome in 'direct' feeding studies due to the low levels of Cry proteins in Bt crops, tests like those developed by Noteborn *et al.* (1995) described above indicate that such studies can be done. The results of less direct methods, such as chemical fingerprinting of metabolic changes that might be detected between Bt crops and their isogenic parents (Noteborn *et al.*, 2000), would appear unnecessary and could easily lead to questionable interpretations. For example, what if the metabolic pool of a mRNA for a certain enzyme is found to be twice as high in a Bt tomato as opposed to its isogenic parent, but all the direct feeding studies show that the tomato is safe? And what if the mRNA for the same enzyme in another tomato variety is shown to be just as high as the mRNA in the Bt tomato. Do any of these results mean the Bt tomato is not safe?

It is important that any new tests developed to evaluate safety be directly relevant to providing data that can serve as a scientific basis for the further assessment of safety. Of course, all methods used need to be subjected to rigorous validation. Studies that result in data of questionable value or that could be subject to misinterpretation should be avoided. Aside from the scientific rationale for taking this approach, a reason for doing this is that many individuals and environmental groups are opposed to Bt crops simply because they object to the use of genetic engineering technology to modify crops. A good example of this sort of thinking comes from Lord Peter Melchett, a former Labour Minister in the United Kingdom, and a leader in Greenpeace's effort to stop the use of biotechnology in agriculture. Lord Melchett opposes the use of science to make decisions on food consumption, preferring that selections be made on an emotional basis (Specter, 2000), stating that 'If its acceptable to choose your car based on emotion and not science, why should it be wrong to choose your food that way?' The public cannot afford to let this sort of thinking dominate or influence views and decisions on what is safe to eat. Therefore, to counter such thinking and avoid the misuse of data, meaningful short- and long-term studies like those carried out by Noteborn et al. (1995) should be the type used to evaluate Bt crop safety. Such studies should detect any adverse effects of natural or modified proteins synthesized by crop plants. In addition, the same rationale is relevant to justifying long-term, i.e. studies of several years, on the effects of Bt crops on non-target organisms, including soil microorganisms to assess any potential ecological effects, be they positive, neutral or negative. The results of initial studies indicate that Bt crops will be much safer for non-target organisms, especially beneficial insects, than chemical insecticides. If long-term studies confirm these initial results, dissemination of this knowledge should improve public confidence in this new and important crop protection technology.

In summary, Bt crops are one of the first practical examples of new and powerful pest-control technologies emerging from the results of decades of basic research and recombinant DNA technology. An overwhelming amount of evidence based on studies of Bt Cry proteins indicates these crops are safe for consumption by humans and other vertebrates, and are much safer for non-target invertebrates and the environment than synthetic chemical insecticides. Based on the scientific evidence supporting these conclusions, the US Environmental Protection Agency will continue to register Bt crops for use in the United States (US EPA, 2001). While additional direct testing of Bt crops is warranted to further assess their safety, there is no reason at present to think that these crops present risks greater than those associated with the consumption of non-Bt crops. In fact, Bt crops may be safer for human consumption than conventional crops because they contain lower levels of mycotoxins and residues of chemical insecticides.

References

Aronson, A.I. (1993) 'The two faces of Bacillus thuringiensis: insecticidal proteins and post-exponnential survival', Molecular Microbiology 7: 489–96.
Baumann, L., Okamoto, K., Unterman, B.M. et al. (1984) 'Phenotypic characterization of Bacillus thuringiensis (Berliner) and B. cereus (Frankland & Frankland)', Journal of Invertebrate Pathology 44: 329–41.
Becker, N. and Margalit, J. (1993) 'Use of Bacillus thuringiensis israelensis against mosquitoes and blackflies', in P.F. Entwistle, J.S. Cory, M.J. Bailey and S. Higgs (eds) Bacillus thuringiensis,

An Environmental Biopesticide: Theory and Practice. Chichester: John Wiley & Sons, pp. 147–70.

Betz, F.S., Forsyth, S.F. and Stewart, W.E. (1990) 'Registration requirements and safety considerations for microbial pest control agents in North America', in M. Laird, L.A. Lacey and E.W. Davidson (eds) *Safety of Microbial Insecticides*. Boca Raton: CRC Press, pp. 3–10.

Brake, J. and Viachos, D. (1998) 'Evaluation of transgenic event 176 Bt-corn in broiler chickens,' *Poultry Science* 77: 648–53.

Butko, P., Huang, F., Pusztai-Carey, M. *et al.* (1996) 'Membrane permeabilization induced by cytolytic delta-endotoxin CytA from *Bacillus thuringiensis* var. *israelensis*', *Biochemistry* 35: 11355–60.

Butko, P., Huang, F., Pusztai-Carey, M. *et al.* (1997) 'Interaction of the delta-endotoxin CytA from *Bacillus thuringiensis* var. *israelensis* with lipid membranes', *Biochemistry* 36: 12862–68.

Chen, X.J., Lee, M.K. and Dean, D.H. (1993) 'Site-directed mutations in a highly conserved region of *Bacillus thuringiensis* δ-endotoxin affect inhibition of short circuit current across *Bombyx mori* midguts', *Proceedings of the National Academy of Sciences of the USA* 90: 9041–45.

Chilcott, C.N. and Ellar, D.J. (1988) 'Comparative toxicity of *Bacillus thuringiensis* var. *israelensis* crystal proteins *in vivo* and *in vitro*', *Journal of General Microbiology* 134: 2551–58.

Chilcott, C.N., Knowles, B.H., Ellar, D.J. *et al.* (1990) 'Mechanism of action of *Bacillus thuringiensis israelensis* parasproal body', in H. de Barjac and D. Sutherland (eds) *Bacterial Control of Mosquitoes and Black Flies*. New Brunswick: Rutgers University Press, pp. 45–65.

Couch, J.A. and Foss, S.S. (1990) 'Potential impact of microbial insecticides on the estaurine and marine environments', in M. Laird, L.A. Lacey and E.W. Davidson (eds) *Safety of Microbial Insecticides*. Boca Raton: CRC Press, pp. 85–97.

Crecchio, C. and Stotzky, G. (1998) 'Insecticidal activity and biodegradation of the toxin from *Bacillus thuringiensis* subsp. *kurstaki* bound to humic acids from soil', *Soil Biology and Biochemistry* 30: 463–70.

Crickmore, N., Bone, E.J., Williams, J.A. *et al.* (1995) 'Contribution of the individual components of the delta-endotoxin crystal to the mosquitocidal activity of *Bacillus thuringiensis* subsp. *israelensis*', *FEMS Microbiology Letters* 131: 249–54.

Crickmore, N., Zeigler, D.R., Feitelson, J. *et al.* (1998) 'Revision of the nomenclature for the *Bacillus thuringiensis* pesticidal crystal proteins', *Microbiology and Molecular Biology Reviews* 62: 807–13.

de Barjac, H. and Frachon, E. (1990) 'Classification of *Bacillus thuringiensis* strains', *Entomophaga* 35: 233–40.

Dejoux, C. and Elouard, J.-M. (1990) 'Potential impact of microbial insecticides on the freshwater environment, with special reference to the WHO/UNDP/World Bank, Onchocerciasis Control Programme', in M. Laird, L.A. Lacey and E.W. Davidson (eds) *Safety of Microbial Insecticides*. Boca Raton: CRC Press, pp. 65–83.

Delecluse, A., Juarez-Perez, V. and Berry, C. (2000) 'Vector-active toxins: structure and diversity', in J.-F. Charles, A. Delecluse and C. Neilsen-LaRoux (eds) *Bacterial Insecticides: from the Laboratory to the Field*. Amsterdam: Kluwer, pp. 101–25.

Estruch, J.J., Warren, G.W., Mullins, M.A. *et al.* (1996) 'Vip3A, a novel *Bacillus thuringiensis* vegetative insecticidal protein with a wide spectrum of activities against epidopteran insects', *Proceedings of the National Academy of Sciences of the USA* 93: 5389–94.

Fares, N.H. and El-Sayed, A.K. (1998) 'Fine structural changes in the ileum of mice fed on δ-endotoxin-treated potatoes and transgenic potatoes', *Natural Toxins* 6: 219–33.

Federici, B.A. (1999) '*Bacillus thuringiensis* in biological control', in T. Bellows, T.W. Fisher and G. Gordh (eds) *Handbook of Biological Control*. San Diego: Academic Press, pp. 575–93.

Federici, B.A. (1993) 'Bacillus thuringiensis: Biology, application, and prospects for further development', in R.J. Akhurst (ed.) *Proceedings of the Second Canberra Meeting on* Bacillus thuringiensis. Canberra: CPN Publications, pp. 1–15.

Federici, B.A. and Bauer, L.S. (1998) 'Cyt1Aa protein of *Bacillus thuringiensis* is toxic to the cottonwood leaf beetle, *Chrysomela scripta*, and suppresses high levels of resistance to Cry3Aa', *Applied and Environmental Microbiology* 64: 4368–71.

Federici, B.A., Luthy, P. and Ibarra, J.E. (1990) 'The parasporal body of BTI: Structure, protein composition, and toxicity', in H. de Barjac and D. Sutherland (eds) *Bacterial Control of Mosquitoes and Black Flies*. New Brunswick: Rutgers University Press, pp. 16–44.

Fisher, R. and Rosner, L. (1959) 'Toxicology of the microbial insecticides, Thuricide', *Agriculture and Food Chemistry* 7: 686–99.

Folmer, J.D., Grant, R.J., Milton, C.T. *et al.* (2000) 'Effect of Bt corn silage on short-term lactational performance and ruminal fermentation in dairy cows', *Journal of Dairy Science* 83: 1182.

Frutos, R., Rang, C. and Royer, M. (1999) 'Managing insect resistance to plants producing *Bacillus thuringiensis* toxins', *Critical Reviews in Biotechnology* 19: 227–76.

Ge, A.Z., Shivarova, N.I. and Dean, D.H. (1989) 'Location of the *Bombyx mori* specificity domain on a *Bacillus thuringiensis* δ-endotoxin protein', *Proceedings of the National Academy of Sciences of the USA* 86: 4037–41.

Gill, S.S., Cowles, E.A. and Pietrantonio, P.V. (1992) 'The mode of action of *Bacillus thuringiensis* endotoxins', *Annual Reviews in Entomology* 37: 615–36.

Glare, T.R. and O'Callaghan, M. (2000) Bacillus thuringiensis: *Biology, Ecology, and Safety*. Chichester, John Wiley & Sons.

Gould, F. (1998) 'Sustainability of transgenic insecticidal cultivars: Integrating pest genetics and ecology', *Annual Reviews in Entomology* 43: 701–26.

Gould, F., Martinez-Ramirez, A., Anderson, A. *et al.* (1992) 'Broad-spectrum resistance to *Bacillus thuringiensis* toxins in *Heliothis virescens*', *Proceedings of the National Academy of Sciences of the USA* 89: 7986–90.

Garczynski, S.F., Crim, J.W. and Adang, M.J. (1991) 'Identification of putative brush border membrane-binding molecules specific to *Bacillus thuringiensis* δ-endotoxin by protein blot analysis', *Applied and Environmental Microbiology* 57: 2816–20.

Halle, I., Aulrich, K. and Flachowsky, G. (1998) 'Einzatz von Maiskornen der sorte cesar und ddes gentechnisch veranderten Bt-hybriden in der broiler mastr', Proc. 5. Tagung. Schweine- und Geflhgelernahrung, 01.03.12.1998, Wittenberg, 265–67.

Hardee, D.D. and Bryan, W.W. (1997) 'Influence of *Bacillus thuringiensis*-transgenic and nectari-less cotton on insect populations with emphasis on the tarnished plant bug (Heteroptera: Miridae)', *Journal of Economic Entomology* 90: 663–68.

Heimpel, A.M. and Angus, T.A. (1963) 'Diseases caused by certain sporeforming bacteria', in E.A. Steinhaus (ed.) *Insect Pathology: An Advanced Treatise*, Vol. 2. New York: Academic Press, pp. 21–73.

Hellmich, R.L., Siegfried, B.D., Sears, M.K. *et al.* (2001) 'Monarch larvae sensitivity *to Bacillus thuringiensis*-purified proteins and pollen', *Proceedings of the National Academy of Sciences of the USA* 98: 11925–30.

Hilbeck, A., Baumgartner, M., Fried, P.M. *et al.* (1998a) 'Effects of transgenic *Bacillus thuringiensis* corn-fed prey on mortality and development time of immature *Chrysoperla carnea* (Neuroptera: Chrysopidae)', *Environmental Entomology* 27: 480–87.

Hilbeck, A., Moar, W.J., Pustai-Carey, M. *et al.* (1998b) 'Toxicity of *Bacillus thuringiensis* Cry1A(b) toxin to the predator *Chrysoperla carnea* (Neuroptera: Chrysopidae)', *Environmental Entomology* 27: 1255–63.

Hofmann, C., Vanderbruggen, H., Hofte, H. *et al.* (1988) 'Specificity of *Bacillus thuringiensis* δ-endotoxins is correlated with the presence of high affinity binding sites in the brush border membrane of target insect midguts', *Proceedings of the National Academy of Sciences of the USA* **85**: 7844–88.

Hofte, H. and Whiteley, H.R. (1989) 'Insecticidal crystal proteins of *Bacillus thuringiensis*', *Microbiological Reviews* **53**: 242–55.

Knowles, D.H. and Ellar, D.J. (1987) 'Colloid-osmotic lysis is a general feature of the mechanisms of *Bacillus thuringiensis* δ-endotoxins with different insect specificity. *Biochimica Biophysica Acta* **924**: 509–18.

Knowles, B.H. and Dow, J.A.T. (1993) 'The crystal δ-endotoxins of *Bacillus thuringiensis*: Models for their mechanism of action on the insect gut', *BioEssays* **15**: 469–76.

Koni, P.A. and Ellar, D.J. (1993) 'Cloning and characterization of a novel *Bacillus thuringiensis* cytolytic toxin δ-endotoxin', *Journal of Molecular Biology* **229**: 319–27.

Larvik, P.B., Bartnicki, D.E., Feldman, H. *et al.* (1995) 'Safety assessment of potatoes resistant to Colorado potato beetle', in K.-H. Engel, G.R. Takeoka and R. Teranishi (eds) *American Chemical Society Series 605*. Washington, DC: ACS, pp. 148–58.

Lacey, L.A. and Siegel, J.P. (2000) 'Safety and ecotoxicology of entomopathogenic bacteria', in J.-F. Charles, A. Delecluse and C. Neilsen-LaRoux (eds) *Bacterial Insecticides: from the Laboratory to the Field*. Amsterdam: Kluwer, pp. 253–73.

Lacey, L.A. and Mulla, M.S. (1990) 'Safety of *Bacillus thuringiensis* ssp. *israelensis* and Bacillus sphaericus to nontarget organisms in the aquatic environment', in M. Laird, L.A. Lacey and E.W. Davidson (eds) *Safety of Microbial Insecticides*. Boca Raton: CRC Press, pp. 169–88.

Lee, M.L., Milne, R.E., Ge, A. *et al.* (1992) 'Location of a *Bombyx mori* receptor binding region on a *Bacillus thuringiensis* δ-endotoxin', *Journal of Biological Chemistry* **267**: 3115–21.

Levine, R. (1991) 'Recognized and possible effects of pesticides in humans', in W.J., Hayes, Jr. and E.R. Laws, Jr. (eds) *Handbook of Pesticide Toxicology*, Vol. 1. San Diego: Academic Press, pp. 275–360.

Li, J., Carroll, J. and Ellar, D.J. (1991) 'Crystal structure of insecticidal δ-endotoxin from *Bacillus thuringiensis* at 2.5 angstrom resolution', *Nature* **353**: 815–21.

Li, J., Koni, P.A. and Ellar, D.J. (1996) 'Structure of the mosquitocidal delta-endotoxin CytB from *Bacillus thuringiensis* sp. *kyushuensis* and implications for membrane pore formation', *Journal of Molecular Biology* **257**: 129–52.

Losey, J.J., Raynor, L. and Cater, M.E. (1999) 'Transgenic pollen harms monarch larvae', *Nature* **399**: 214.

Lozzia, G.C., Furlanis, C., Manachini, B. *et al.* (1998a) 'Effects of Bt corn on *Rhodopalosiphum padi* (Rhynchota, Aphidae) and on its predator *Chrysoperla carnea* Stephen (Neuroptera, Chrysopidae). *Bolletino Zool. Agr. Bachic. Ser. II* **30**: 153–64.

Lozzia, G.C., Rigamonti, I.E. and Agosti, M. (1998b) 'Metodi di valutazione degli effetti de Mais transgenico sugli artropodi non beresaglio', *Notiz. Protez. Plante* **8**: 27–39.

Luthy, P. and Ebersold, H.R. (1981) '*Bacillus thuringiensis* delta-endotoxin: histopathology and molecular mode of action', in E.W. Davidson (ed.) *Pathogenesis of Invertebrate Microbial Diseases*. Totowa: Allanheld, Osman & Co., pp. 235–67.

Masoero, F., Moschini, M., Rossi, F. *et al.* (1999) 'Nutritive value, mycotoxin contamination and in vitro rumen fermentation of normal and genetically modified corn (CryIAb) grown in northern Italy', *Maydica* **44**: 205–209.

McClintock, J.T., Schaffer, C.R. and Sjoblad, R.D. (1995) 'A comparatiove review of the mammalian toxicity of *Bacillus thuringiensis*-based pesticides', *Pesticide Science* **45**: 95–105.

Meadows, M.P. (1993) '*Bacillus thuringiensis* in the environment: ecology and risk assessment', in P.F. Entwistle, J.S. Cory, M.J. Bailey and S. Higgs (eds) Bacillus thuringiensis, *An Environmental Biopesticide: Theory and Practice*. Chichester: John Wiley & Sons, pp. 193–220.

Melin, B.E. and Cozzi, E.M. (1990) 'Safety to non-target invertebrates of lepidopteran strains of *Bacillus thuringiensis*', in M. Laird, L.A. Lacey and E.W. Davidson (eds) *Safety of Microbial Insecticides*. Boca Raton: CRC Press, pp. 149–67.

Mulla, M.S. (1990) 'Activity, field efficacy, and use of *Bacillus thuringiensis israelensis* against mosquitoes', in de Barjac and Sutherland (eds) *Bacterial Control of Mosquitoes and Black Flies*. New Brunswick: Rutgers University Press, pp. 134–60.

Munkvold, G.P., Hellmich, R.K. and Showers, W.B. (1997) 'Reduced fusarium ear rot and symptomless infection in kernals of maize genetically engineered for European corn borer resistance', *Phytopathology* 87: 1071–77.

Munkvold, G.P., Hellmich, R.L. and Rice, L.G. (1999) 'Comparison of fumonisn concentrations in kernals of transgenic Bt maize hybrids and non transgenic hybrids', *Plant Diseases* 83: 130–38.

Orr, D.B. and Landis, D.A. (1997) 'Oviposition of European corn borer (Lepidoptera: Pyralidae) and impact of natural enemy populations in transgenic versus isogenic corn', *Journal of Economic Entomology* 90: 905–909.

Noteborn, H.P.J.M., Bienenmann-Ploum, M.E., van den Berg, J.H.J. *et al.* (1995) 'Safety assessment of the *Bacillus thuringiensis* insecticidal crystal protein Cry1A(b) expressed in transgenic tomatoes', in K.-H. Engel, G.R. Takeoka and R. Teranishi (eds) *American Chemical Society Series 605*. Washington, DC: ACS, pp. 134–47.

Noteborn, H.P.J.M., Lommen, A., van der Jagt, R.C. *et al.* (2000) 'Chemical fingerprinting for the evaluation of unintended secondary metabolic changes in transgenic food crops', *Journal of Biotechnology* 77: 103–14.

Palm, C.J., Schaller, D.L., Donegan, K.K. *et al.* (1996) 'Persistence in soil of transgenic plant produced *Bacillus thuringiensis* var. *kurstaki* δ-endotoxin', *Canadian Journal of Microbiology* 42: 1258–62.

Perlak, F.J., Fuchs, R.L., Dean, D.A. *et al.* (1991) 'Modification of the coding sequence enhances plant expression of insect control protein genes', *Proceedings of the National Academy of Sciences of the USA* 88: 3324–28.

Pilcher, C.D., Obrycki, J.J., Rice, M.E. *et al.* (1997) 'Preimaginal development, survival, and field abundance of insect predators on transgenic *Bacillus thuringiensis* corn', *Environmental Entomology* 26: 446–54.

Pleasants, J.M., Hellmich, R.L., Dively, G.P. *et al.* (2001) 'Corn pollen deposition on milkweeds in and near cornfields', *Proceedings of the National Academy of Sciences of the USA* 98: 11919–24.

Saik, J.E., Lacey, L.A. and Lacey, C.M. (1990) 'Safety of microbial insecticides to vertebrates – domestic animals and wildlife', in M. Laird, L.A. Lacey and E.W. Davidson (eds) *Safety of Microbial Insecticides*. Boca Raton: CRC Press, pp. 115–32.

Sanders, P.R., Lee, T.C., Groth, M.E. *et al.* (1998) 'Safety assessment of insect-protected corn', in Thomas, J.A. (ed.) *Biotechnology and Safety Assessment*, 2nd edition. London: Taylor & Francis, pp. 241–56.

Sangadala, S., Walters, F.W., English, L.H. *et al.* (1994) 'A mixture of *Manduca sexta* aminopeptidase and alkaline phosphatase enhances *Bacillus thuringiensis* insecticidal CryIA(c) toxin binding and $^{86}Rb^+$-K^+ efflux *in vitro*', *Journal of Biological Chemistry* 269: 10088–92.

Schnepf, E., Crickmore, N., Van Rie, J. *et al.* (1998) '*Bacillus thuringiensis* and its pesticidal crystal proteins', *Microbiology and Molecular Biology Reviews* 62: 775–806.

Schuler, T.H., Poppy, G.M., Kerry, B.R. *et al.* (1998) 'Insect-resistant transgenic plants', *Trends in Biotechnology* 16: 168–75.

Sears, M.K., Hellmich, R.L., Stanley-Horn, D.E. *et al.* (2001) 'Impact of Bt corn pollen on monarch butterfly populations: a risk assessment', *Proceedings of the National Academy of Sciences of the USA* 98: 11937–42.

Shelton, A.M., Zhao, J.Z. and Roush, R.T. (2002) 'Economic, ecological, food safety, and social consequences of the development of Bt transgenic plants', *Annual Reviews in Entomology* 47: 845–81.

Siegel, J.P. (1997) 'Testing the pathogenicity and infectivity of entomopathogens to mammals', in Lacey, L. (ed.) *Manual of Techniques in Insect Pathology*. San Diego: Academic Press, pp. 325–36.

Siegel, J.P. (2001) 'The mammalian safety of Bacillus thuringiensis-based insecticides', *Journal of Invertebrate Pathology* 77: 13–21.

Siegel, J.P. and Shadduck, J.A. (1990a) 'Safety of microbial insecticides to vertebrates – humans', in M. Laird, L.A. Lacey and E.W. Davidson (eds) *Safety of Microbial Insecticides*. Boca Raton: CRC Press, pp. 101–13.

Siegel, J.P. and Shadduck, J.A. (1990b) 'Mammalian safety of *Bacillus thuringiensis israelensis*', in de Barjac and Sutherland (eds) *Bacterial Control of Mosquitoes and Black Flies*. New Brunswick: Rutgers University Press, pp. 202–17.

Specter, M. (2000) 'The pharmageddon riddle', *The New Yorker*, April 10, pp. 58–71.

Stanley-Horn, D.E., Dively, G.P., Hellmich, R.L. *et al.* (2001) 'Assessing the impact of Cry1Ab-expressing corn pollen on monarch butterfly larvae in field studies', *Proceedings of the National Academy of Sciences of the USA* 98: 11931–36.

Tapp, H. and Stotzky, G. (1998) 'Persistence of the insecticidal toxin from Bacillus thuringiensis subsp. kurstaki in soil', *Soil Biology and Biochemistry* 30: 471–76.

Thayer, A.M. (1999) 'Ag biotech food: risky or risk free?', *Chemical Engineering News* November 1, pp. 11–20.

Thiery, I. and Frachon, E. (1997) 'Bacteria: identification, isolation, culture and preservation of entomopathogenic bacteria', in L. Lacey (ed.) *Manual of Techniques in Insect Pathology*. San Diego: Academic Press, pp. 55–77.

Thomas, W.E. and Ellar, D.J. (1983) 'Mechanism of action of *Bacillus thuringiensis* var. *israelensis* insecticidal δ-endotoxin', *FEBS Letters* 154: 362–68.

US Environmental Protection Agency (2001) 'Biopesticides Registration Action Document – *Bacillus thuringiensis* (Bt) plant-incorporated protectants', epa.gov/pesticides/biopesticides/reds/brad_bt_pip2.htm

Van Rie, J., Jansens, S., Hofte, H. *et al.* (1989) 'Specificity of *Bacillus thuringiensis* δ-endotoxins: importance of specific receptors on the brush border membrane of the mid-gut of target insects,' *European Journal of Biochemistry* 186: 239–47.

Vinson, S.B. (1990) 'Potential impact of microbial insecticides on beneficial arthropods in the terrestrial environment', in M. Laird, L.A. Lacey and E.W. Davidson (eds) *Safety of Microbial Insecticides*. Boca Raton: CRC Press, pp. 43–64.

Wolfersberger, M.G. (1990) 'The toxicity of two *Bacillus thuringiensis* delta δ-endotoxins to gypsy moth larvae is inversely related to the affinity binding cites on the brush border membranes for toxins', *Experentia* 46: 475–77.

Wu, D. and Aronson, A.I. (1992) 'Localized mutagenesis defines regions of the *Bacillus thuringiensis* δ-endotoxin involved in toxicity and specificity', *Journal of Biological Chemistry* 267: 2311–17.

Wu, D. and Federici, B.A. (1993) 'A 20-kilodalton protein preserves cell viability and enhances CytA crystal formation during sporulation in *Bacillus thuringiensis*', *Journal of Bacteriology* 175: 5276–80.

Wu, D., Johnson, J.J. and Federici, B.A. (1994) 'Synergism in mosquitocidal toxicity between CytA and CryIVD proteins using inclusions produced from cloned genes of *Bacillus thuringiensis*', *Molecular Microbiology* 13: 965–72.

Yu, L., Berry, R.R. and Croft, B.A. (1997) 'Effects of *Bacillus thuringiensis* toxins in transgenic cotton and potato on *Folsomia candida* (Collembola: Isotomidae) and *Oppia nitens* (Acari: Orbatidae)', *Ecotoxicology* 90: 113–18.

Chapter 9

Case study: recombinant baculoviruses as microbial pesticidal agents

Ivan E. Gard, Michael F. Treacy and John J. Wrubel

Introduction

Baculoviruses are a group of viral pathogens that cause fatal disease in arthropods. More specifically, these baculoviruses infect primarily only members of the class Insecta and even within this class limit themselves to only a few orders, i.e. Lepidoptera, Diptera, Hymenoptera and Coleoptera (Granados and Federici, 1986). Baculoviruses have been studied extensively dating back to the 1960s. The literature is filled with a plethora of information on their biology and safety. The wild-type baculoviruses have been evaluated for use as insecticides numerous times, especially during the late 1960s and up into the early 1970s. The first baculoviral product introduced in the commercial arena was Elcar™ (*Helicoverpa zea* nucleopolyhedrovirus [NPV], HzNPV). Three other non-commercial preparations produced and used by the US Forest Service for control of forestry pests were developed and registered under the names of Gyp-Chek™ (*Lymantria dispar* NPV, LdMNPV), EM Biocontrol–1™ (*Orgyia pseudotsugata* NPV, OpMNPV) and NeoCheck-S™ (*Neodiprion sertifer* NPV, NsMNPV). These products have received only limited use. However, recently the Elcar™ product was reborn under the trade name of GemStar™ and is being sold in various countries around the world. As of 1995, a total of eight baculoviruses were registered for insect control by the US Environmental Protection Agency (EPA).

Intricate evolutionary relationships have developed between the baculoviruses and the arthropods that they exploit. This close evolutionary relationship has resulted in the restriction of host range (Black *et al.*, 1997). Over the past 40 years, extensive studies evaluating the safety of baculoviruses against vertebrate species have been carried out. A summary of the data indicates that over 26 different insect baculoviruses have been tested for pathogenicity against 10 different mammalian species, including rats, dogs, mice, guinea-pigs, monkeys and humans. The methods of administration in these tests varied, including oral, intravenous injection, intercerebral injection, intramuscular injection and topical application. In no case was there any indication of toxicity or allergenicity. These results have been summarized in a number of reviews (Ignoffo, 1973, 1975; Burges *et al.*, 1980; Doller, 1985).

Specificity is also a major factor in making baculoviruses safe to non-target insects. Baculoviruses generally only infect insects in a few families. In some cases only one species in a genus may be susceptible (e.g. LdMNPV). Those insect species that are considered beneficial (control pest insect species) are generally in families other than those targeted by baculoviruses. There are a few exceptions to this rule, however, as they can also infect some desirable lepidopterans. This insect specificity makes baculoviruses excellent candidates for use in integrated pest management (IPM) programs as they are compatible with other insect control technologies (bioinsecticides, chemical insecticides and transgenic plants), and have no effect on any non-target insects such as honeybees.

If baculoviruses have all of these attractive parameters, why are they not widely used as insecticides? Baculoviruses are unable to rapidly kill a susceptible insect. It may take 5–15 days to kill a host larva following infection. This period is affected by such factors as temperature, dosage and larval age. Younger larvae are generally more susceptible to infection and die sooner, but even though they are infected they continue to eat, usually up until the day they die. Obviously a crop with a low economic threshold cannot tolerate this continued feeding damage until the insect dies. Growers want to see dead insects soon after the application of any insecticide.

Fortunately, recombinant technology offers a new life to the old concept of utilizing baculoviruses as insecticides. It is now possible to genetically engineer a baculovirus with a toxin gene, giving it improved insecticidal properties and increasing the speed of action. These toxin genes usually code for an insect-specific, proteinaceous toxin. The genes can be isolated from the venom of certain arthropods (e.g. scorpions or spiders). Typically, when a baculovirus containing a toxin gene infects an insect, the feeding is rapidly halted and causes a quick death. Insertion of the insecticidal gene does not affect the LD_{50} of the baculovirus but does significantly hasten the killing speed through the pharmokinetic interaction of the toxin within the insect. Even before actual death, the toxin causes cessation of feeding resulting in less plant damage, then paralysis, and eventually death. Diet-overlay assays using an LD_{99} quantity of recombinant NPV have shown that, when second-instars of susceptible insect species are exposed to the wild-type virus versus the genetically engineered virus, at 10 days the LD_{50} is the same for both groups; however, the wild-type infected larvae characteristically take as long as 5.1 days to die versus 2.6 days for larvae infected with genetically engineered virus (Treacy and All, 1996).

Greenhouse studies have shown that this speed of kill translates to significantly better plant protection. In one greenhouse study, cotton plants sprayed with a genetically engineered virus containing a scorpion toxin gene were compared with plants treated with a chemical standard, esfenvalerate. The plants were then artificially infested with target insect species and then evaluated 4–5 days after treatment. The cotton incurred 18.9% versus 13.0% damaged squares for the two treatments, respectively. Statistically, both treatments were equal (Treacy and All, 1996). Numerous US field trials have been conducted, again showing that the speed of kill is influential in reducing the damage. One study conducted in North Carolina (USA) on tobacco against the tobacco budworm, showed that the genetically engineered virus performed as well against this insect as did the standard, acephate (Treacy, 1997). Additionally, toxin production is limited only to the body of the insect (i.e. there is no release of toxicant into the environment as with conventional insecticides). This makes the genetically engineered baculovirus a prime candidate for insecticidal use, and in some situations makes it competitive with chemical insecticides. Block and Zhihong (1997) have given a good summary of the progression of the technology in engineering of baculoviruses, culminating with the actual field release of a toxin gene carrying baculovirus by a commercial company in 1995.

Public relations activity

These genetically manipulated baculoviruses obviously need approval by regulatory agencies prior to any field release and eventual registration. This entails the development of significant safety data which will be discussed later in this chapter.

Using the scientific process to demonstrate safety of a genetically engineered organism to the public is only part of the 'proof of safety' process. This information has to be transmitted to the public to establish the true level of safety in their perception. Biotechnology has come under public scrutiny to varying degrees depending on where one is in the world. Environmental activists have attacked biotechnology from scientific, ethical and personal viewpoints. These activities have created doubts, which must be addressed, no matter how trivial or unscientific they may appear. Obviously the development of a plan is necessary. Assuming that approval of regulatory agencies is adequate (to make a release) could be a major mistake, as demonstrated by the experience of the National Environment Research Council (NERC) Institute of Virology, Oxford, England. All the necessary regulatory clearances were obtained for a release, but no effort had been made to inform the public around the application site. The end result was that the public felt that 'a fast one was being pulled'. The researchers then had to backtrack and attempt to educate an irate public.

American Cyanamid Company conducted field releases in 1995, 1996 and 1998 of baculoviruses containing an insect-specific toxin gene from either the scorpion *Androctonus australis* or the straw-itch mite *Pyemotes tritici*. Before any release was attempted, American Cyanamid Company scientists developed a safety database (extracted from the literature) and through carefully conducted experimentation. This database was summarized and written into a brochure. Also, a video was produced, which explained the engineering process and the resultant recombinant baculovirus' impact on the susceptible host. Both the brochure and video were made available for public distribution. This information was utilized at numerous open meetings with representatives from many areas, academia, regulatory, mass media, environmental groups, politicians and the general public. Thus communication lines were opened and the dialog that developed served to mitigate any concerns.

Early on, American Cyanamid Company scientists opened communication lines with responsible national environmental groups to explain the technology. Gaining the support of these groups inevitably helped ease any concerns that may have been present. Using the basic elements of risk communication, a policy was adopted that all the safety information available would be made public. This policy was followed throughout the public education process and in all US EPA submissions (i.e. all safety data supplied to the US EPA were available to anyone for the asking). This strategy resulted in the generation of very little negative response from the aforementioned groups.

Risk assessment

In addition to evaluating the insecticidal properties of baculoviruses against targeted pest species, it is essential also to assess the potential environmental impact of such gene-inserted pathogens prior to, as well as during, initial small-scale field trials. In their review of ecological considerations pertaining to recombinant baculoviruses,

Richards *et al.* (1998) noted that an environmental impact evaluation must address the effect of the recombinant entity on non-target species and populations, as well as the impact of the genetic modification on the ability of the virus to persist in the environment, relative to that of the wild-type or feral counterpart. It is important to determine whether the addition of a foreign gene alters the host-range of the virus, and whether the insecticidal protein produced in the virus-infected insect will present a hazard to other species. Additionally, comparative testing of recombinant and feral baculoviruses for physical characteristics, genetic fitness and dispersal properties can provide insight as to relative stability in the environment.

Recombinant nucleopolyhedroviruses

Nucleopolyhedroviruses isolated from alfalfa looper, *Autographa californica* (AcMNPV), and cotton bollworm, *Helicoverpa zea* (HzNPV), have been engineered as vectors for carrying and expressing insecticidal genes into the body cavities of pestiferous insect species. AcMNPV is known to infect about 39 species of Lepidoptera (Entwistle and Evans, 1985). Larvae of tobacco budworm, *Heliothis virescens* (F.), and cabbage looper, *Trichoplusia ni* (Hubner), are among the most permissive to infection by AcMNPV, whereas others such as *H. zea* are only moderately sensitive, or semipermissive, to this virus (Vail *et al.*, 1978; Possee *et al.*, 1993; Huang *et al.*, 1997). Conversely, HzNPV is highly virulent against larvae of the entire *Heliothine* spp. complex, i.e. *H. virescens* and *Helicoverpa* spp. (Ignoffo and Garcia, 1992; Grewal *et al.*, 1998).

A number of insect-selective toxin genes have been inserted into the genomes of AcMNPV and HzNPV in an attempt to make them more effective as biocontrol agents in cropping systems. Two such genes, *AaIT* and *LqhIT*, which were isolated from venoms of the Algerian scorpion, *Androctonus australis* (Hector), and Israeli yellow scorpion, *Leiurus quinquestriatus hebraeus* Birula, respectively, are known to inappropriately modulate neuronal sodium channels. The protein encoded by *AaIT* acts as an excitatory sodium channel agonist, causing repetitive firing of the insect's motor nerves and overstimulation of skeletal muscle (Walther *et al.*, 1976; Zlotkin *et al.*, 1985). Symptoms exhibited by insects intoxicated by AaIT include cessation of feeding, paralysis and death. Conversely, LqhIT is a depressant neurotoxin, causing a blockade to sodium conductance and suppression of axonal depolarization and neuronal action potentials (Zlotkin *et al.*, 1993). LqhIT blocks neuromuscular transmission and induces progressive paralysis in affected insects. A third invertebrate toxin gene which has been incorporated into selected baculoviral genomes is *tox34*, which was cloned from the genomic array encoding venom of the straw-itch mite, *Pyemotes tritici* (Tomalski *et al.*, 1989; Tomalski and Miller, 1991). In assays conducted on primary cultures of motor neurons isolated from selected lepidopteran species, it has been shown that the product of the *tox34* gene, a 33-kDa protein designated as TxP-I, causes a blockade of presynaptic inward calcium currents, which likely results in reduced secretion of the neurotransmitter acetylcholine (Torruellas *et al.*, 1998).

The recombinant baculoviruses AcMNPV-AaIT, AcMNPV-LqhIT, AcMNPV-tox34, HzNPV-AaIT and HzNPV-LqhIT have been evaluated for insecticidal efficacy in open field trial settings (Treacy and All, 1996; All and Treacy, 1997; American Cyanamid

Company, 1998; Heinz *et al.*, 1999; Treacy *et al.*, 1999). Considerable research has been devoted to assessing mammalian and environmental safety of these aforementioned baculoviruses, some of which will be discussed in this section of the chapter.

Selectivity of isolated toxins

Studies conducted to define the pharmacology of purified AaIT and LqhIT have shown these toxins to possess remarkable specificity. Although both of these toxins are very potent against insects, they appear to have no effect on mammals even at very high dosages.

Using hemocoelic injection methodology, DeDianous *et al.* (1987) determined that LD_{50} values for AaIT against selected species of Diptera and Orthoptera ranged from 20 to 3770 ng/kg. Conversely, these same researchers found that subcutaneous injection of AaIT, at dosages as high as 50 g/kg, caused no adverse effects in mice. In separate oral and inhalation dosing studies conducted on laboratory mice, Possee *et al.* (1993) showed that AaIT at 1 µg/animal had no effect on the animals' body weight, dietary intake, survival or appearance of internal organs at necropsy. Experiments *in vitro* and *in situ* have demonstrated that AaIT causes convulsive firing and overstimulation of insect motor neurons and skeletal muscle (Loret *et al.*, 1991), whereas it caused no effects in neuromuscular preparations isolated from mammals and selected species of Crustacea and Arachnida (Tintpulver *et al.*, 1976; Rathmayer *et al.*, 1977; Ruhland *et al.*, 1977). Further, in experiments conducted on isolated nervous tissue, radiolabeled AaIT readily bound to such tissue from insects but not to nervous tissue from crustaceans or mammals (Gordon *et al.*, 1985; Teitelbaum *et al.*, 1979).

LqhIT has also demonstrated a high affinity for insect neuronal receptors and a high degree of potency against species within this class, but no detectable toxicity to mammalian or crustacean species (Zlotkin *et al.*, 1991; Moskowitz *et al.*, 1994). Researchers at DuPont Agricultural Products (1996) intravenously injected individual mice with LqhIT at a dosage of 0.5 mg/kg and then observed them for 14 days. None of the mice exhibited any clinical signs of toxicity over the duration of the study, and necropsies performed on the animals at the conclusion of the test revealed no gross lesions on tissues and organs as a result of exposure to LqhIT.

Compared with AaIT and LqhIT, studies pertaining to pharmacology of TxP-I are limited. However, Tomalski *et al.* (1989) reported that the PD_{50} (paralysis dosage) of this straw-itch mite toxin in larvae of wax moth, *Galleria mellonella*, was about 0.5 mg/kg, whereas TxP-I had no effect on laboratory mice at dosages as high as 50 mg/kg.

Vertebrate safety of recombinant NPVs

It is intuitive to expect that recombinant NPVs would have no impact on mammals, birds or fish, because the NPVs would have to infect cells of these vertebrate species in order to produce toxin. As noted earlier in this chapter, it is well documented in the scientific literature that NPVs are highly specific for insect hosts, with AcMNPV and HzNPV able to infect only certain species of Lepidoptera. However, it has been considered prudent to repeat such vertebrate pathogenicity/toxicity studies with the recombinant forms of NPVs.

Possee *et al.* (1993) conducted a series of studies with AcMNPV-AaIT to determine its acute oral, dermal and subcutaneous toxicity or pathogenicity to rats and guinea-pigs. Over a 14-day period following a single dosing of each animal with 1×10^6 occlusion bodies (OBs) of recombinant NPV, all rats and guinea-pigs remained healthy and showed no abnormalities in weight gain, dietary intake, physical appearance or motor skills. Additionally, necropsies performed at the end of the study showed that no gross changes occurred in organs of treated animals.

When used as a bioinsecticide, a recombinant NPV would be formulated and sprayed onto crops in the form of occlusion bodies. Since toxin would not be present in OBs of a recombinant NPV, the only route of exposure to the toxin by vertebrates would be through ingestion of infected insect larvae or herbivory of foliage contaminated with insect cadavers. In an attempt to mimic this route of exposure, researchers at American Cyanamid Company (1998) fed laboratory mice a diet which consisted partly of *H. virescens* larvae that had been previously infected with AcMNPV-tox34. Prior to use in the study, *H. virescens* were exposed to 10 times the larval LD_{95} of AcMNPV-tox34. Three days after exposure, larvae that exhibited paralysis (i.e. indicative of TxP-I production and intoxication) were homogenized and fed to mice by gavage. Each mouse was fed the equivalent of four infected *H. virescens* larvae on a daily basis for five consecutive days. No clinical signs of toxicity or effects on survival were seen in treated mice during the study period. Dietary intake and body weight gains for treated and non-treated mice were equal throughout the five-day study. Macroscopic evaluation of tissues and organs in mice which were sacrificed at the end of the five days of dosing revealed no treatment-related pathological changes or abnormalities.

American Cyanamid Company evaluated AcMNPV-AaIT in a series of studies which were established by the US EPA to ascertain the safety of microbial pesticides (i.e. Code of Federal Regulations, Title 40, Part 158, Section 740). In an acute oral toxicity-limit test, AcMNPV-AaIT had no impact on survival of laboratory rats, nor did it impact internal organs of the animals. The budded form of the recombinant virus was also evaluated for its ability to infect mammalian cells (i.e. diploid human lung cell line MRC-5). AcMNPV-AaIT did not alter cell doubling time or cell morphology, and as expected, there was no evidence of viral replication or expression of AaIT. Mallard ducks, bobwhite quail and rainbow trout were fed diets consisting of AcMNPV-AaIT OBs as well as *H. virescens* larvae infected with AcMNPV-AaIT, and no adverse effects were observed in any of these three vertebrate species. Finally, no abnormalities were observed in the aquatic invertebrate, *Daphnia magna*, when it was maintained in aqueous suspensions of AcMNPV-AaIT.

Researchers at DuPont Agricultural Products (1994) compared feral AcMNPV and AcMNPV-LqhIT for infectious and pathogenic behavior in laboratory rats using oral, intravenous and inhalation routes of exposure. Neither virus was infectious or toxic to rats by any of the aforementioned means of exposure at dosages ranging from 1×10^7 to 1×10^8 OB/animal. DuPont researchers also confirmed that AcMNPV-LqhIT was not infective or cytotoxic to human liver (Chang liver), intestine (strain 407) and lung (strain W138) cell lines. Additional studies showed that bobwhite quail receiving oral and intraperitoneal exposure to OBs, and trout and grass shrimp exposed to the NPV via food and/or aquatic environment, were unaffected by AcMNPV-LqhIT.

Invertebrate selectivity of recombinant NPVs

In order for toxin to be expressed by a recombinant baculovirus, the virus must first establish a productive systemic infection within the host (particularly if the toxin gene is under transcriptional control of a late promoter). If an organism is non-permissive to infection by the vectoring agent, the foreign gene will not be transcribed, i.e. host-range is determined by viral gene products rather than by those of the introduced gene. Using either hemocoelic injection or a *per os* inoculation methodology, Huang *et al.* (1996) tested budded and pre-occluded forms of the following viruses for their abilities to establish an infection and replicate within selected non-lepidopteran insects: AcMNPV, LdMNPV, OpMNPV and *Bombyx mori* NPV (BmNPV). Each of the aforementioned viruses was engineered to express an inserted reporter gene (e.g. luciferase, β-galactosidase), so that even symptomless infections could be identified within the insects (i.e. viral replication could be based on detection of reporter gene product). Huang *et al.* demonstrated that individuals from 13 species of insects, representing Blattodea, Coleoptera, Diptera, Hemiptera, Homoptera, Neuroptera and Orthoptera, did not support detectable replication of any of the four viruses included in the study.

Maximum challenge laboratory studies have been conducted to determine the impact of certain recombinant NPVs on growth, behavior and survival of numerous species of non-lepidopteran invertebrates. Depending on species or growth stage tested in these laboratory assays, animals were exposed to AcMNPV-AaIT, AcMNPV-LqhIT, AcMNPV-tox34 and/or HzNPV-AaIT via ingestion of contaminated artificial diet or foliage (OBs), consumption of virus-infected prey (OBs, budded virus and expressed toxin), and hemocoelic injection (budded virus) (Bishop *et al.*, 1995; Heinz *et al.*, 1995; McNitt *et al.*, 1995; McCutchen *et al.*, 1996; American Cyanamid Company, 1997; Treacy *et al.*, 1997; DuPont Agricultural Products, 1998). Results from these laboratory studies showed that none of the aforementioned recombinant NPVs directly caused adverse effects in individuals of the following predatory, parasitic, herbivorous or saprophytic invertebrate species: green lacewing (*Chrysopa carnea* Stephen), insidious flower bug (*Orius insidiosus* (Say)), ground beetle (*Pterostichus modidus*), honeybee (*Apis mellifera*), hymenopterous parasitoid (*Microplitis croceipes*), twospotted spider mite (*Tetranychus urticae* Koch), boll weevil (*Anthonomus grandis grandis* Boheman), western corn rootworm (*Diabrotica virgifera virgifera* LeConte), Japanese beetle (*Popillia japonica* Newman), earthworm (*Lumbricus* spp.), Chinese mantid (*Tenodera aridfolia*), funnel web spider (*Ixeuticus* spp.), subterranean termite (*Reticulotermes flavipes* Kollar), convergent lady beetle (*Hippodamia convergens*), bigeyed bug (*Geocoris* spp.), red imported fire ant (*Solenopsis invicta* Buren), the social wasp (*Polistes metricus* (Say)), German cockroach (*Blatella germanica*) and common fruitfly (*Drosophila melanogaster*).

In a series of small-scale field trials conducted on cotton, tobacco and cabbage, it was determined that weekly foliar sprays of AcMNPV-AaIT, at rates as high as 2.0×10^{12} OB/ha, had no effects on diversity or density of non-target arthropods, relative to arthropod populations in non-treated plots of each crop (Treacy and All, 1996; All and Treacy, 1997; Treacy, 1997). At least 45 species, representing 26 families of insects, spiders and mites, were found in collections taken from these small-plot field tests. Heinz *et al.* (1999) found that plots of field-grown cotton which were sprayed with either HzNPV-LqhIT or AcMNPV-LqhIT had similar populations of non-target predatory arthropods as those plots treated with feral forms of the viruses. Further, cotton treated with the aforementioned LqhIT-inserted viruses had significantly greater numbers of beneficial arthropods than plots treated with conventional insecticides, such as the pyrethroid esfenvalerate and the semi-synthetic macrolide emamectin benzoate.

Published and non-published studies have also shown that even within Lepidoptera, recombinant NPVs exhibit a host-range similar to that of their feral counterparts. For example, in a series of dose–response assays conducted on larvae of 52 different species of Lepidoptera, Bishop *et al.* (1995) found that AcMNPV-AaIT exhibited potency or virulence against each species at a level similar to that imparted by feral AcMNPV.

There have been results from time-limited assays which would appear to suggest that AcMNPV-semipermissive *H. zea* might be more sensitive to gene-inserted forms of AcMNPV than the feral form of this virus (Treacy and All, 1996). However, studies such as those reported by Treacy and All were designed to compare recombinant and feral viruses for speed of insecticidal activity against lepidopteran pest species, and thus focused on levels of larval mortality during the initial 8–10 days of viral exposure. Although such studies can demonstrate potential of a recombinant virus to be an effective insecticide, they do not provide insight as to the overall impact of such a gene-inserted NPV, relative to is feral counterpart, on the life cycle of a semipermissive lepidopteran species. Therefore, entomologists at American Cyanamid's Agricultural Research Center in Princeton, New Jersey (USA) recently conducted a laboratory study to compare wild-type AcMNPV, AcMNPV-AaIT and AcMNPV-tox34 for effects on survival of AcMNPV-semipermissive *H. zea* at selected points in time throughout its life cycle, i.e. middle and late larval development and adult eclosion (or 5, 10 and 35 days after placing second-instars on viral contaminated diet). Based on temporal LD_{50} values and presence of non-overlapping fiducial limits, the two aforementioned recombinant AcMNPVs were about 800 and 25 times more lethal to *H. zea* than feral AcMNPV at 5 and 10 days after initial exposure to the viruses, respectively (Table 9.1). However, when LD_{50} values were calculated from final test-population responses to treatments (i.e. 35 days after test initiation), it was found that AcMNPV-AaIT, AcMNPV-tox34 and AcMNPV were statistically equal in virulence against *H. zea*. Results from this later study indicated that when the feral and recombinant forms of AcMNPV were allowed to complete their infection cycles in AcMNPV-semipermissive *H. zea*, there were little to no difference among the NPVs in virulence. The differential *H. zea* LD_{50} values between the two recombinant viruses and wild-type AcMNPV seen during the first 10 days for the study were likely due to pharmacology of the expressed toxins (per the recombinant NPVs), and not a result of any genome-related changes in viral infectivity.

Environmental stability of feral and recombinant NPV

The potential for recombinant NPVs to displace their feral counterparts in the environment would be dependent upon expression of an advantage in fitness by the genetically altered virus. Fitness of a virus is determined by its physical characteristics and capacity for production of active progeny. Evidence gathered to date indicates

Table 9.1 Virulence of feral and recombinant forms of AcMNPV against the AcMNPV-semipermissive lepidopteran pest species, *Helicoverpa zea*

Treatment[a,b]	LD_{50} values (95% FLs) in OBs/16 cm^2 at selected time-points during the assay		
	5 days	10 days	35 days
AcMNPV	6.4×10^7	7.4×10^5	3.8×10^3
	(1.0–10.8)	(3.2–23.2)	(1.7–7.2)
AcMNPV-AaIT	6.3×10^4	2.3×10^4	9.4×10^2
	(3.8–10.9)	(1.3–4.0)	(2.1–25.8)
AcMNPV-tox34	9.9×10^4	3.8×10^4	5.9×10^2
	(5.5–18.5)	(2.1–6.7)	(2.1–17.1)

[a] Viral treatments were suspended in water and 0.01% SDS; 0.4 mL of viral suspension was deposited onto the surface of each diet-coated 16 cm^2 arena.
[b] Each viral dosage was replicated 4 times (one 3-day-old 2nd instar was placed in each arena, 16 insects per treatment-dosage); insect mortality was rated 5, 10 and 35 days after infestation of treated arenas.

that NPVs which have been engineered for enhanced insecticidal speed without significant alteration in host-range are likely to be less competitive in the environment than corresponding feral viruses.

Insertion of a gene encoding an insecticidal protein does not impact the physical or biochemical properties of viral occlusion bodies. Thus, such recombinant and feral NPVs are equally sensitive to abiotic and biotic degradation factors. Photolysis and photo-mediated oxidation are the primary mechanisms for inactivation of baculoviruses (Ignoffo and Garcia, 1994). It is thought that ultraviolet (UV) sunlight in the UV-B spectrum can directly cause strand breakage in viral DNA, or lead to production of highly reactive radicals (e.g. peroxides) which, in turn, degrade viral particles. Field studies have shown that selected feral NPVs which were applied to cotton foliage lost 70–80% of their initial biological activity in less than four days (Ignoffo and Baxter, 1971; Jones and McKinley, 1987). Similarly, following spray-application of a wettable powder formulation of AcMNPV onto field-grown cotton, American Cyanamid Company researchers found that 3-hour-old foliar residues of the virus caused 98% mortality in populations of *H. virescens* larvae, whereas larval mortality was only 5% on cotton leaves that had been aged in the field for two days. Although NPVs are rapidly inactivated when exposed to sunlight, their persistence when buried in soil can be for months or years. In their review of viral persistence in soil, England *et al.* (1998) highlighted the following as factors that can impact stability of viral OBs and DNA: (a) proteolytic microorganisms (e.g. bacterium); (b) soil pH; (c) soil temperature (e.g. freezing and thawing); (d) soil-binding capacity; (e) soil moisture; and (f) predation by soil-dwelling protozoa.

Regarding the genetic fitness of feral and gene-inserted viruses described in this chapter, a characteristic that limits the survivability of recombinant NPVs is the reduced ability to produce viable progeny. The yield of progeny occlusion bodies from a baculovirus infection is a key factor controlling viral population dynamics in the field. The slow killing speed imparted by a feral NPV on its host is strategic, because the virus can continue to replicate and maximize the progeny numbers while the infected larva stays alive. Conversely, recombinant NPVs containing genes that encode insecticidal protein rapidly kill the infected insect, thereby reducing the amount of time available to produce occlusion bodies. In a laboratory study conducted at American Cyanamid Company's Agricultural Research Center, third-instar *H. virescens* were dosed with OBs of feral NPV or any of seven different AcMNPV-AaIT clones (differing in promoter-signal sequences within the chimeric toxin-gene cassette). On average, each of the AcMNPV-AaIT clones killed *H. virescens* larvae in half the time required by wild AcMNPV, and when cadavers were weighed and assessed for quantity of viral progeny, larvae infected with any of the AcMNPV-AaIT isolates had about five-fold less body mass and contained approximately eight-fold fewer OBs than larvae that were infected with feral NPV. Numerous published studies have indicated that production of progeny by AcMNPV-AaIT and AcMNPV-LqhIT clones in larvae of *H. virescens* and *T. ni* is markedly reduced, in some cases by more than 10-fold, compared with the number of OBs produced by feral AcMNPV in these same host species (Tomalski and Miller, 1991; Cory *et al.*, 1994; DuPont Agricultural Products, 1996; Ignoffo and Garcia, 1996; Kunimi *et al.*, 1996).

Fuxa *et al.* (1998) conducted a series of laboratory and greenhouse tests to compare wild and recombinant forms of AcMNPV for their abilities to propagate and disperse following infection and death of insect hosts. Similar to observations made by other researchers noted above, Fuxa *et al.* found that *T. ni* survival time and viral reproduction was significantly reduced for insects infected with AcMNPV-AaIT than for those infected with feral AcMNPV. Further, one week after placement of virus-infected *T. ni* onto non-treated collard plants in the greenhouse, it was found that larvae infected with wild AcMNPV deposited about 25 times more OBs onto collard foliage than *T. ni* infected with AcMNPV-AaIT. This difference in release of OBs onto crop foliage was attributed to (a) fewer OBs being produced in AcMNPV-AaIT infected larvae and (b) reduced incidence of cadaver disintegration by AcMNPV-AaIT diseased hosts and a subsequent decrease in liberation of OBs onto collard leaves. Based upon their findings, Fuxa *et al.* suggested that AcMNPV-AaIT would not be as capable as feral AcMNPV in establishing infections in subsequent hosts, and would therefore be selected against in competition with its feral counterpart in the environment. Indeed, in a caged-plot field trial, Cory *et al.* (1994) determined that transmission of virus between successive cohorts of *T. ni* was significantly lower on cabbage sprayed with AcMNPV-AaIT than in plots of cabbage treated with feral AcMNPV.

A recent laboratory experiment conducted by entomologists and molecular biologists at American Cyanamid Company showed that AcMNPV-AaIT was not competitive with feral AcMNPV in co-infected *H. virescens* larvae. In this experiment, individual larvae were initially exposed to a mixture (LD_{95}) of AcMNPV-AaIT and AcMNPV at a 10:1 numeric ratio of OBs, respectively. Viral OBs produced from this first larval population were extracted from cadavers, pooled and fed to a second cohort of *H. virescens*. This process of passing viral progeny onto subsequent host-insects continued through six cohort sequences. Restriction enzyme digestion and quantitative DNA blot hybridization were used to determine relative abundance of AaIT-containing viral genomes extracted from each larval passage. Analyses of virus extracted from the initial larval passage showed that 78% of the viral DNA

Table 9.2 Proportion of viral DNA molecules containing the AaIT gene following sequential passage of a binary mixture of AcMNPV-AaIT and feral AcMNPV through cohorts of *Heliothis virescens* larvae

Larval passage[a]	Mean per cent of extracted viral DNA molecules containing AaIT gene
1	78
2	55
3	45
4	15
5	3
6	nd[b]

[a] Initial larval cohort was infected with a 10:1 OB ratio of AcMNPV-AaIT plus feral AcMNPV; subsequent larval cohorts were infected with viral progeny from the previous cohort.
[b] Not detectable.

molecules contained the AaIT gene (Table 9.2). However, the viral DNA population collected from the sixth serial passage consisted entirely of feral AcMNPV genotype, i.e. AaIT-containing DNA molecules were not detected in the samples. Such results again suggest that AcMNPV-AaIT is at a competitive disadvantage with feral AcMNPV. Thus, such changes made to the viral genome to make it a more effective insecticide may also be responsible for increasing the likelihood of extinction of the recombinant NPV in the environment.

Regulatory affairs

The US EPA is charged with the regulation of pesticides under both the Federal Insecticide Fungicide and Rodenticide Act (FIFRA) and the Federal Food, Drug and Cosmetic Act as they were amended by the Food Quality Protection Act (FQPA) of 1996. This function is managed by the Agency's Office of Pesticide Programs (OPP) which is tasked with assuring that the use of a pesticide does not cause unreasonable adverse effects to humans or the environment (under FIFRA) and that there is a reasonable certainty no harm will result from the aggregate exposure to the pesticide residue (under FQPA). A pesticide, as defined by FIFRA, is 'any substance . . . intended for preventing, destroying, repelling, or mitigating any pest'. This broad definition includes baculoviruses as well as other microbial products being used for control of insect pests.

A separate group within OPP was established in 1994 to specifically facilitate the registration of biopesticides such as the baculoviruses and encourage their use. This group, called the Biopesticides and Pollution Prevention Division (BPPD) performs risk/benefit and risk management functions for microbial pesticides among other responsibilities centered around the use of safer pesticides and IPM programs. Within BPPD, responsibility is divided among a Microbial Pesticides Branch (which includes plant-pesticides), a Biochemicals Pesticides Branch, and a Pollution Prevention Staff.

In order to register a pesticide for use under FIFRA, certain data requirements must be met which allow the Agency to assess the risk to humans and the environment from the use of the product. In general, biopesticides pose fewer risks than conventional pesticides; thus, EPA's data requirements for registration are typically less than those required to register a conventional product. Since less data need be reviewed by EPA, the time from submission to registration of a biopesticide also is typically less than that needed to register a conventional product. Nonetheless, the Agency is tasked with the conduct of a thorough evaluation of data to ascertain the potential risks.

In addition, any pesticide used on food or feed crops in the US must have an established tolerance or, due to the nature of the pesticide, be exempted from the requirement of a tolerance. A tolerance is the legal maximum residue concentration of a pesticide allowed on food or feed. If a crop is found to have pesticide residues above the legal limit or if pesticidal residues are found on a crop which is not supported by a tolerance, the crop may be considered adulterated and be seized by Federal or State enforcement agencies. Again, since biopesticides are generally considered inherently less harmful than conventional pesticides, tolerance exemptions are typically granted by EPA for microbial products.

The data requirements for the registration of microbial pesticide products can be found in the Code of Federal Regulations, Title 40, Part 158, Section 740a–d (40 CFR §158.740). Data required for registration depends upon the proposed use of the microbial pesticide products (i.e. terrestrial food use, forestry use, indoor use, etc.) and includes, but is not limited to, the areas of product analysis, residues, mammalian toxicology, non-target organism toxicology, and environmental expression. Data requirements for an Experimental Use Permit (EUP) under FIFRA §5 are typically a subset of what is required for a full registration under FIFRA §3. Examples of tolerance exemption regulations for microbial products can be found under 40 CFR §180.

Prior to making a development decision and generating data for registration or EUP purposes, the efficacy of a microbial product is typically ascertained under field use conditions. According to 40 CFR §172 Subpart C, field efficacy testing can be accomplished on a small scale (i.e. less than 10 acres for terrestrial crops) without EPA approval for naturally occurring, indigenous microbial products as long as the treated crop is destroyed at the completion of the trial. However, for certain genetically modified microbial products, the Agency requires the submission of a notification of the intent to conduct field trials. Notification is required for small-scale field testing of microbial pesticides 'whose pesticidal properties have been imparted or enhanced by the introduction of genetic material that has been deliberately modified' (i.e. gene-inserted constructs) and for 'non-indigenous microbial pesticides that have not been acted upon by the US Department of Agriculture (i.e. either by issuing or denying a permit or determining that a permit is unnecessary; or a permit is not pending with USDA)'. Although previously not exempt from the requirement of a notification under 40 CFR §172 Subpart C, a notification is no longer required for field testing genetically modified 'microbial pesticides resulting from deletions or rearrangements within a single genome' (i.e. gene-deleted constructs).

The notification itself must include information and data on the microbial pesticide including but not limited to its taxonomy and natural habitat, means and limits of detection of specific analytical methods, techniques used to genetically modify the microbe, the identity and location of the gene segments inserted or deleted, its physical, chemical and biological features, its host-range and survivability in the environment, along with detailed information on the field trial program such as rates, specific trial locations and application methods, target pests, site security measures, methods of sanitation of equipment, methods used to monitor movement of the microorganism, and a statement of composition of the experimental formulation. In addition, the means of detecting potential adverse effects must be submitted along with methods of controlling the microbial pesticide if detected outside the test area. Basically, this information and data are a subset of what would be required for a full registration or EUP for the genetically modified microbial pesticide.

The Agency has established a mechanism in the notification process for protecting data claimed to be confidential business information (CBI). EPA also allows for the petitioning for exemption from the requirement of a notification under 40 CFR §172.52.

Once the notification is submitted, the Agency has established a timeframe of 90 days to review the data and respond back to the submitter; however, approval of the notification by EPA must be granted before the field trials can take place. In addition, the EPA announces receipt of the notification in *the Federal Register* and thereby opens a public comment period.

Upon review of the notification and public comment(s) the Agency can: (1) approve the proposed test; (2) approve the proposed test with certain modifications to the program; (3) request additional information from the submitter; (4) require an EUP be submitted under FIFRA §5; or (5) disapprove the proposed test and any EUP application.

Following Federal approval of the field testing of the microbial pesticide, individual states (i.e. California) may have their own regulatory approval processes in place and the appropriate State regulatory authorities should be consulted prior to conducting field trials.

As part of its program to develop recombinant baculovirus biopesticide, American Cyanamid Company first submitted a notification to conduct small-scale field testing of its *egt* gene-deleted *Autographa californica* MNPV construct in 1993. Although this gene-deleted construct does not require a notification under the current regulations, at the time of submission, notifications were indeed required for gene modifications of this type.

Based on information submitted with the notification in conjunction with literature from the public domain, the Agency concluded that 'no unreasonable adverse effects on human health and environmental effects from the conduct of this small-scale field test are foreseen'. EPA approved the conduct of the field trial program with a soil-monitoring requirement and feeding of the sampled soil to the highly permissive *H. virescens* larvae. The Agency required that any larvae showing signs of viral infection after being exposed to the treated soil must be examined for occlusion bodies (OBs) and any OBs found must be analyzed by polymerase chain reaction assay (PCR) to determine if recombinant DNA is involved.

The following year, American Cyanamid Company submitted a notification of intent to conduct small-scale field testing of its AaIT gene-inserted *Autographa californica* MNPV (AcMNPV-AaIT) construct. Upon review of the submitted data and in response to public comment, the Agency requested more information concerning the environmental expression (i.e. persistence in the environment) and host-range of the gene-inserted baculovirus. After several rounds of meetings and correspondence with EPA personnel aimed at addressing their concerns, the initial notification for field testing was ultimately approved by the Agency, albeit with substantially increased mitigation (i.e. limited acreage, site disinfection procedures, soil barriers, 'wild-type' baculovirus oversprays), increased monitoring (i.e. soil analyses for OBs, non-target insect surveys), and decreased acreage requirements in comparison to the gene-deleted construct. The mitigation and monitoring requirements were aimed at addressing the questions on persistence and host-range and more fully characterizing any potential risk.

As the initial field trials were underway, scientists at American Cyanamid Company continued to conduct laboratory studies focused on the host-range determination by looking at various species of lepidopteran and non-lepidopteran insects. Additionally, a 'competition' study was conducted which proved that the ecologically disadvantaged modified baculovirus (i.e. recombinant produces less OB progeny than the wild-type virus) would not be able to overcome the naturally occurring virus in the environment. These studies, in conjunction with the results from two years of field monitoring work, provided the Agency with sufficient data to ultimately reduce the mitigation and monitoring requirements in follow-on notifications. By 1997, the Agency had monitoring data collected from the 1995 and 1996 trials which, along with numerous laboratory bioassays, showed that AcMNPV-AaIT had no effect on non-target arthropods (at least 26 families of Insecta and Arachnida representing 45 species were encountered in monitoring surveys with no ill-effect observed). In addition, soil monitoring data from the earlier trials showed that the recombinant 'does not vary in its environmental fate (i.e. persistence) profile in comparison to the naturally occurring *Autographa californica* MNPV'. The Agency subsequently approved a small-scale field program for the 1997 season and removed the requirements for soil/non-target insect monitoring and site disinfection. In fact, EPA no longer requires notification for field testing of AcMNPV-AaIT as long as certain minimal mitigation criteria are followed.

Also in 1997, American Cyanamid Company submitted a notification for small-scale, crop-destruct field testing of an AaIT gene-inserted construct based on a *H. zea* NPV baculovirus. This notification built on the experience gained from the modified AcMNPV field work and included laboratory-derived results which once again showed that neither the host-range nor the infectivity of the recombinant *H. zea* NPV changed in comparison with the wild-type baculovirus.

The Agency approved the small-scale field trial program and stated that no further notifications were necessary for similar constructs of AaIT gene-inserted *H. zea* NPV 'if there is no indication of increased host-range' as long as the tests are not conducted in areas adjacent to potentially susceptible endangered insects and that wild-type *H. zea* NPV baculovirus is introduced to the site as part of a standard treatment or overspray. According to EPA, this overspray of the trial site is a valid way to provide a source of wild-type baculovirus to compete with the genetically engineered baculovirus.

In a notification submitted in 1998, American Cyanamid Company requested approval of small-scale field testing of a *tox34* gene-inserted AcMNPV (AcMNPV-tox34). Expression of the *tox34* gene in recombinant AcMNPV significantly improved the efficacy of the baculovirus in the same fashion as expression of the *AaIT* gene in the earlier AcMNPV constructs. This submission once again provided EPA with information on the host-range and infectivity of the construct. As this *tox34* gene-insert construct introduced an insect-specific toxin not previously reviewed by EPA, laboratory data were submitted to ascertain the mammalian toxicity characteristics of the toxin expressed in infected target insect larvae. No adverse toxicological effects were seen. The notification was approved (and further notifications of the *tox34* construct waived) with the same wild-type baculovirus overspray restriction and infectivity testing as the AaIT gene-insert *H. zea* NPV construct.

Overall, EPA concerns with small-scale field testing programs of genetically modified baculoviruses focused on potential risks of an increase in host-range. Because of the novel technology and after a thorough review of the submitted host-range and environmental safety data, EPA approved the initial small-scale field programs of the original recombinant baculoviruses with extensive risk mitigation criteria. As the database on host-range, infectivity and environmental expression increased, the Agency was able to rely on past studies and extensive monitoring as a foundation to support newer construct introductions. This increase in familiarity with the science ultimately led to the approval of small-scale field testing of several different recombinant constructs with lessened mitigation requirements and laid the groundwork for further development work.

Summary

Baculovirus safety in conjunction with recombinant technology should be perceived as having two levels: level 1, species specificity and level 2, target specificity.

The safety of feral baculoviruses has been clearly established through several decades of high-quality research. The first level of safety is dictated by the fact that infection can occur only in the limited number of species susceptible to any baculovirus. In the recombinant form of the virus, a second level of safety is added, in that the inserted gene encodes a toxin which binds only to insect targets.

Research on the recombinant viruses is somewhat limited since their availability is a recent development. The constructs currently available meet the two-level safety thesis. The limited database does show that these constructs do not infect species outside their established host-range and, in fact the addition of the toxin gene does not even alter the LD_{50}. The major benefit is a faster kill and better crop protection (desirable changes). Very sophisticated cellular level experiments have shown that there is no replication in non-target hosts (vertebrate and invertebrate). Finally, the recombinant NPV is clearly at a competitive disadvantage to the feral NPV due to its limited ability to propagate.

The growing database has been well received by rational environmental groups, academics and growers. USA regulatory agencies are becoming convinced about the safety of these recombinants and see them taking their place in the grower's arsenal of insect control tools.

Acknowledgement

The authors would like to acknowledge the assistance of Tina Holsten in the typing and proofing of this manuscript. Her considerable efforts have contributed significantly to the accuracy and consistency of this work.

References

All, J.N. and Treacy, M.F. (1997) 'Improved control of *Heliothis virescens* and *Helicoverpa zea* with a recombinant form of *Autographa californica* nuclear polyhedrosis virus and interaction with Bollgard™ cotton', in P. Duggar and D.A. Richter (eds) *Proceedings Beltwide Cotton Conferences*. Memphis, Tennessee: National Cotton Council, pp. 1294–96.

American Cyanamid Company (1997) 'Notification to conduct small-scale testing of a genetically engineered microbial pesticide', in *OPP Public Docket*. Washington, DC: United States Environmental Protection Agency, 241NMP-L.

American Cyanamid Company (1998) 'Notification to conduct small-scale testing of a genetically engineered microbial pesticide', in *OPP Public Docket*. Washington, DC: United States Environmental Protection Agency, 241NMP-A.

Bishop, D.H.L., Hirst, M.L., Possee, R.D. *et al.* (1995) 'Genetic engineering of microbes: virus insecticides – a case study', in P.A. Hunter, G.K. Darby and N.J. Russell (eds) *Fifty Years of Antimicrobials: Past Perspectives and Future Trends*. Cambridge, UK: Cambridge University Press, pp. 249–77.

Black, B.C., Brennan, L.A., Dierks, P.M. *et al.* (1997) 'Commercialization of baculoviral insecticides', in L.K. Miller (ed.) *The Baculoviruses*. New York: Plenum Press, pp. 341–87.

Burges, H.D., Croizier, G. and Huber, J. (1980) 'Review of safety tests on baculoviruses', *Entomophaga* 25: 329–40.

Cory, J.S., Hirst, M.L., Williams, T. *et al.* (1994) 'Field trial of a genetically improved baculovirus insecticide', *Nature* 370: 138–40.

DeDianous, S., Hoarau, F. and Rochat, H. (1987) 'Reexamination of the specificity of the scorpion *Androctonus australis* Hector insect toxin toward arthropods', *Toxicon* 25: 411–17.

Doller, G. (1985) 'The safety of insect viruses as biological control agents', in *Viral Insecticides for Biological Control*. New York: Academic Press, pp. 399–439.

DuPont Agricultural Products (1996) 'Notification to conduct small-scale testing of a genetically engineered microbial pesticide', in *OPP Public Docket*. Washington, DC: United States Environmental Protection Agency, 352NMP-4.

DuPont Agricultural Products (1998) 'Notification to conduct small-scale testing of a genetically engineered microbial pesticide', in *OPP Public Docket*. Washington, DC: United States Environmental Protection Agency, 352NMP-A.

England, L.S., Holmes, S.B. and Trevors, J.T. (1998) 'Review: persistence of viruses and DNA in soil', *World Journal of Microbiology and Biotechnology* 14: 163–69.

Entwistle, P.F. and Evans, H.F. (1985) 'Viral control', in L.I. Gilbert and G.A. Kerkut (eds) *Comprehensive Insect Physiology, Biochemistry and Pharmacology*. Oxford: Pergamon Press, pp. 347–412.

Fuxa, J.A., Fuxa, J.R. and Richter, A.R. (1998) 'Host-insect survival times and disintegration in relation to population density and dispersion of recombinant and wild-type nucleo-polyhedroviruses', *Biological Control* 12: 143–50.

Gordon, D., Zlotkin, E. and Catterall, W.A. (1985) 'The binding of an insect-selective neurotoxin and saxitoxin to insect neuronal membranes', *Biochimica et Biophysica Acta* 82: 130–36.

Grewal, P.S., Webb, K., VanBeek, N.A.M. *et al.* (1998) 'Virulence of *Anagrapha falcifera* nuclear polyhedrosis virus to economically significant Lepidoptera', *Journal of Economic Entomology* 91: 1302–306.

Heinz, K.M., McCutchen, B.F., Hermann, R. *et al.* (1995) 'Direct effects of recombinant nuclear polyhedrosis viruses on selected nontarget organisms', *Journal of Economic Entomology* 88: 259–64.

Heinz, K.M., Smith C., Minzenmayer, R. and Flexner, J.L. (1999) 'Utilization of recombinant viral pesticides within a cotton IPM program', in P. Duggar and D.A. Richter (eds) *Proceedings Beltwide Cotton Conferences*. Memphis, Tennessee: National Cotton Council (in press).

Huang, X.P., Davis, T.R., Hughes, P. *et al.* (1997) 'Potential replication of recombinant baculoviruses in nontarget insect species: reporter gene products as indicators of infection', *Journal of Invertebrate Pathology* 69: 234–45.

Ignoffo, C.M. (1973) 'Effects of entomopathogens on vertebrates', *Annals of New York Academy of Sciences* 217: 141–64.

Ignoffo, C.M. (1975) 'Entomopathogens as insecticides', *Environmental Letters* 8: 23–40.

Ignoffo, C.M. and Baxter, O.F. (1971) 'Microencapsulation and ultraviolet protectants to increase sunlight stability of an insect virus', *Journal of Economic Entomology* 64: 850–53.

Ignoffo, C.M. and Garcia, C. (1992) 'Combinations of environmental factors and simulated sunlight affecting activity of inclusion bodies of the *Heliothis* (Lepidoptera:Noctuidae) nucleopolyhedrosis virus', *Environmental Entomology* 21: 210–13.

Ignoffo, C.M. and Garcia, C. (1994) 'Antioxidant and oxidative enzyme effects on the inactivation of inclusion bodies of the *Heliothis* baculovirus by simulated sunlight-UV', *Environmental Entomology* 23: 1025–29.

Ignoffo, C.M. and Garcia, C. (1996) 'Rate of larval lysis and yield and activity of inclusion bodies harvested from *Trichoplusia ni* larvae fed on wild or recombinant strain of the nuclear polyhedrosis virus of *Autographa californica*', *Journal of Invertebrate Pathology* 68: 196–98.

Jones, K.A. and McKinley, D.J. (1987) 'Persistence of *Spodoptera littoralis* nuclear polyhedrovirus on cotton in Egypt', *Aspects of Applied Biology* 14: 323–34.

Kunimi, Y., Fuxa, J.R. and Hammock, B.D. (1996) 'Comparison of wild type and genetically engineered nuclear polyhedrosis viruses of *Autographa californica* for mortality, virus replication and polyhedra production in *Trichoplusia ni* larvae', *Entomology Experimental and Applied* 81: 251–57.

Loret, E.P., Martin-Eauclaire, M.F., Mansuelle, P. *et al.* (1991) 'An anti-insect toxin purified from the scorpion *Androctonus australis* Hector also acts on the alpha- and beta-sites of the mammalian sodium channel sequence and circular dichoism studies', *Biochemistry* 30: 63–640.

McCutchen, B.F., Hermann, R., Heinz, K.M. *et al.* (1996) 'Effects of recombinant baculoviruses on a nontarget endoparasitoid of *Heliothis virescens*', *Biological Control* 6: 45–50.

McNitt, L., Espelie, K.E. and Miller, L.K. (1995) 'Assessing the safety of toxin-producing baculovirus biopesticides to a nontarget predator, the social wasp *Polistes metricus* Say', *Biological Control* 5: 267–78.

Possee, D., Hirst, M., Jones, L.D. *et al.* (1993) 'Field tests of genetically engineered baculoviruses', in *Opportunities for Molecular Biology in Crop Production*. British Crop Protection Council, monograph 55, pp. 23–36.

Rathmayer, M., Walther, C. and Zlotkin, E. (1977) 'The effect of different toxins from scorpion venom on neuromuscular transmission and nerve action potential in crayfish', *Comprehensive Biochemistry and Physiology* 56: 35–38.

Richards, A., Mathews, M. and Christian, P. (1998) 'Ecological considerations for the environmental impact evaluation of recombinant baculovirus insecticides', *Annual Reviews in Entomology* 43: 493–517.

Ruhland, M., Zlotkin, E. and Rathmeyer, W. (1977) 'The effect of toxins from the venom of the scorpion *Androctonus australis* on a spider nerve-muscle preparation', *Toxicon* 15: 157–60.

Teitelbaum, Z., Lazarovici, P. and Zlotkin, E. (1979) 'Selective binding of the scorpion venom insect toxin to insect nervous tissue', *Insect Biochemistry* 9: 343–46.

Tintpulver, M., Zerachia, T. and Zlotkin, E. (1976) 'The actions of toxins derived from scorpion venom on the ileal smooth muscle preparation', *Toxicon* 14: 371–77.

Tomalski, M. and Miller, L.K. (1991) 'Expression of a paralytic neurotoxin gene to improve insect baculoviruses as biopesticides', *Biotechnology* 10: 545–49.

Tomalski, M.D. and Miller, L.K. (1991) 'Insect paralysis by baculovirus mediated expression of a mite neurotoxin gene', *Nature* 352: 82–85.

Tomalski, M.D., Kutney, R., Bruce, W.A. et al. (1989) 'Purification and characterization of insect toxins derived from the mite *Pyemotes tritici*', *Toxicon* **10**: 1151–67.

Torruellas, J., Kumar, L., Oland, L.A. et al. (1998) 'Blockade of insect calcium channels expressed in primary neuronal cultures of *Manduca sexta* with straw-itch mite toxin', *Society of Neuroscience Abstracts* **24**: 1080.

Treacy, M.F. (1997) 'Efficacy and nontarget arthropod safety of an AaIT gene-inserted baculovirus: results from field and laboratory studies conducted during 1995–1996', in A. Parkinson (ed.) *Biopesticides and Transgenic Plants*. Southborough, Massachusetts: International Business Communications, pp. 3.1.1–3.1.17.

Treacy, M.F. and All, J.N. (1996) 'Impact of insect-specific AaIT gene insertion on inherent bioactivity of a baculovirus against tobacco budworm and cabbage looper', in P. Duggar and D.A. Richter (eds) *Proceedings Beltwide Cotton Conferences*. Memphis, Tennessee: National Cotton Council, pp. 911–17.

Treacy, M.F., All, J.N. and Kukel, C.F. (1997) 'Invertebrate selectivity of a recombinant baculovirus: case study on AaHIT gene-inserted *Autographa californica* nuclear polyhedrosis virus', in K. Bondari (ed.) *New Developments in Entomology*. Trevandrum: Research Signpost, pp. 57–68.

Treacy, M.F., Rensner, P.E. and All, J.N. (1999) 'Differential insecticidal properties exhibited against heliothine species by two viral vectors encoding a similar chimeric toxin gene', in P. Duggar and D.A. Richter (eds) *Proceedings Beltwide Cotton Conferences*. Memphis, Tennessee: National Cotton Council (in press).

Vail, P.V., Jay, D.L., Stewart, F.D. et al. (1978) 'Comparative susceptibility of *Heliothis virescens* and *Heliothis zea* to the nuclear polyhedrosis virus isolated from *Autographa californica*', *Journal of Economic Entomology* **71**: 293–96.

Walther, C., Zlotkin, E. and Rathmeyer, W. (1976) 'Action of different toxins from scorpion *Androctonus australis* on locust nerve-muscle preparation', *Journal of Insect Physiology* **22**: 1187–94.

Zlotkin, E., Kadouri, D., Gordon, D. et al. (1985) 'An excitatory and depressant insect toxin from scorpion venom both affect sodium conductance and possess a common binding site', *Archives of Biochemistry and Biophysics* **240**: 877–87.

Zlotkin, E., Gurevitz, M., Fowler, E. et al. (1993) 'Depressant insect selective neurotoxins from scorpion venom: chemistry, action and gene cloning', *Archives of Insect Biochemistry and Physiology* **22**: 55–73.

Chapter 10

Case study: virus-resistant crops

Hector Quemada

Squash and virus disease

Agricultural importance of cucurbits and squash

Squash (*Cucurbita pepo* L.) belongs to the Cucurbitaceae, a family of crops that is an important source of food throughout the world. While not as important as the cereals or oilseed crops, such as soybean, nor other vegetable crops such as potatoes and tomatoes, cucurbits are nevertheless grown on significant acreages worldwide. In 1998, cantaloupes and other melons were grown on 1 044 672 hectares, producing 17 764 188 metric tonnes of food; cucumbers and gherkins were grown on 1 567 389 hectares, producing 26 673 943 metric tonnes; while pumpkins, squash, and gourds were grown on 1 136 083 hectares, producing 14 169 983 metric tonnes (FAO, 1998). Cucurbit crops, including squash, are used for food in all regions of the world, but are also used for oil and other non-food uses, such as musical instruments or household implements.

Impact of virus disease on squash

Squash, like the other major cucurbit crops, is susceptible to infection by several viruses. The most damaging are cucumber mosaic virus (CMV), papaya ringspot virus watermelon strain (PRSV-W), watermelon mosaic virus 2 (WMV2) and zucchini yellow mosaic virus (ZYMV) (Provvidenti, 1990; Zitter *et al.*, 1996; Fuchs *et al.*, 1998). While the extent to which viruses cause economic damage in cucurbits, and squash in particular, is not well documented, the problem is extensive enough that diligent effort to breed squash resistant to CMV, PRSV-W, WMV2 and ZYMV has been a goal of breeders for many decades (Provvidenti, 1990).

Disease breeding in squash

Sources of genes for resistance or tolerance to viruses have been reported in various relatives of squash and in *Cucurbita pepo* itself. However, introducing useful levels of resistance to more than one virus into commercially acceptable cultivars has been difficult to achieve (Provvidenti, 1990; Table 10.1). While releases of zucchini-type and English marrow-type varieties carrying resistance to a single virus (CMV) has been achieved (Anonymous, 1992), breeding for multiple virus resistance has only

Table 10.1 Traits for disease resistance reported in *Cucurbita pepo* (as of 1994)

Trait	Report	Reference
CMV resistance	*C. pepo*, *C. pepo* 'Delicata', *C. pepo* 'Cinderella', *C. pepo* 'Tiger Cross', *C. pepo* 'Supremo', *C. pepo* (from *C. martinezii*)	Anonymous, 1992; Enzie, 1940; Kyle *et al.*, 1993; Martin, 1960; Walkey & Pink, 1984; Zitter *et al.*, 1991; USDA (Germplasm Resources Information Network), North Central Region Plant Introduction Center, Ames, IA
PRSV-W	*C. pepo*, *C. pepo* XPHT1815 and XPHT1817	USDA (Germplasm Resources Information Network), North Central Region Plant Introduction Center, Ames, IA; Schultheis and Walters, 1998
WMV2	*C. pepo*	USDA (Germplasm Resources Information Network), North Central Region Plant Introduction Center, Ames, IA
Scab	*C. pepo*, *C. pepo* (from *C. martinezii*)	USDA (Germplasm Resources Information Network), North Central Region Plant Introduction Center, Ames, IA; Kyle *et al.*, 1993
Powdery mildew	*C. pepo*, *C. pepo* (from *C. martinezii*)	USDA (Germplasm Resources Information Network), North Central Region Plant Introduction Center, Ames, IA; D. Groff, personal communication, 1993
CMV+WMV2+ ZYMV+PRSV-W	*C. pepo* (from *C. ecuadorensis*, *C. martinezii* and *C. moschata*)	Gilbertalbertini *et al.*, 1993
Downy mildew	*C. pepo*	USDA (Germplasm Resources Information Network), North Central Region Plant Introduction Center, Ames, IA; D. Groff, personal communication, 1993

recently been successful. A summer squash cultivar, Whitaker, has recently been developed by researchers at Cornell University (Anonymous, 1999). This cultivar is resistant to multiple pathogens, including viruses and fungi. The difficulty of this accomplishment is emphasized by the fact that Whitaker was the result of work conducted by a team of researchers for more than a decade. Because the sources of resistance were related species not normally sexually compatible with *C. pepo* (*C. ecuadorensis*, *C. okeechobeensis* ssp. *martinezii* and *C. moschata*), these researchers had to employ a commonly used breeding tool, embryo rescue, to overcome the barriers to crossing between species.

Squash as a species

Cucurbita pepo is a crop native to North America. It consists of several subspecies and varieties that are completely interfertile (Andres, 1987; Wilson, 1990), and are therefore presently considered part of a single species (Decker, 1988; Decker-Walters *et al.*, 1993). *C. pepo* ssp. *pepo* includes cultivated pumpkins, marrows and a few ornamental gourds; *C. pepo* ssp. *ovifera* var. *ovifera* includes cultivated crookneck, scallop and acorn squashes, and most ornamental gourd cultivars; *C. pepo* ssp. *ovifera* var. *texana* includes free-living populations found in Texas; *C. pepo* ssp. *ovifera* var. *ozarkana* includes free-living populations found in the Mississippi Valley and the Ozark Plateau; *C. pepo* ssp. *fraterna* includes non-cultivated populations found in northeastern Mexico.

Populations of *C. pepo* exist today under non-cultivated conditions in the United States. These free-living populations range from northeastern Mexico and Texas, to Alabama, and through the Mississippi Valley to Illinois (Smith *et al.*, 1992; Cowan and Smith, 1993). Some free-living populations, especially in Arkansas, Louisiana and Mississippi, occur as weeds in soybean and cotton fields, and are difficult to control (Oliver *et al.*, 1983; Boyette *et al.*, 1984; Smith *et al.*, 1992; Bryson and Byrd, 1998).

Gene exchange between free-living and cultivated populations has been documented (Kirkpatrick and Wilson, 1988; Smith *et al.*, 1992; Decker-Walters *et al.*, 1993). Furthermore, some free-living populations may have originated as escapes from cultivation, which might have undergone subsequent introgression with other nearby cultivated, or other free-living populations (Decker and Wilson, 1987; Wilson, 1990; Asch and Asch, 1992; Cowan and Smith, 1993; Decker-Walters *et al.*, 1993).

Genetic engineering of squash

Because of the need to obtain useful virus resistance in a commercially acceptable cultivar, an alternative to breeding was attempted: the direct introduction of genes conferring resistance into already known commercial genotypes. The genetic engineering approach was to introduce the coat protein genes of the viruses in question, a strategy first successful in tobacco (Powell-Abel *et al.*, 1986), and subsequently shown to be effective in cucurbit species (Gonsalves *et al.*, 1992; Fang and Grumet, 1993).

Constructs containing viral coat protein genes were introduced into proprietary inbred lines of *C. pepo* ssp. *ovifera* var. *ovifera* via *Agrobacterium*-mediated transformation. The engineering of the constructs used in the transformations, as well as the transformation procedure, are described in Tricoli *et al.* (1995). Based on greenhouse testing and small-scale field tests, two lead lines, ZW20 and CZW3, were chosen as parents of commercial hybrids. The characteristics of these lines are described in detail by Tricoli *et al.* (1995).

ZW20

Line ZW20 was transformed with disarmed *Agrobacterium* harboring a binary Ti-plasmid (pPRBN-ZYMV72/WMBN22; Tricoli *et al.*, 1995). This plasmid contained expression cassettes consisting of the coat protein genes of zucchini yellow mosaic virus (ZYMV) and watermelon mosaic virus 2 (WMV2) under the control of the cauliflower mosaic virus 35S promoter. The T-DNA also contained, as a selectable marker, a neomycin phosphotransferase II (NPTII) gene from *E. coli* transposon Tn5, under the control of the cauliflower mosaic virus 35S promoter. The transformation event contained multiple insertion events, some of which were structurally complex. Among these insertion events were two loci containing genes encoding both the ZYMV and WMV2 coat proteins, but not the NPTII gene. One locus provided a moderate level of resistance to the two target viruses, while the other provided a high level of resistance. Advanced breeding lines were made homozygous for the high-resistance locus, but were segregating for the moderate-resistance locus. During this process, the NPTII marker was also eliminated by selection of lines that not only were homozygous for the high-resistance locus, but had also lost the selectable marker through segregation.

CZW3

Line CZW3 was transformed with the plasmid pPRBN-CMV73/ZYMV72/WMBN22 (Tricoli *et al.*, 1995), containing expression cassettes for the coat protein genes for ZYMV, WMV2 and cucumber mosaic virus (CMV), as well as the NPTII selectable marker. Line CZW3 was a simpler transformant, consisting of only one insertion event, containing all three coat protein genes, as well as the selectable marker. Advanced breeding lines were homozygous for the transformed locus.

Field results

Field trials at a number of locations showed that the lines ZW20 and CZW3 were effective in controlling infections by ZYMV, WMV2, and – in the case of CZW3 – CMV as well. Tricoli *et al.* (1995) mechanically inoculated two different ZW20 sublines, including one homozygous for the high-resistance locus. This subline showed only 20% infection after inoculation with WMV2, and no infection at all after ZYMV infection, compared with 100% infection of non-transgenic controls. The same researchers studied the resistance of CZW3 after mechanical inoculation with CMV, ZYMV and WMV2 individually, and in a mixed infection. Segregating inbred lines were tested, as well as a hybrid in which the segregating inbred was used as a parent. After individual and mixed inoculation of transgenic segregants, only one plant out of a total of 146 showed mild symptoms, while all of the 103 non-transgenic segregants as well as all of the 259 non-transgenic control plants became severely symptomatic. Further tests were carried out in the field, comparing virus resistance in these lines after mechanical and aphid inoculation (Fuchs and Gonsalves, 1995). Other tests studied the economic benefit of the resistance provided by lines ZW20 and CZW3 either as an inbred line or as the parent to a hybrid (Arce-Ochoa *et al.*, 1995; Clough and Hamm, 1995; Fuchs *et al.*, 1998). These tests confirmed that virus resistance conferred by the coat protein genes was able to dramatically increase crop yield and economic return under the pressure of naturally vectored disease in the field.

USDA exemption

The determination of efficacy against viral disease under field conditions led to the preparation of a petition for a determination of non-regulated status from the United States Department of Agriculture, Animal and Plant Health Inspection Service (USDA-APHIS). USDA-APHIS is the primary agency responsible for regulating work, in the greenhouse and field of transgenic plants. It also has the authority to grant non-regulated status to transgenic plants that in its opinion no longer present a plant pest risk.

The main responsibility for conducting an assessment of safety rests with USDA-APHIS. The assessment relies upon a review of the information package supplied to the agency by the petitioner, but the agency can and does seek out information on its own, and relies on the expertise of the reviewers on its staff. All relevant information is used by the agency to conduct an environmental assessment, and – if the assessment allows – grant non-regulated status.

One guiding principle followed by the USDA-APHIS in assessing risk is to compare the characteristics of the transgenic plant with the characteristics of plants that are produced by traditional plant breeding. Plants produced by traditional breeding are not regulated, because history provides sufficient assurance of safety. Thus, if the transgenic plant were determined to pose a risk different from that posed by plants already existing in nature or plants that would be produced by traditional breeding, then non-regulated status would be withheld. If, on the other hand, the risks presented by the transgenic plant are not different from those that are presented by traditionally bred varieties, then non-regulated status can be granted.

Since USDA-APHIS relies heavily (but not exclusively) on the information contained in the petition for non-regulated status, information in such a petition should include the following (see 7CFR340, Section 6 of the United States Code of Federal Regulations):

1   A description of the biology of the plant prior to the transformation. The biological description should include detailed taxonomic information.
2   Relevant experimental data and publications relating to the work conducted on the transgenic plant.
3   A description of the differences in genotype between the transgenic plant and the corresponding non-transgenic plant. Information must be provided on the organisms from which the transgenes or other control elements were obtained, the method of transformation and the vectors used during the transformation process, as well as the inserted genetic material and its product(s). USDA-APHIS, the Canadian Food Inspection Agency, and Health Canada, have standardized the types of molecular and expression data submitted in a petition for non-regulated status and the criteria by which they should be judged. This information has been published and is available at the following URLs:
    http://www.inspection. gc.ca/english/plaveg/pbo/usdaø3e.shtml
    http://www.inspection.gc.ca/english/plaveg/pbo/usdaø4e.shtml
4   A description of the phenotype of the transgenic plant, including any known and potential differences from the corresponding non-transgenic plant, in order to substantiate the claim that the transgenic plant is unlikely to pose a greater plant pest risk than the non-transgenic plant from which it was derived. This information should include the following: plant pest risk characteristics, disease and pest susceptibilities, expression of the gene product, new enzymes, or changes to plant metabolism, weediness of the transgenic plant, impact on the weediness of any other plant which is sexually compatible with the transgenic plant, agricultural or cultivation practices, effects of the transgenic plant on non-target organisms, indirect plant pest effects on other agricultural products, transfer of genetic information to organisms with which it cannot interbreed, and any other information which might be relevant. Any information indicating that a transgenic plant may

pose a greater plant pest risk than the corresponding non-transgenic plant must be included.

5   Field test reports for all trials of the transgenic line in question that were conducted under USDA-APHIS permit or notification. The field test reports should include any observations of deleterious effects on the environment caused by the transgenic plant.

Based on information received by USDA-APHIS from various sources including the petition, the Agency granted non-regulated status to ZW20 and CZW3 (Medley, 1994a; Acord, 1996), and published their support for this decision (Medley, 1994b, c; Payne, 1996). In keeping with their goal of considering all information that might be available to them, USDA-APHIS made use not only of data supplied by the petitioners, but also relied on the literature, external expert opinion, and comparison with traits that were already familiar to agriculture. This approach acknowledged that scientifically valid risk assessment could be conducted by relying on information obtained from a wide variety of sources.

Weediness

Of particular concern in determining the plant pest risk of transgenic squash was the potential for the crop itself to become a weed, or for new weeds to arise as a consequence of gene flow from the crop to sexually compatible wild populations. This question was especially important because the transgenic squash would be released in a region that was known to be the center of origin for *C. pepo*. After evaluating all the information available, USDA-APHIS concluded that ZW20 and CZW3 were unlikely to cause a weed problem or a threat to genetic diversity of free-living squash if released from regulation (Medley, 1994b, c; Payne, 1996). With regard to causing a weed problem, the agency concluded that the transgenic lines were unlikely to become weeds themselves, and that the virus-resistance traits were unlikely to cause free-living squash to become weeds.

The conclusion that the transgenic lines themselves were unlikely to become weeds was based on three reasons. First, the inbred lines used in the transformations belonged to the yellow crookneck squash type (*C. pepo* ssp. *ovifera* var. *ovifera*), which did not exhibit weedy characteristics and was not considered to be a weed in standard texts and weed lists. Second, introduction of CMV (for CZW3) as well as ZYMV and WMV2 resistance into the crop was not different from other disease- or pest-resistance traits already present in the *C. pepo* gene pool or introduced into it by traditional breeding. These traits had already been introduced into squash, and even into some commercial cultivars (see Table 10.1), yet these cultivars had not become weeds. Finally, the two transgenic lines did not exhibit any other characteristics that were more 'weedy' than those exhibited by traditionally bred squash cultivars.

Aside from considering the potential problem caused by the transgenic lines themselves, USDA-APHIS also evaluated the possibility that traits transferred from ZW20 and CZW3 to free-living relatives would cause those free-living plants to become more 'weedy'. USDA-APHIS concluded that this problem was unlikely to happen. The agency assumed that the transgenic lines and their cultivated progeny would be

Table 10.2 Survey of free-living *C. pepo* populations for evidence of virus infection

Sample	State collected	Detection method					Distance to soybean	Distance to squash
		Visual symptoms	Double diffusion serology	Host plant indexing	ELISA	Electron microscopy		
1-AR	Arkansas	Slight chlorosis	–	–	–	–	25 feet	Unknown
2-AR	Arkansas	–	–	–	–	–	0.25 miles	0.5 miles
1-LA	Louisiana	–	–	–	–	–	Within the field	0.5 miles
2-LA	Louisiana	–	–	–	–	–	75 feet	>6 miles
3-LA	Louisiana	Powdery mildew?	–	–	–	–	30 feet	>6 miles
4-LA	Louisiana	–	–	–	–	–	0.1 mile	0.1 mile
1-MS	Mississippi	–	–	–	–	–	>2 miles	>2 miles
2-MS	Mississippi	–	–	–	–	–	>2 miles	>2 miles
3-MS	Mississippi	–	–	–	–	–	>2 miles	>2 miles

– = virus absent.

able to pollinate free-living squash plants and transmit the virus resistance traits to them. This undoubtedly had occurred in the past with other disease- or pest-resistance traits that had been introduced into *C. pepo* (see Table 10.1). However, despite this opportunity, no weedy hybrid progeny had emerged. Since there was no evidence to suggest that the transgenic traits would behave differently from the traditionally bred traits, the historical lack of weed problems provided a sufficient level of assurance that crosses between transgenic and free-living squash were unlikely to create a weed problem as well. Furthermore, the cultivated squash genotypes that were used for transformation possessed characteristics that placed those plants at a selective disadvantage in a non-cultivated habitat. The transmission of these traits to free-living relatives along with the transgene would place recipient plants at a disadvantage, further reducing their weedy potential. Finally, surveys of free-living squash populations taken by the applicant failed to detect infections of CMV, WMV2 or ZYMV (Table 10.2). Therefore, there was no evidence that viruses limited the populations of free-living squash. Consequently, it was not likely that virus resistance would release these populations from control, nor was it likely that virus resistance conferred a significant selective advantage in free-living habitats.

The lack of selective pressure exerted by viruses was also cited by USDA-APHIS as the reason for concluding that genetic diversity in free-living populations would not be reduced. Without selection pressure, there was no reason to expect selection of a subset of genotypes into which the virus resistance would have been crossed.

Transcapsidation/recombination

The commercial introduction of viral sequences via transgenic plants – and in particular viral genes encoding viral coat proteins – present the potential for interactions between the transgenes or their products, and the viruses to which the transgenic squash might remain susceptible. These interactions include transencapsidation of heterologous viral RNA by coat proteins expressed in the plants, phenotypic mixing between viral coat protein subunits expressed by the transgenic plant or by the transgenic plant and infecting viruses, and recombination between transgenic viral components and other viruses infecting the transgenic plants. DeZoeten (1991) pointed out that these phenomena have the potential to produce new viruses.

USDA-APHIS concluded that the risk of transencapsidation and phenotypic mixing in transgenic virus-resistant plants would be no greater than their likelihood of occurrence in non-transgenic plants already in agriculture. Data submitted by the petitioners, and other information gathered by the agency, showed that the amounts of viral coat protein produced in transgenic plants resistant to CMV, ZYMV and WMV2 were less than the amounts found in susceptible, virus-infected plants. Table 10.3 illustrates this fact with data obtained from inoculations of ZW20. With respect to this set of data, USDA-APHIS noted that ZYMV and WMV2 coat protein levels increase when ZW20 plants were inoculated with ZYMV and WMV2 singly or combined (Medley, 1994c). The agency speculated that this increase might be due to limited replication of the viruses in inoculated ZW20 plants, or because the coat proteins produced by the transgenes were stabilized by limited replication of the inoculated viruses. However this was a phenomenon reported to occur already in other potyviruses (Farnelli et al., 1992). USDA-APHIS also relied on data summarized

Table 10.3 Amount of viral coat protein in transgenic line ZW20, either inoculated or uninoculated

Genotype (inoculum)	Mean amount of coat protein measured (standard deviation)			
	ZYMV	WMV2	PRSV-W	CMV
Nt (uninoculated)	B	B	B	B
Nt (CMV)	≤	≤	≤	0.451 (0.0267)
Nt (PRV)	≤	≤	+++	≤
Nt (WMV2)	≤	1.130 (0.107)	≤	≤
Nt (ZYMV)	1.348 (0.127)	0.006 (0.003)[a]	≤	≤
ZW20 (uninoculated)	≤	0.693 (0.003)	≤	≤
ZW20 (CMV)	≤	0.319 (0.051)	≤	0.694 (0.0272)
ZW20 (WMV2)	0.092 (0.006)	0.775 (0.020)	≤	≤
ZW20 (ZYMV)	≤	0.577 (0.046)	≤	≤
ZW20 (WMV2+ZYMV)	0.053 (0.005)	0.999 (0.067)	≤	≤
ZW20 (PRV) Plant 1	≤	1.956 (0.060)	+++	0.0025[b]
ZW20 (PRV) Plant 2	0.097 (0.019)	0.214 (0.619)	+++	≤
ZW20 (PRV) Plant 3	0.101 (0.013)	2.248 (0.040)	+++	≤
ZW20 (PRV) Plant 4	≤	0.930 (0.155)	+++	≤
ZW20 (PRV) Plant 5	≤	0.108 (0.034)	+++	≤
ZW20 (PRV) Plant 6	0.122 (0.053)	2.906 (0.044)	+++	≤

Data are shown for ZW20 plants possessing the high resistance locus. Data are pooled from six plants, assayed in triplicate, except PRSV-W. Because of variability of symptom severity among PRSV-W inoculated plants, they are given separately. Nt, nontransgenic; B, baseline; ≤, below or equal to baseline; +++, readings too high for quantitation.

[a] Only two samples > baseline.
[b] Only one sample > baseline.

in Table 10.3 to conclude that contrary to the reported ability of ZYMV to enhance CMV replication in double infections (Poolpol and Inouye, 1986), the ZYMV coat protein expression in transgenic plants did not have the same effect on infecting CMV. These data, considered along with the fact that multiple infections of ZYMV, WMV2 and PRSV-W are usual in squash fields in the United States, led USDA-APHIS to conclude that if there were indeed risks posed by transencapsidation in transgenic plants expressing viral coat proteins, they were the same as those already posed by naturally occurring multiple infections of non-transgenic plants.

USDA-APHIS also addressed the question of whether commercial release of ZW20 and CZW3 was likely to increase the likelihood of appearance of new viruses through recombination between RNA encoded by the transgenes and RNA from other viruses infecting the transgenic plants. For their decision, USDA-APHIS relied primarily on the literature and the opinions of experts (Medley, 1994c; Payne, 1996). It was acknowledged that recombination could occur between viruses infecting a plant, or between transcripts in transgenic plants and infecting viruses. However, the known examples occurred in virus groups other than those which were in question, and occurred under experimental conditions where selection for recombinants was high (Bujarski and Kaesberg, 1986; Robinson et al., 1987; Angenent et al., 1989; Allison et al., 1990; Falk and Bruening, 1994; Greene and Allison, 1994). Moreover, at least ZYMV and WMV2 did not show evidence of being prone to recombination during its evolutionary history (Robaglia et al., 1989; Shukla and Ward, 1989; Quemada et al., 1990; Sudarsono et al., 1993).

Data concerning the amount of RNA in transgenic plants versus infected non-transgenic plants was important in assessment by USDA-APHIS of the risk of recombination posed by ZW20 and CZW3. An example of the data regarding RNA levels in these plants is presented in Figure 10.1. This figure shows the levels of CMV, ZYMV, WMV2, and PRSV-W RNA in total RNA extracted from transgenic CZW3 and non-transgenic plants of the original inbred line, when infected with CMV, ZYMV, WMV2, PRSV-W, CMV+ZYMV+WMV2, or all four viruses. All inoculations show that in the cases where viral coat protein transgenes are present, the levels of viral RNA produced in the infected non-transgenic plants are significantly greater than the RNA transcribed from the transgenes. This is true even in the case of apparent low-level replication of WMV2 (Figure 10.1c, although the symptoms were severe). In the case where the transgenic and non-transgenic plants were, as expected, equally susceptible to PRSV-W, the amount of coat protein RNA was the same in both types of plants.

Even though it recognized that the full implications of these data were not clear, the Agency took the position that the significantly lower levels of RNA in transgenic plants gave some assurance of safety (Payne, 1996). While it was impossible to quantify the exact level of potential recombination, the agency concluded that it would be acceptably low, that it probably did not differ from the potential for such a phenomenon to occur already in nature, and that natural selection would reduce the chances that such recombinants would survive. Finally, USDA-APHIS concluded that even if recombinants were to survive, they would not significantly alter the already-present risk of producing recombinant viruses, and that the recombinant viruses could be managed in the same way that new recombinant viruses are managed in agriculture today (Payne, 1996).

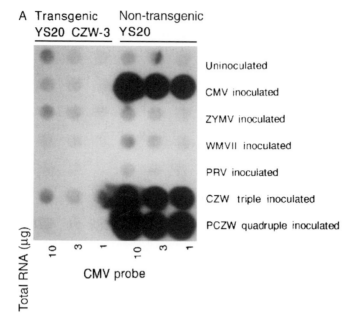

A Transgenic   Non-transgenic
   YS20 CZW-3   YS20

Uninoculated

CMV inoculated

ZYMV inoculated

WMVII inoculated

PRV inoculated

CZW triple inoculated

PCZW quadruple inoculated

Total RNA (μg)

10   3   1   10   3   1

CMV probe

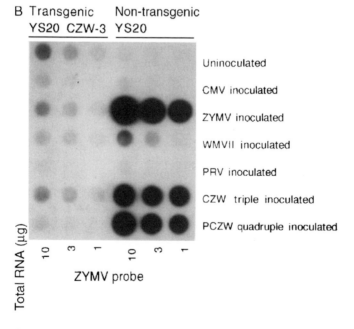

B Transgenic   Non-transgenic
   YS20 CZW-3   YS20

Uninoculated

CMV inoculated

ZYMV inoculated

WMVII inoculated

PRV inoculated

CZW triple inoculated

PCZW quadruple inoculated

Total RNA (μg)

10   3   1   10   3   1

ZYMV probe

Figure 10.1 Dot blots of RNA extracted from transgenic (YS20 CZW3) and non-transgenic (YS20) squash after inoculation with viruses either singly or in combination. Ten, three and one μg of each RNA was blotted and probed with radiolabelled clones of the coat protein genes of (A) CMV, (B) ZYMV, (C) WMV2 and (D) PRSV-W.

C Transgenic Non-transgenic
YS20 CZW-3 YS20

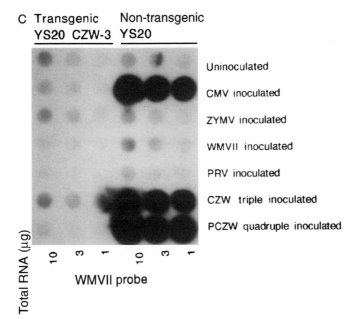

Uninoculated

CMV inoculated

ZYMV inoculated

WMVII inoculated

PRV inoculated

CZW triple inoculated

PCZW quadruple inoculated

Total RNA (µg)

10 3 1 10 3 1

WMVII probe

D Transgenic Non-transgenic
YS20 CZW-3 YS20

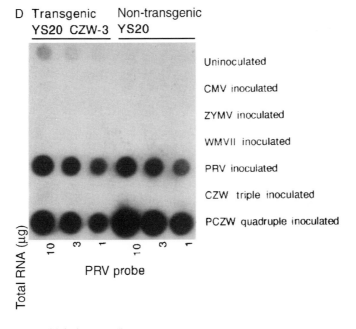

Uninoculated

CMV inoculated

ZYMV inoculated

WMVII inoculated

PRV inoculated

CZW triple inoculated

PCZW quadruple inoculated

Total RNA (µg)

10 3 1 10 3 1

PRV probe

Figure 10.1 (continued)

EPA exemption from tolerance

As is the case with USDA-APHIS, the US Environmental Protection Agency (EPA) also has the primary responsibility of conducting a safety assessment, based on information supplied by the developer of a transgenic crop and other information that the agency might obtain, including that of external and internal experts. When the first transgenic squash line (ZW20) was ready for commercialization, the EPA did not have a final rule covering transgenic plants. However, consultation with the EPA took place during the time the agency was formulating its proposed rules. These rules (Browner, 1994) extended the agency's jurisdiction to all plants expressing disease- or pest-resistance traits, since they could be viewed as producing a substance that was pesticidal – that is, they mitigated the effect of a pest (Browner, 1994). While the agency claimed jurisdiction over these plants, it recognized that certain broad categories of plants could be exempted from regulation. For example, plants produced through traditional breeding, or plants that repelled pests or disease through the modification of some structural feature could be exempted since they usually presented no new pesticidal exposure to humans consuming those plants. Furthermore, the EPA proposed to exempt coat protein produced in transgenic plants from regulation under the Federal Insecticide, Fungicide, and Rodenticide Act (FIFRA) and the Federal Food, Drug, and Cosmetic Act (FFDCA), again because they presented no new exposures to humans consuming them (Browner, 1994).

Even though the rules were only proposed, consultation with the EPA led to the decision to submit a petition for exemption from tolerance (the legal limit for a pesticide chemical residue in or on a food), as required by the FFDCA prior to the commercialization of any pesticide. In order to obtain this exemption, it was important to show that the establishment of a tolerance level was not required to protect the public health. This goal was achieved for line ZW20 by measuring the amount of viral coat protein in samples of various melon and squash fruit on sale at major grocery stores in one representative locality in the United States. Results showed that levels of viral coat protein in transgenic fruit were within the range found for most of the fruit in the sample. For some samples, the fruit on the market had many-fold higher levels of coat protein than in the transgenic squash (Table 10.4).

These data were used to support the claim that viral coat proteins in general, and ZYMV and WMV2 coat proteins in particular, were not substances that fitted the two main criteria which the EPA planned to use to determine whether a pesticidal substance would require review. First, in order to trigger EPA review the pesticidal substance would be one that was not derived from a known food source. Second, the substance would present the following changes in pesticidal exposure to humans: it would be found in significantly higher levels in the transgenic plant than that normally found in food, it was being transferred to a different crop than that in which it was already found, it was not already found in edible tissue, and it was not already found in unprocessed food.

Other arguments advanced in the petition in support of the exemption from the requirement of a tolerance were based on the premise that a demonstrated history of safe use was sufficient to assess safety. Examples from the literature showed that humans were already exposed to viral coat proteins in general, not only through consumption of naturally infected produce, but also through ingestion of food obtained

Table 10.4 Amounts of viral coat protein in transgenic ZW20 compared with levels of viral coat protein found in melons and squash already consumed by humans

Fruit	Viral coat protein (µg/kg fruit)			
	CMV	PRSV-W	WMV2	ZYMV
ZW20	ND	ND	68.4	430.6
CZW	8.28	a	a	a
C-251	355 200	18 000	14 400	10 320
C-252	130 464	5 472	10 944	115 488
C-253	ND	252 000	28 800	720
C-255	ND	ND	864	ND
CJ-1	>2 400 000	1 200	8 400	ND
CJ-2	>3 216 000	ND	14 000	ND
CJ-3	>3 216 000	ND	12 864	ND
H-1	ND	7 200	9 840	ND
H-2	ND	6 840	1 800	ND
H-3	ND	ND	2 200	ND
H-4	359	4 752	3 888	173
H-5	269	3 168	3 168	260
H-6	238	ND	2 592	ND
H-7	ND	5 928	1 824	137
H-8	664	13 272	1 896	190
H-9	82	960	24	24
H-10	ND	ND	250	ND
H-11	ND	ND	1 560	ND
H-12	ND	ND	480	ND
H-13	ND	ND	2 200	ND
H-14	ND	3 120	720	ND
H-15	ND	10 080	1 700	ND
H-16	ND	ND	3 100	ND
Y-1	ND	ND	11 424	ND
Y-2	ND	ND	ND	ND
Y-3	ND	ND	1 152	ND
Y-4	ND	ND	13 056	ND
Z-1	ND	ND	140	ND
Z-2	ND	ND	ND	ND
Z-3	ND	ND	454	ND
Z-4	ND	ND	ND	ND
Z-5	ND	ND	ND	ND
Z-6	ND	ND	576	ND
Z-7	43	ND	2 592	ND
Z-8	14	ND	2 900	ND

C and CJ, cantaloupe samples; H, honeydew melon; Y, yellow crookneck squash; Z, zucchini squash; ND, no virus detected.

[a] Not measured for CZW3, since the ZYMV and WMV2 coat proteins had already received an exemption from tolerance through the ZW20 petition, and no PRV coat protein gene was introduced.

from crops that had been subjected to cross protection (the intentional inoculation of a crop with a mild strain of a virus in order to protect the crop from subsequent infection by a more virulent strain of the virus). The petition also argued that the data in Table 10.4 supported the claim that expression of ZYMV and WMV2 in ZW20 would not present any changed exposure of humans to allergens. Because ZYMV and WMV2 coat proteins are commonly present in cucurbit fruit, anyone who might be allergic to viral coat protein in squash (a phenomenon for which no documented cases were known) would have the same reaction to the transgenic fruit as to non-transgenic fruit they had already consumed. The petition also addressed exposures due to dermal contact and inhalation. The petition argued that a history of safety had been established through the lack of reports of toxicity or allergenicity from researchers who had undoubtedly been exposed to these viruses by both the dermal and inhalation routes. Finally, the petition pointed out that viral coat proteins were ubiquitous in the environment, and therefore a natural component of it. Infected cucurbit fruit that are not marketable are routinely plowed back into the soil, thus adding large amounts of viral coat protein to the environment. Furthermore, studies in the literature have reported the recovery of plant viruses from ground water runoff in Europe and North America. The EPA considered the data and the arguments presented in the petition and granted an exemption from tolerance for the ZYMV and WMV2 coat proteins produced within ZW20 plants (Barolo, 1994).

Subsequently, the EPA granted a more categorical exemption from tolerance for ZYMV and WMV2 coat proteins in all food commodities. The decisions and supporting reasoning are summarized in the following paragraphs, which are based on the published decision (Johnson, 1997a). A detailed review of this decision is especially instructive because it provides insight into the issues that by law must now be considered by the EPA. A detailed review also provides another example of the reliance of US regulatory agencies not only on data submitted by petitioners, but also on the literature and other sources of information.

In granting the broader exemption from tolerance, the EPA considered several issues that by law they were obligated to address, especially as a result of the passage of the Food Quality Protection Act of 1996 (FPQA), which amended the FFDCA. The amended FFDCA allowed EPA to establish an exemption from the requirement for a tolerance only if EPA determines that the tolerance is 'safe'. By law, 'safe' meant that there was 'a reasonable certainty that no harm will result from aggregate exposure to the pesticide chemical residue, including all anticipated dietary exposures and all other exposures for which there is reliable information' (Johnson, 1997a). This includes exposure through drinking water and in residential settings, but does not include exposure due to its use during application or other occupational activities. In deciding on exemptions from tolerance the EPA is also required to pay special attention to exposure of infants and children (Johnson, 1997a).

The EPA examined two areas of concern in determining the risks from aggregate exposure to pesticide residues. First, the agency examined the toxicity of the 'pesticides' (ZYMV and WMV2 coat proteins). Second, it examined exposure to the pesticide through food, drinking water, and through other exposures that occur as a result of pesticide use in residential settings.

In considering toxicity, petitioners are required by law to submit data regarding acute toxicity, genotoxicity, reproductive and developmental toxicity, as well as subchronic and chronic toxicity. However the petitioner requested, and EPA granted, waivers from these data requirements. The waivers were based on the long history of mammalian consumption of the entire plant virus particle in foods, without causing any deleterious human health effects. The EPA once again acknowledged that virus-infected plants currently are – and have always been – a part of both the human and domestic animal food supply, and that there is no evidence that plant viruses are toxic to humans and other vertebrates. Furthermore, the EPA pointed out that plant viruses are unable to replicate in mammals or other vertebrates, thereby eliminating the possibility of human infection. As additional assurance of safety, only the coat protein would be expressed in the plant, and this component alone would be incapable of forming infectious particles.

The EPA also had to consider the toxicity of the introduced DNA (coding and regulatory regions) directing the synthesis of the coat proteins. The EPA acknowledged that DNA is common to all forms of plant and animal life and that it was consumed as a safe component of food. The specific nucleic acids that were used to transform the squash plants were characterized by the petitioners, and a review of the nucleic acid sequence led EPA to conclude that no mammalian toxicity was anticipated from the genes encoding the coat proteins of ZYMV and WMV2.

In considering exposures, the EPA had to consider available information concerning exposures from the coat protein in food and all other non-occupational exposures, including drinking water from groundwater or surface water, and exposure through its use in gardens, lawns or buildings. With respect to dietary exposure in food, the EPA concluded, as they did before, that the use of viral coat protein in squash would not result in any new dietary exposure to plant viruses, because (1) entire infectious particles of ZYMV and WMV2 are found in the fruit, leaves and stems of most plants, (2) viruses are ubiquitous in the agricultural environment at levels higher than will be present in transgenic plants, and (3) virus-infected food plants have historically been a part of the human (including that of infants and children) and domestic animal food supply, with no known adverse effects to human or animal health.

Moreover, the EPA concluded that there was no evidence to indicate that harmful effects occurred as a result of exposure to viral coat proteins through other means, such as non-food oral, dermal and inhalation routes. They arrived at this conclusion by relying not only on data submitted by the petitioners, but on the expert opinion of its Scientific Advisory Panel, which concluded that the levels of virus in the agricultural environment are much higher than those present in transgenic plants, and that the existing presence of viral coat proteins in the food supply provided a scientific basis for exempting transgenic plants that express viral coat proteins from the requirement of a tolerance.

With respect to drinking water exposure, the EPA concluded that potential exposures through this route would be negligible. The agency reasoned that viral coat proteins produced in plants are an integral part of the living tissue of the plant, and are therefore subject to rapid degradation and decay. Consequently, viral coat proteins would not accumulate in the environment or in the food chain. Therefore, because of their presumed transience, these proteins would present only negligible exposure in drinking water obtained from surface or groundwater sources. The EPA used similar reasoning to conclude that exposure of engineered coat proteins as a consequence of their use in such areas as around homes, parks, recreation areas, athletic fields and golf courses, would be negligible.

The EPA concluded that the likelihood of exposure to plant viral coat proteins, including ZYMV and WMV2 coat proteins, through any dermal or respiratory routes would be unlikely. Although physical contact with the plant or raw agricultural food obtained from the plant may present some opportunity for dermal exposure, the potential amounts involved in this exposure were judged to be negligible compared with exposure through the diet. Furthermore, the EPA reasoned that the skin would provide a barrier that would not be crossed by viral coat proteins. Similarly, exposure through inhalation was judged to be negligible compared with potential exposure through the diet. The most likely opportunity for respiratory exposure through inhalation was through pollen. However, the EPA concluded that it was again unlikely that the amount of exposure to the coat proteins in the pollen would be significant. Viral coat proteins, when present in pollen, would be part of the pollen grain and would not cross the barrier presented by the mucous membrane of the respiratory tract. Therefore, this route would not add any additional exposure to that presented by the dietary route.

A new requirement imposed by the Food Quality Protection Act involved the assessment of available information, indicating that there might be special concerns regarding exposure of infants and children to viral coat proteins. The EPA concluded that viral coat proteins were present in all foods, including those foods consumed by infants and children. It had no evidence to indicate that the amount of exposure was different between adults and children, nor that the effect of viral coat proteins differed between the two groups. Most importantly, the EPA had no evidence suggesting that exposure of adults or infants and children was in any way harmful.

Finally, the EPA had to consider whether ZYMV and WMV2 coat proteins, or the DNA needed to produce them in plants, would have an effect on the immune and endocrine systems. Because the agency had no information indicating that such an effect could be expected, it did not require that the petitioner submit information regarding that question.

The decision to grant an exemption from tolerance for ZYMV and WMV2 coat proteins in ZW20 and subsequently in all raw commodities also cleared the way for commercialization of CZW3. EPA then granted an exemption from tolerance for the coat protein of CMV in all raw commodities, based on the same safety assessment it conducted for the other two coat proteins (Johnson, 1997b). Line CZW was included in this exemption and was therefore cleared for commercialization.

FDA food safety analysis

In contrast to USDA-APHIS and EPA, the United States Food and Drug Administration (FDA) makes the safety assessment of new plant varieties, and the foods made from them, the responsibility of the developer or manufacturer. In 1992, the FDA published guidelines to clarify the obligations of any producer of new plant varieties, whether produced through traditional plant breeding, or through techniques of recombinant DNA. The principle governing the guidelines is that assessment of the safety of a product (in this case food produced from a new plant variety) should focus on the product's characteristics rather than the method by which it was made.

The authority of FDA to regulate food is a post-market authority (Kessler, 1992). Thus, a food producer can market a food containing a new component that it has assessed to be generally recognized as safe (GRAS). If this component is not GRAS, then a food additive petition must be filed and granted before food containing that new component can be marketed (Kessler, 1992). The 1992 guidelines were meant to help producers in determining whether a food safety concern existed. If a concern existed, a consultation with FDA would take place to determine whether a new component could be considered GRAS, or whether a food additive petition was required. In practice, consultation with FDA regarding new transgenic plants that will become food usually occurred regardless of safety concerns. The consultation assures producers that FDA is aware of the developer's safety assessment process, and serves to identify those parts of the safety analysis that might not be acceptable to the Agency. This process enables the producer to proceed toward commercialization with confidence that no unanticipated safety concerns will be encountered at late stages of product development. Recently, the FDA has published its intention to make this consultation process mandatory (Henney and Shalala, 2000).

The food safety assessment conducted for ZW20 has been presented elsewhere (Quemada, 1996) and will therefore only be summarized in this section. The assessment applies equally well to CZW3. In general, the safety assessment specified by the guidelines examines the host plant (the plant which has been modified by introduction of foreign DNA), the organisms from which the foreign DNA has been obtained, and the new compounds produced as a result of the introduction of the foreign DNA.

First, the guidelines direct the developer to determine whether there is a history of safe consumption of the species that was transformed. In the case of squash, the species has had a history of safe consumption for at least 3000 years (Heiser, 1989; Smith *et al.*, 1992), and is consumed in many forms throughout the world. The particular genotypes transformed to produce ZW20 and CZW3 have been in use commercially for several years.

The guidelines also ask the developer to consider whether characteristics of the host species, related species, or the lines used in transformation warrant analytical or toxicological tests. If so, they also ask whether the results of the tests provide evidence that the toxicant levels in the new plant variety do not present a safety concern. Squash indeed has characteristics warranting analytical testing, since the plant is known to produce cucurbitacins, which are extremely bitter toxic alkaloids. Squash breeders are aware that cucurbitacins must be eliminated from commercial varieties. Therefore, standard breeding practices take into consideration the potential presence of this compound, especially if genes from wild or free-living relatives are crossed into cultivated squash.

The FDA guidelines indicate that the types of testing currently used to screen for known toxicants have been effective in managing the risk of such compounds, and are therefore appropriate to apply to the case of transgenic crops. For the cucurbitacins, tasting fruits for bitterness has been an effective test method for the detection of this compound. The documented sensitivity of the human sense of taste for this compound is as low as 1–10 parts per billion, depending upon the particular form of cucurbitacin (Metcalf *et al.*, 1980; Rymal *et al.*, 1984). At 10 parts per billion, this detection level is 34 000 times less than the reported oral LD_{50} in mice (Merck, 1989). Therefore, tasting fruit is an appropriate analytical test for the presence of cucurbitacin, and the presence of these compounds can be detected long before toxic levels are reached. The non-bitter taste of ZW20 and CZW3 indicates the absence of cucurbitacins in these lines, as they were in the original lines before transformation. This result

Table 10.5 Compositional analysis of transgenic squash, the corresponding commercial variety, and values for important nutrients reported in the literature

Component	Transgenic range	Non-transgenic range	Literature values (Pennington, 1989)
Protein (g/100 g)	0.8–1.2	0.8–1.4	0.9
Moisture (g/100 g)	93.6–94.5	93.7–94.8	94.2
Fat (g/100 g)	<0.1–0.1	<0.1	0.3
Ash (g/100 g)	0.4–0.7	0.5–0.8	
Total dietary fiber (g/100 g)	1.0–1.2	1.0–1.1	1.1
Carbohydrates (g/100 g)	4.1–4.3	3.9–4.4	4.0
Calories (Calories/100 g)	16.4–18.9	14.4–18.4	18.5
Fructose (g/100 g)	0.9–1.2	1.1–1.3	
Glucose (g/100 g)	0.8–1.1	1.0–1.2	
Sucrose (g/100 g)	<0.2	<0.2–0.2	
Lactose (g/100 g)	<0.2	<0.2	
Vitamin C (mg/100 g)	15.1–22.4	14.1–23.2	7.7
β-carotene (mg/100 g)	<0.03–0.05	<0.03–0.04	
Vitamin A (IU/100 g)	<50–80	<50–70	338
Calcium (mg/100 g)	13.3–29.4	15.7–29.7	21.5
Iron (mg/100 g)	0.287–0.367	0.372–0.478	0.477
Sodium (mg/100 g)	<2.50–3.98	<2.50–2.88	

Samples were obtained from three separate field locations, grown under typical agricultural practices.

therefore provides evidence that the toxicant levels in the transgenic lines (if any are present at all) do not present a concern that would warrant toxicology testing.

The final consideration with respect to the host plant is the question of whether the concentration and bioavailability of important nutrients in the transformed line are within the range normally seen in the non-transformed plant. Nutritional analyses of the transgenic lines compared with their non-transformed counterparts were conducted (Table 10.5). These data demonstrate that no change in the important nutrients has occurred in the transgenic lines.

The guidelines then call for consideration of the organisms from which the foreign DNA has been obtained. The question of whether the donor organisms are known to be allergenic is considered, as is the question of whether characteristics of the donor organisms warrant toxicological testing. Neither ZYMV, WMV2 nor CMV are known to be allergenic, nor is there any evidence to suggest that these organisms are toxic. Therefore, the nature of the donor organisms does not raise safety concerns.

Finally, for new proteins produced by the introduced foreign DNA, the FDA guidelines ask whether the new protein is present in food derived from the plant; whether the protein is derived from a food source, or is similar to an edible protein; whether food made from the donor is known to be allergenic; whether the introduced protein is known to be toxic; whether ingestion of the new protein will be comparable to the intake of this protein or a similar one in other food; and whether the new protein is likely to become a major constituent of the human diet. As shown in Table 10.4, the CMV, ZYMV and WMV2 coat proteins are already present in squash and other cucurbits. While not food themselves, plant viruses are found in squash, and the coat proteins can therefore be considered as being derived from a food. Furthermore, the

CMV, ZYMV and WMV2 coat proteins are almost identical to the coat proteins (presumably edible) found in cucurbit fruit. Squash containing viruses is not known to be allergenic, nor is it known to be toxic. Intake of viral coat proteins from the fruit of transgenic plants will be comparable to the intake of these proteins in non-transgenic infected fruit. Finally, the amounts of viral coat protein expressed in transgenic ZW20 and CZW3 are too low to present a possibility of becoming a major dietary constituent.

These data were presented to the FDA in order to inform the Agency of the justification for the judgement that no food additive petition was warranted for ZW20 and CZW3 (they were therefore GRAS). Since no objections to these analyses were raised by the Agency, commercialization proceeded.

Conclusion

The assessment of safety of transgenic crops continues to be a topic of scientific and public concern. A review of the issues that surround the debate over the safety of these crops (Butler et al., 1999) reveals that the issues addressed seven years ago during the commercialization of virus-resistant squash – which was the first commercialized virus-resistant crop – are still unresolved.

The appropriate perspective on safety is one that was expressed well before transgenic crops were at issue (Lowrance, 1976):

> A thing is safe if its risks are judged to be acceptable. By its preciseness and connotative power, this definition contrasts sharply with simplistic dictionary definitions that have 'safe' meaning something like 'free from risk.' Nothing can be absolutely free of risk . . . Because nothing can be absolutely free of risk, nothing can be said to be absolutely safe. There are degrees of risk, and consequently there are degrees of safety.

This is the perspective that is taken by the agencies responsible for regulating transgenic crops in the US. They all judge the level of risk of these crops by comparing them to the existing level of risk, such as risks presented by crops developed by traditional breeding or risks presented by natural phenomena. Furthermore, even though the risk presented by these crops might be greater than those presented by current unregulated activities, the acceptability of that risk also determines whether these crops are safe enough to commercialize and use. It is this latter issue that adds the dimension of public perception and acceptance to the debate. Clearly, it is this dimension that explains the vastly different views of safety of transgenic crops between the public in North America, particularly the United States, and the public in Europe.

The deployment of transgenic crops will most likely continue and increase with time. The potential benefits to the consumer, the farmer, as well as to the companies developing these crops will serve as a potent force for the exploitation of this technology. As this continues, scientific inquiry into the potential risks should continue. These investigations will help to further clarify areas of real concern. By focusing attention on these areas, which will probably be very few, safe use of this technology will be further improved.

Acknowledgements

The author thanks Dr Deena S. Decker-Walters, The Cucurbit Network, for allowing the use of an unpublished description of the species, *Cucurbita pepo* in section III. Tables 10.1 and 10.2 are reprinted with permission – Food Safety Evaluation Copyright OECD, 1996.

References

Acord, B. (1996) 'Asgrow Seed Co.; availability of determination of nonregulated status for squash line genetically engineered for virus resistance', *Federal Register* **61**: 33484–85.

Allison, R., Thompson, C. and Ahlquist, P. (1990) 'Regeneration of a functional RNA virus genome by recombination between deletion mutants, and requirement for cowpea chlorotic mottle virus 3a and coat protein genes for systemic movement', *Proceedings of the National Academy of Sciences of the USA* **87**: 1820–24.

Andres, T. (1987) '*Cucurbita fraterna*, the closest wild relative and progenitor of *C. pepo*', *Cucurbit Genetics Cooperative Report* **10**: 69–71.

Angenent, G., Postthumus, E., Brederose, F. *et al.* (1989) 'Genome structure of tobacco rattle virus strain PLB: further evidence on the occurrence of RNA recombination among tobraviruses', *Virology* **171**: 271–74.

Anonymous (1992) *Thompson and Morgan Seed Catalog, 1992.* Jackson, New Jersey: Thompson and Morgan, Inc.

Anonymous (1999) 'A new multi-disease resistant summer squash', *The Cucurbit Network News* **6**: 3.

Arce-Ochoa, J., Dainello, F., Pike, L. *et al.* (1995) 'Field performance comparison of two transgenic summer squash hybrids to their parental hybrid line', *HortScience* **30**: 492–93.

Asch, D.L. and Asch, N.B. (1992) 'Archaeobotany', in R. Stafford (ed.) *Geoarchaeology of the Ambrose Flick Site.* Kampsville, Illinois: Center for American Archaeology, pp. 177–268.

Barolo, D. (1994) 'Watermelon mosaic virus-2 coat protein, zucchini yellow mosaic virus coat protein, and the genetic material necessary for the production of these protein in transgenic squash plants; tolerance exemption', *Federal Register* **59**: 54824–25.

Boyette, G., Templeton, E. and Oliver, L. (1984) 'Texas gourd (*Cucurbita texana*) control', *Weed Science* **32**: 649–55.

Browner, C. (1994) 'Plant-pesticides subject to the Federal Insecticide, Fungicide, and Rodenticide Act and the Federal Food, Drug, and Cosmetic Act', *Federal Register* **59**: 60496–518.

Butler, D., Reichhardt, T., Abbott, A. *et al.* (1999) 'Long-term effect of GM crops serves up food for thought', *Nature* **398**: 651–56.

Bryson, C. and Byrd, J. Jr. (1998) 'Texas gourd (*Cucurbita texana*) management in cotton', paper presented at the Southern Weed Science Society, January 26–28, 1998.

Clough, G. and Hamm, P. (1995) 'Coat protein transgenic resistance to watermelon mosaic and zucchini yellow mosaic virus in squash and cantaloupe', *Plant Disease* **79**: 1107–109.

Bujarski, J. and Kaesberg, P. (1986) 'Genetic recombination between RNA components of a multipartite plant virus', *Nature (London)* **321**: 528–31.

Cowan, C. and Smith, B. (1993) 'New perspectives on a wild gourd in eastern North America', *Journal of Ethnobiology* **13**: 17–54.

Decker, D. (1988) 'Origin(s), evolution, and systematics of *Cucurbita pepo* (Cucurbitaceae)', *Economic Botany* **42**: 4–15.

Decker, D. and Wilson, H. (1987) 'Allozyme variation in the *Cucurbita pepo* complex: *C. pepo* var. *ovifera* vs. *C. texana*', *Systematic Botany* **12**: 263–73.

Decker-Walters, D.S., Walters, T.W., Cowan, C.W. *et al.* (1993) 'Isozymic characterization of wild populations of *Cucurbita pepo*', *Journal of Ethnobiology* **13**: 55–72.

DeZoeten, G. (1991) 'Risk assessment – do we let history repeat itself?', *Phytopathology* 81: 585–86.

Enzie, W. (1940) 'The Geneva Delicata squash', *Farm Research* 6: 12.

Falk, B. and Bruening, G. (1994) 'Will transgenic crops generate new viruses and new diseases?', *Science* 263: 1395–996.

Fang, G. and Grumet, R. (1993) 'Genetic engineering of potyvirus resistance using constructs derived from the zucchini yellow mosaic virus coat protein gene', *Molecular Plant Microbe Interactions* 6: 358–67.

FAO (1998) FAOStat Database. Food and Agricultural Organization of the United Nations, http://apps.fao.org/cgi-bin/nph.db.pl?subset=agriculture

Fuchs, M. and Gonsalves, D. (1995) 'Resistance of transgenic hybrid squash ZW20 expressing the coat protein genes of zucchini yellow mosaic virus and watermelon mosaic virus 2 to mixed infections by both potyviruses', *Bio/Technology* 13: 1466–73.

Fuchs, M., Tricoli, D., Carney, K. *et al.* (1998) 'Comparative virus resistance and fruit yield of transgenic squash with single and multiple coat protein genes', *Plant Disease* 82: 1350–56.

Gilbertalbertini, T., Lecoq, H., Pitrat, M. *et al.* (1993) 'Resistance of *Cucurbita moschata* to watermelon mosaic virus type 2 and its genetic relationship to resistance to zucchini yellow mosaic virus', *Euphytica* 69: 231–37.

Gonsalves, D., Chee, P., Provvidenti, R. *et al.* (1992) 'Comparison of coat protein-mediated and genetically derived resistance in cucumber to infection by cucumber mosaic virus under field conditions with natural challenge inoculations by vectors', *Bio/Technology* 10: 1562–70.

Greene, A. and Allison, R. (1994) 'Recombination between viral RNA and transgenic plant transcripts', *Science* 263: 1423–25.

Heiser, C. (1989) 'Domestication of the Cucurbitaceae: *Cucurbita* and *Lagenaria*', in D. Harris and G. Hillman (eds) *Foraging and Farming*. London: Unwin Hyman.

Henney, J. and Shalala, D. (2001) 'Premarket notice concerning bioengineered foods', *Federal Register* 66: 4706–38.

Johnson, S. (1997a) 'Coat proteins of watermelon mosaic virus–2 and zucchini yellow mosaic virus and the genetic material necessary for its production; exemption from the requirement of a tolerance', *Federal Register* 62: 44579–82.

Johnson, S. (1997b) 'Coat protein of cucumber mosaic virus and the genetic material necessary for its production; exemption from the requirement of a tolerance', *Federal Register* 62: 44575–79.

Kessler, D. (1992) 'Statement of policy: foods derived from new plant varieties; notice', *Federal Register* 57: 22984–3005.

Kirkpatrick, K.J. and Wilson, H.D. (1988) 'Interspecific gene flow in *Cucurbita*: *C. texana* vs. *C. pepo*', *American Journal of Botany* 75: 519–27.

Kyle, M., Moriarty, G. and Munger, H. (1993) 'Cucumber, melon, and squash germplasm from the Cornell collection', *Cucurbit Genetics Cooperative Report* 16: 90–91.

Lowrance, W. (1976) *Of Acceptable Risk*. Los Altos, CA: William Kaufmann.

Martin, M. (1960) 'Inheritance and nature of cucumber mosaic virus resistance in squash', *Dissertation Abstracts* 20: 3462.

Medley, T. (1994a) 'Availability of determination of nonregulated status for virus resistant squash; notice', *Federal Register* 239: 64187–89.

Medley, T. (1994b) 'APHIS/USDA petition 92-204-01 for determination of nonregulated status for ZW-20 squash: environmental assessment and finding of no significant impact', United States Department of Agriculture document, December 7, 1994.

Medley, T. (1994c) 'Response to the Upjohn Company/Asgrow Seed Company petition 9-204-01 for determination of nonregulated status for ZW-20 squash', United States Department of Agriculture document, December 7, 1994.

Merck (1989) *Merck Index*, 11th edition. Rahway, NJ: Merck and Co.

Metcalf, R., Metcalf, R. and Rhodes, A. (1980) 'Cucurbitacins as keiromones for diabroticite beetles', *Proceedings of the National Academy of Sciences of the USA* 77: 3769–72.

Oliver, L., Harrison, S. and McClelland, M. (1983) 'Germination of Texas gourd (*Cucurbita texana*) and its control in soybeans (*Glycine max*)', *Weed Science* 31: 700–706.

Payne, J. (1996) 'USDA/APHIS petition 95-352-01P for determination of nonregulated status for CZW–3 squash: environmental assessment and finding of no significant impact', United States Department of Agriculture document, June 14, 1996.

Pennington, J. (1989) *Food Values of Portions Commonly Used*, 15th edition. Philadelphia, Pennsylvania: J.B. Lippincott.

Poolpol, P. and Inouye, T. (1986) 'Enhancement of cucumber mosaic virus multiplication by zucchini yellow mosaic virus in doubly infected cucumber plants', *Annals of the Phytopathology Society of Japan* 52: 22–30.

Powell-Abel, P., Nelson, R., De, B. *et al.* (1986) 'Delay of disease development in transgenic plants that express the tobacco mosaic virus coat protein gene', *Science* 232: 738–43.

Provvidenti, R. (1990) 'Viral diseases and genetic sources of resistance in *Cucurbita* species', in D. Bates, R. Robinson and C. Jeffrey (eds) *Biology and Utilization of the Cucurbitaceae*. Ithaca, NY: Cornell University Press, pp. 427–35.

Quemada, H. (1996) 'Food safety evaluation of a transgenic squash', in *Food Safety Evaluation*. Paris: OECD, pp. 71–79.

Quemada, H., Sieu, L., Siemieniak, D. *et al.* (1990) 'Watermelon mosaic virus II and zucchini yellow mosaic virus: cloning of 3′ terminal regions, nucleotide sequences, and phylogenetic comparisons', *Journal of General Virology* 71: 1451–60.

Robaglia, C., Durand-Tardif, M., Tronchet, M. *et al.* (1989) 'Nucleotide sequence of potato virus Y (N strain) genomic RNA', *Journal of General Virology* 70: 935–47.

Robinson, D., Hamilton, W., Harrison, B. *et al.* (1987) 'Two anomalous tobravirus isolates: evidence for RNA recombination in nature', *Journal of General Virology* 68: 2551–61.

Rymal, K., Chambliss, O., Bond, M. *et al.* (1984) 'Squash containing toxic cucurbitacin compounds occuring in California and Alabama', *Journal of Food Protection* 47: 270–71.

Schultheis, J. and Walters, S. (1998) 'Yield and virus resistance of summer squash cultivars and breeding lines in North Carolina', *HortTechnology* 8: 31–39.

Shukla, D. and Ward, C. (1989) 'Identification and classification of potyviruses on the basis of coat protein sequence data and serology', *Archives of Virology* 106: 171–200.

Smith, B.D., Cowan, C.W. and Hoffman, M.P. (1992) 'Is it an indigene or a foreigner?', in B.D. Smith (ed.) *Rivers of Change: Essays on the Origins of Agriculture in Eastern North America*. Washington, DC: Smithsonian Institution Press, pp. 67–100.

Sudarsono, Woloshuk, S., Xiong, Z. *et al.* (1993) 'Nucleotide sequence of the capsid cistrons from six potato virus Y (PVY) isolates infecting tobacco', *Archives of Virology* 132: 161–70.

Tricoli, D., Carney, K., Russell, P. *et al.* (1995) 'Field evaluation of transgenic squash containing single or multiple virus coat protein gene constructs for resistance to cucumber mosaic virus, watermelon mosaic virus 2, and zucchini yellow mosaic virus', *Bio/Technology* 13: 1458–65.

Walkey, D. and Pink, D. (1984) 'Resistance in vegetable marrow and other *Cucurbita* spp. to two British strains of cucumber mosaic virus', *Journal of Agricultural Science, Cambridge* 104: 325–29.

Wilson, H. (1990) 'Gene flow in squash species', *Bioscience* 40: 449–55.

Zitter, T., Pink, D. and Walkey, D. (1991) 'Reaction of *Cucurbita pepo* L. cv. Cinderella to strains of cucumber mosaic virus', *Cucurbit Genetics Cooperative Report* 14: 125–28.

Zitter, T., Hopkins, D. and Thomas, C. (1996) *Compendium of Cucurbit Diseases*. St. Paul, Minnesota: American Phytopathological Society.

Index

Note: page numbers in *italics* refer to figures and tables

AaIT toxin gene 204
AaIT toxin specificity 205
Acinetobacter calcoacetinus 112
Advisory Committee for Agricultural
    Biotechnology (USDA) 29
Advisory Committee on Novel Foods and
    Processes (UK) 45, 49, 57
  transparency 60
agricultural biotechnology, US federal
    regulations 5–6
Agrobacterium tumefaciens 138, 139
agronomic traits 1
  improvements vii
algae, plant pests *23*
allergens 94, 102–4
  analytical data requirements of FDA 12,
    232, 236
  avoidance 98
  Brazil nut 94
  expression in recipient organism 106
  glycans 81
  labelling of food 19
  pollen 81
  risk protection 104–6
  safety evaluation 10, 11
  sequence similarity studies for transferred
    genes 105
  upregulation in recipient organism 106
  see also food allergenicity; food allergies
allergic reactions 98
amino acid sequence, of transferred genes
    104–5
amino acids
  aromatic 146
  canola seed composition 146, *147*
amino peptidase 173
aminomethylphosphonic acid (AMPA) 138,
    139
ampicillin resistance gene transfer 112

anaphylaxis 95, *95, 96*
  food allergens 102
  treatment 99
Androctonus australis 203, 204
angioedema 102
Animal and Plant Health Inspection Service
    (APHIS) of USDA 21–2
  genetically modified squash regulation
    exemption 222–7, *228–9, 230, 231,*
    232–4
  new organisms 24
  non-regulated status determination 24–5,
    26
  notification process 25–6
    petitioning 27–8
  permits 24
  risk assessment 223
  safety test assessment 223
animal studies, safety assessment 57–8,
    76
anti-carcinogenic compounds 78
anti-nutritional substances
  analytical data requirements of FDA 12
  oilseed rape 140–1
anti-oxidative compounds 78
antibacterial compounds, genes encoding
    119
antibiotic resistance 111
  determinants 117, 120
    mutation 122
    transfer 111, 122
  gene transfer 16, 112
  marker genes 16
  safety-in-use 12
antibiotics, plasmid transfer effects 122,
    124
aphids, Bt maize trials 188–9
Apis mellifera 181
Archaea conjugation 117

arthropods, fatal disease induction by
    baculoviruses 201
Asilomar Conference 2
aspiration 97
asthma
    bronchial 100
    food allergens 102
    ingested foods 101, *102*
    mouse model 106
atopic dermatitis *95*, 96
atopic disease 100
Autographa californica
        nucleopolyhedrovirus (AcMNPV)
        204, 206
    environmental stability 210–11
    host-range determination 214
    notification for field testing 213–15
    selectivity 207–8
    virulence 208, *209*

B cell epitopes 103
Bacillus cereus 168
Bacillus thuringiensis (Bt) 31
    bacterial insecticides 175–7
    biology 165–75
    endotoxins 165, 167, 193
        crystals 176
        mode of action 171
        production 168
    δ-endotoxins 166, 168, 176
    β-exotoxins 166
    IgE induction 94
    insect-resistant tomatoes 75
    insecticidal proteins 168–70
    parasporal bodies 167, 168, 174, 175
    protein toxins 165, 166, 167, 168–9
        binding proteins 173
        insecticidal 165
    safety 176–7
        mammalian 183–5
        non-target invertebrates 179–83
        testing 178–9
    spore preparations 168
    sporulating cells 165, *166*
    subspecies diversity 167–8
    toxin synergists 166
    transmissible plasmids 168
    see also Bt *entries;* Cry proteins; Cyt
        proteins
Bacillus thuringiensis (Bt) modified corn 1
bacteria 98
    antibiotic-resistant in food 122
    conjugation 115–19, 122
        frequency 119
    conjugative transposon transfer 117
    DNA transfer 111

    to bacteriophage 113–15
    human gut models 122, *123*
    DNA transformation 111–12
    human gut microflora 120
    marker gene non-expression in novel
        hosts 120
    plant pests *23*
    plasmid maintenance 119
    probiotic 110, 111
    retrotransfer 118
bacterial products 98
bacteriophage
    attack 110
    DNA transfer 113–15
    genome 114
    host range 114
    human gut 114–15
    particles 114
baculoviruses
    evolutionary relationship with arthropods
        201
    feral 204, 208–11
    inactivation mechanisms 209
    progeny occlusion body yield from
        infection 209–10
    specificity 201–2
baculoviruses, recombinant 201–15
    comparative testing with feral 204
    environmental impact evaluation 204
    environmental stability 208–11
    insecticidal efficacy evaluation 204–5
    public relations activity 203
    recombinant technology 202
    regulation 211–15
    risk assessment 203–11
    speed of kill 202
    toxin genes 202, 203
    toxin specificity 205
basophils 95
biological control programs with Bt crops
    194
biosafety marker genes *see* marker genes
blackfly control 181, 182
Bombyx mori recombinant
        nucleopolyhedrovirus (BmNPV)
        207
Brassica napus 138, 139
Brazil nut allergen 94
breast feeding 98
bronchodilators 98–9
Bt crops 164–95
    advantages 164
    aquatic organism testing 187–8
    biological control programs 194
    Cry proteins 175
    extent of use 164

feeding studies with mammals 194
integrated pest management programs
    194
non-target organisms 165, 187–91, 194,
    195
safety 193–4
    human 165
    mammals 191–3
    non-mammalian vertebrates 191
    non-target invertebrates 187–91
    non-target mammals 194
    non-target organisms 195
safety testing 185–7, 194–5
    validation 195
soil microorganisms 190–1
terrestrial organism testing 187–8
transgenic 175–7
Bt formulations 175
    tolerance requirement exemption 179
Bt genes 80
    nomenclature 170
Bt insecticides 193–4
    aquatic invertebrate tests 181
    caterpillar pest control 183
    comparison with alternative control
        technologies 182–3
    crop spraying prior to harvest 185
    endangered species 180
    field condition evaluation 181–2
    moth larvae 180
    safety
        human tests 184–5
        to mammals 183–5
        to non-target invertebrates 179–83
        standard tests 183
    specificity 179, 180
    tolerance requirement exemption 179
Bt proteins
    degradation 177
    levels in Bt maize 190
    post-translational modifications 187
    soil particle binding 191
Bt toxin 31

cadherin-like proteins 173
Canada
    guidelines on safety of genetically
        modified plants 139
    registration of canola term 70
cancer, glucosinolate effects 141–2
canola
    breeding programmes and development
        69–70
    erucic acid/glucosinolate reduced levels
        70, 141
    fatty acid content changes 71, 72

fatty acid structural changes 71
glucosinolates adverse effect 69–70
herbicide-tolerant 6, 70–1
high erucic acid levels 69
high oleic/low linolenic 71
low erucic acid varieties 69, 141
    application of substantial equivalence
        70–2
    compositional modification 71
    LEAR oil see LEAR oil
    see also GT73 canola
meal (feed supplement) 69
    processed 69, 141
    safety assessment 71
    uses 69–70
registration of term (Canada) 70
substantial equivalence application 67
substantial equivalence case study see
    substantial equivalence
tolerant to Roundup® herbicide 138–59
see also GT73 canola; Westar canola
carbohydrates
    malabsorption 99
    modified 11
        pre-market review 12
CAT-Tox(L) assay 87, 88
catfish, channel 191
cattle, Bt crop safety 193
cDNA libraries 79–80
cDNA microarrays 79–80
celiac disease 100–1, 103
chalasia 97–8
chemical fingerprinting 82, 83, 84–5, 89
chicken
    Bt crop safety studies 191
    food protein gastroenteropathy 99
children
    exposure to viral coat proteins 234
    see also infants
cholelithiasis 98
cholesterol
    Cyt protein affinity 174
    reduction 48, 59
Chrysoperla carnea 188, 189, 190
ciguatera poisoning 98
Codex Ad Hoc Intergovernmental Task
    Force on Foods Derived from
    Biotechnology 66, 72
Codex Alimentarius Commission 66
commercial planting, federal regulation 6
commercialization
    environmental safety concerns 41
    FDA requirements 9–16
    genetically modified crops 74–5
    human health concerns 41
    USDA requirements 24–9

communication, public relations for
    baculoviruses 203
competence 111–12
compositional analysis of plant 13
concept of familiarity 5
contact dermatitis 102
contact reactions 102
Coordinated Framework for Regulation of
    Biotechnology (US, 1986) 3, 4, 6
  EPA role 30
  FDA role 7
  USDA role 22
corn *see* maize
cotton, Bt 32
cotton, insecticide application 32
cotton bollworm 204
cough, chronic 97
Council of Ministers (EC) 51
Council on Agricultural Science and
    Technology (CAST) 40
Council on Environmental Quality (US)
    3–4
cow's milk allergy 95
cow's milk protein 99
CP4 5-enolpyruvylshikimate-3-phosphate
    synthase (CP4 EPSPS) 138, 146
  safety assessment 139
crop consumption, limitation of substantial
    equivalence 68
cross-pollination of GM plant with wild
    plants 28
cruciferous vegetables, anti-cancer effects
    141–2
crustacea, allergy 104
cry genes 170
  sequence analysis 172
Cry proteins *166*, 168
  activation 177
  binding proteins 173
  Bt transgenic crops 175
  Cyt protein synergy 174, 175
  degradation after soil incorporation 191
  efficacy 193
  genetically engineered crops 185–7
  insecticidal 174
  insecticidal process at molecular level 171
  insertion 171
  mode of action 170–1
  modification by plants 187
  nomenclature 169–70
  non-target organism testing 188–9, 190–1
  post-inertional processing 171
  safety evaluation 178–9
  safety testing in mammals 191–3
  structure 172–4
  transgenic plant 187

Cry1Ab toxin 189, 191
  mammalian safety testing 192–3
Cry3A molecule structure 172, *173*
Cry3A protein 193
cucumber mosaic virus (CMV) 219, *220*
  resistance introduction into squash
    221–2, 224–5, 226–7, *228–9*
  viral coat protein *231*, 236–7
    tolerance exemption 234
Cucurbita pepo 219, 220–1
cucurbitacin E 74
cucurbitacins 235–6
cystic fibrosis 97
cyt genes 170
Cyt proteins 168
  affinity for lipid portion of membrane 174
  Cry protein toxicity synergy 174, 175
  membrane protein disruption 175
  mode of action 174–5
  nomenclature 169–70
  specificity 177
  structure 174–5
  toxicity 175
cytotoxicity assays 86

Danaus plexippus 189
dapsone 101
dermatitis herpetiformis 101, 103
Desulfovibrio desulfricans 115
determination of non-regulated status
    (DONRS) 24–5, 26, 27
  agency response 28
  notices of petitions 28
  petition process 27–8
diatoms, phytoplanktonic 98
diet
  elimination 98, 100, 101
  food protein-induced gastroenteropathy
    99
dietary supplements, Novel Foods
    Regulation (EC) 48, 49
dinoflagellates 98
disease resistance vii
  breeding in squash 219–20
  introduction into squash 221–2, 224–5,
    226–7, *228–9*
DNA
  naked 112–13
  requirements for competence by bacteria
    112
  toxicity of introduced 233
  transduction 113–15
  transfer
    Archaea 117
    between bacteria in human gut models
      122, *123*

during bacterial conjugation 115–19
  by conjugation in natural environments
    118–19
  dietary factors 128–9
  from GMOs to human gut microflora
    111
  from GMOs to human gut microflora
    monitoring 119–22, *123*, 124–9
  Lactococcus lactis modified strains 125
  microbial interactions 125
transformation 111–12
  factors affecting in natural
    environments 112–13
  transposition intermediate 117
see also plasmids
DNA fragments, gastrointestinal tract wall
  113
DNA from genetic modification, presence in
  food 55, 56, 57
DNA microarray 79–80, 88
  gene expression profiling 87–8
  technology 89
domoic acid 98
drugs 98
dyes 98

eczema 96, 100
  occupational 102
egg, food protein gastroenteropathy 99
Elcar™ 201
EM Biocontrol-1™ 201
embryo rescue 220
endangered species 180
Enterococcus, plasmid initiation of
  conjugation 116
enterocolitic syndromes 102–3
environment
  antibiotic resistance gene transfer 16
  APHIS permits for releases 24
  impact of rDNA technology 4
  monitoring of impacts 5
  National Research Council report on
    hazards 4
  plant-incorporated protectants impact
    36–7
  risk assessment *35*
  safety concerns with commercialization
    of new crop varieties 41
  US regulatory approach to safety 6
environmental impact assessment
  recombinant baculoviruses 204
  USDA 25
Environmental Protection Agency (US) 3
  authority 29–30
  baculovirus registration 201, 211–15
  Bt crop registration 195

exemptions 33, 230, *231*, 232–4
federal regulation 6
herbicide-tolerant crops 33–4
introduced DNA toxicity assessment 233
mandate 30–1
microbial pesticide registration 212–15
notification of field trials 212–15
Office of Pesticide Programs (OPP) 211
pesticide residue aggregate exposure
  232–3
public availability of submissions 203
registration requirements for PIPS 34–8
  development 39–40
  international criticism 40
role 30
squash exemption from tolerance 230,
  231, 232–4
enzyme deficiencies 96–7
eosinophilia 99, 100
epinephrine 98–9
erucic acid 140, 141
  GT73 canola 149
  high levels in canola 69
  low levels in canola *see* canola
Erwinia chrysanthemi 112
Escherichia coli
  competence induction 111–12
  plasmid transfer 128–9
eukaryotic stress gene assay 87
European Academy of Allergy and Clinical
  Immunology 94
European Commission
  Group of Advisers on the Ethical
    Implications of Biotechnology 54
  novel food application submission 50, 51
  Scientific Committee for Food 57
European Community
  Common Catalogue for seeds 51, 53
  control of genetically modified organisms
    46–7
  Directive 90/220/EEC 46–7, 51
  Directive 97/618/EC 75
  Directive 2001/18/EC 47
  experimental release of genetically
    modified organisms 46–7
  genetically modified organisms
    traceability regulations 57
  labelling of foodstuffs 54
  legislation 46–61
  marketing of genetically modified
    organisms 47
  Novel Foods Regulation (1997a) 45, 46,
    47–51, *52*, 53–61
    approval process 49–51, *52*
    substantial equivalence 76
  registration of GM crops vii

Regulation 49/2000 56
Regulation 50/2000 57
Regulation 1139/98 55, 56, 57, 76
experimental use permit (EUP) 34

familiarity concept 5
FAO/WHO Expert Consultation on
    Allergenicity of Foods Derived from
    Biotechnology (2001) 15–16
FAO/WHO Expert Consultation on
    Biotechnology and Food Safety
    (1996) 65
  application of substantial equivalence 67
FAO/WHO Expert Consultation on Foods
    Derived from Biotechnology (2000)
    65
fat replacers, low-calorie 55
fats, genetically modified 11
  canola seed composition 145–6
fatty acids
  low-erucic acid rapeseed (canola) oil 71,
    72
  newly introduced
    GT73 canola 148–51
    pre-market review 12
  structural changes in canola
    modifications 71
Federal Food, Drug, and Cosmetic Act
    (FFDCA) 5, 6–7, 8
  adulteration provisions 8
  coat protein in transgenic plant
    exemption 230
  FDA mandate 8
  food additive provisions 8
  labelling of food 17
  pesticide tolerance levels 30, 232
  recombinant baculovirus regulation 211
Federal Insecticide, Fungicide, and
    Rodenticide Act (FIFRA) 6
  exemptions 33
    coat protein in transgenic plant 230
  pesticide registration 29, 211–12
  rDNA technology 30
  recombinant baculovirus regulation
    211–12
Federal Plant Pest Act (FPPA) 6, 22
  regulated articles 22, 24
federal regulatory agencies 3, 4
field trials
  APHIS permits 24
  design 58
  federal regulation 6, 212–13
  genetically modified microbial pesticides
    212–15
  new rDNA crop varieties 63
  performance standards of USDA 26

performance studies 27–8
  sites 58
fish allergies 98
  food protein gastroenteropathy 99
fish poisons 98
Flavr Savr™ tomato, commercial planting 6
fluorescent in situ hybridization (FISH)
    120–1
food
  adventitious contamination of non-GM
    material 56
  adverse reaction classification 94
  antibiotic-resistant bacteria 122
  fermented 110, 111
    GMO ingestion 124
    GMOs in starter cultures 119
  hypersensitivity 95–102
  labelling of not bioengineered 20
  non-toxic reactions 94, 95
  sold to mass caterers 56
food, bioengineered/novel
  analysis of intended and unintended
    effects 66–7
  approval procedures 49–51
  case study presentations (OECD Working
    Group) 64
  classes 59–60
  compositional analysis 58
  containing/consisting of genetically
    modified organisms 51
  definitions 9, 47–9
  DNA from genetic modification 55, 56,
    57
  European regulatory requirements 45–62
  field trials 58
  gastrointestinal tract effects 86
  key nutrients 58
  labelling 76
    requirements 19
    voluntary 17, 19–20
  LEAR oil (low-erucic acid rapeseed oil)
    as 70
  likely intake 58–9
  natural toxicants 58
  no longer equivalent to traditional 76
  post-marketing monitoring 61
  pre-market approval 75
  pre-market consultation process 9
  proteins from genetic modification 55,
    56, 57
  safe upper limits 58
  safety 57–8
    comparative approach to assessment
      58, 223
    substantial equivalence approach see
      substantial equivalence

scope 9
substantial equivalence 77
    see also substantial equivalence
surveillance programme with product
    launch 59
transgenic plant comparison with
    traditional plant 58, 223
wholesomeness 76
food additives 98
    labelling 57
    petitions 235
food allergenicity viii, 13–16, 94
    animal models 106
    delayed reaction 99–101
    demographics 95
    donor organisms 236
    FDA requirements 13–16
    glycans 81
    immediate reactions 95–9
    plant-incorporated protectants 38
    reaction types 95–102
    technical approaches 105–6
    transgenic squash 232
    see also allergens
food allergies 94
    diagnostic factors 97
    IgE-mediated 95, 96, 99
    patient education 99
    self-treatment 99
    treatment 98–9
Food and Agriculture Organization (FAO)
    76
    see also entries beginning FAO
Food and Drug Adminstration (FDA; USA)
    3
    allergenicity assessment 13–16, 232, 236
    authority 6–7
    completed consultations for new plant
        varieties 18
    consultation process 16–17
    determination on safety 13
    federal regulation 6
    filing process 16–17
    food additive petition 235
    food safety analysis 234–7
    labelling of food 17, 19–20
    mandate 8–9
    presubmission consultation 20–1
    requirements for commercialization
        9–16
    role 7–8
food antigens 96
food challenge studies 105
food crops
    agronomical traits 74
    natural variation 58

naturally occurring genotypic variation
    58
new varieties 40–1
quality traits 74
transgenic
    analysis of unintended effects 74–90
    food safety 75–6
food flavourings, labelling 57
food processing
    plant-incorporated protectants stability
        38
    rDNA release from genetically modified
        microorganisms 119
Food Quality Protection Act (FQPA) 38,
    211, 232
    viral coat proteins 234
food safety vii
    analysis requirements of FDA 234–7
    genetically modified crop plants 75–6
    strategies for assessing 76
    substantial equivalence assessment 64–6
        effectiveness 65–6
        see also substantial equivalence
    Working Group on Food Safety and
        Biotechnology assessments 64
    see also food, bioengineered/novel
Freedom of Information Act (FOIA) 21
fumonisins 193
fungae, plant pests 23
furanocoumarins 74

galactose-4-epimerase deficiency 97
galactosemia 97
gastroenteritis, eosinophilic 99–100
gastroenteropathy
    food protein-induced 99
    protein-losing 100
gastrointestinal inflammatory diseases 96–7
gastrointestinal mucosal disease 100
gastrointestinal tract, mammalian 111
GemStar™ 201
gene(s)
    differential expression 79–80
    ethically sensitive 55
    expression profiling 87–8
    intestine-specific 88
    see also marker genes
gene products
    comparison for sequence homology
        15–16
    novel 75
gene sequences
    evaluation 12
    homology 15–16
    interaction with other genes 12
gene-splicing see rDNA technology

gene transfer
  allergenic source 104–6
  amino acid sequence 104–5
  reactivity to specific IgEs 106
genetic engineering 110
  baculoviruses 202
  RNA levels in transgenic plants 227, 228–9
  transgenic plant comparison with traditional plant 223
  viral genes encoding viral coat protein introduction 225–6
genetic modification
  allergenic potential of foods 104
  compositional changes in tomatoes 84–5
  newly expressed allergens 104
  nutrient changes 77, 78
  potential of unintended effects 77
  risk assessment in food products 76
  toxicant changes 77, 78
  unintended side effects 78
  see also rDNA technology
genetically modified crop plants
  agronomic traits 74
  analysis of unintended effects 74–90
  Canadian guidelines 139
  FDA food safety analysis 234–7
  metabolic capabilities 77
  quality traits 74
  risk assessment 76
  screening for human health effects 88
  single gene modification 77
  substantially equivalent to analogous food products 77
  tiered safety evaluation 85–6
  toxicological profiling 85–8
  unintended side effects 78
genetically modified microorganisms 110
  rDNA release in food processing 119
  release into natural environments 118
genetically modified organisms (GMOs)
  biosafety assessment 121
  chemical fingerprints 84–5
  colonization of gut 126–8
  control of deliberate release into environment 46
  design 126–8
  DNA transfer to human gut microflora 120
  monitoring 119–22, 123, 124–9
  experimental release into environment 46–7
  fermented food starter cultures 119
  legal framework for market introduction 75
  marketing 47

metabolic perturbations in foods 75, 76
persistence in human gut 124–5
rDNA stability 124, 128
rDNA transfer 111
  to human gut microflora 129
recombinant baculoviruses 203
safety evaluation 76
traceability regulations 57
genomics 78, 90
genotoxic potency 87
genotoxicity endpoints 86
gliadin 100, 101, 103
glucosinolates 140, 141–2
  alkyl 149–50, 151, 156, 158
  beneficial health effects 141–2
  cancer effects 141–2
  canola rat feeding study 144
  GT73 canola 149–50, 151
    backcrossed lines 150, 151
  reduced levels 141
gluten challenge 101
gluten-sensitive enteropathy 100–1
N-glycans 80–1
glycoconjugates 173
glycoproteins, water-soluble 103
glycosylation 80
glyphosate 138
  soya tolerance 48
  see also Roundup® herbicide, canola tolerance
glyphosate oxidoreductase (GOX) 138
  safety assessment 139
Gonyaulax 98
Grain Inspection Packers and Stockyards Administration (GIPSA) 29
Gram-negative bacteria, conjugation 116
Gram-positive bacteria, conjugation 116–17
green fluorescent protein 120
Group of National Experts on Safety in Biotechnology (GNE) (OECD) 63
GT73 canola 138, 139
  agronomic properties 155, 156, 157
  processed meal 143–5, 151, 152, 153, 154, 155
    equivalence with other varieties 158
    fatty acids 153, 155
    glucosinolates 150
    refined oil 153, 155
    toasted 151, 152, 153, 154
  processing 142
  rat feeding study 143–5, 154, 155, 156, 157
    equivalence to other varieties 158
    organ weights 155, 156, 157, 158
    statistical analysis 144–5, 158

seed composition 140
seed compositional analysis 142, *143*,
     145–6, *147*, 148–51
   antinutrients/nutrients 158
   fat 145–6
   fatty acid composition 148–51
   glucosinolates 149–50, *151*, 156,
      158
   protein 145–6
   sinapines 151, *152*
   statistical analysis 143
   substantial equivalence 156, 158
   substantial equivalence testing 139–40,
      156, 158
   see also substantial equivalence
Gyp-Chek™ 201
gypsy moth control 180

Halobacterium volcanii 117
Helicoverpa zea, AcMNPV virulence 208,
      209
Helicoverpa zea nucleopolyhedrovirus
      (HzNPV) 201, 204, 207, 208
   field trials 214–15
Heliothine spp. complex 204
Heliothis virescens, AcMNPV infection
      204, 206, 209–11
herbicide resistance vii
herbicide tolerance *see* tolerance levels for
      pesticides
hives, food allergen reaction 102
HLA-DR3DQ2 haplotype 100
honeybee 181
horticultural plants, non-food 6
human flora-associated (HFA) animals
      122
human gut
   animal models 121–2
   antibiotic resistance gene transfer 16
   DNA transduction 114–15
   DNA transfer between bacteria 111
      in models 122, *123*
   GMOs
      colonization 126–8
      persistence 124–5
   naked DNA entry 112–13
   novel food effect 86
   transconjugants 124
   in vitro/in vivo models 121
   wall exposure to DNA fragments 113
human gut microflora 111, 120
   barrier to persistence of *L. lactis* 125–6
   glucosinolate break down 141
   plasmids
      mobilization 128
      transfer 121, 122

rDNA persistence 128
rDNA transfer from GMOs 111, 120,
      124, 129
   monitoring 119–22, *123*, 124–9
human health
   Bt insecticides 184–5
   concerns with commercialization of new
      crop varieties 41
   genetic modification effects 75
   GM crop plant screening 88
   monitoring of impacts 5
   National Research Council report 4
   plant-incorporated protectants 38
   post-marketing monitoring of novel
      foods 61
   US regulatory approach to safety 6
human pathogens, novel 111
hypersensitivity
   delayed reaction 99–101, 102
   immediate 95–9
hypersensitivity pneumonitis (HP) 101,
      102
hypoalbuminemia 100
hypogammaglobulinemia 100

identity preservation systems 56
immunoglobulin A (IgA) granular depositis
      101
immunoglobulin E (IgE)
   allergen-specific 102
   antigen-specific 96
   binding 103
   contact reactions 102
   food allergies 95
   food-induced gastroenteropathy
      100
   induction by Bt 94
   transferred gene reactivity to 106
immunologic analysis of plants 14
infants
   exposure to viral coat proteins 234
   food allergens 104
   food hypersensitivity reactions 98
insecticidal protein expression in food
      crops 6
insecticides
   bacterial 175–7
      activated toxins 186
      safety testing 177–9
   Bt-based *176*
   chemical application
      comparison with Bt maize 190
      conparison of effects with Bt 182–3
   chemical application reduction 31, 32,
      164–5
      Bt crops 193

insects
  beneficial 181
    plant-incorporated protectants impact
      36–7
  fatal disease induction by baculoviruses
    201
  midgut membrane 171
    Bt toxin activation 177
    microvilli 171, 173, 177
  natural enemies 165
  neuronal receptor affinity of baculovirus
    toxins 205
  plant pests 23
  recombinant nucleopolyhedrovirus
    selectivity 207–8
    specificity 205–6
  resistance vii, 1
integrated pest management programs
  baculoviruses 202
  Bt crops 194
international guidance, safety assessment of
  GMO-derived foods 68, 69
intestinal cell systems 86
iron-deficiency anemia 100

kanamycin-resistance gene *nptII* 112

labelling of food 17, 19–20
  absence of bioengineered 20
  additives 57
  allergenic potential 104
  de minimis threshold 56
  ethical aspects 54, 55
  FDA requirements 19
  flavourings 57
  identity preservation systems 56
  misleading statements 20
  no longer equivalent to traditional 76
  non-prepacked foods 57
  not bioengineered 20
  Novel Foods Regulation (EC) 48, 54–7
  voluntary 17, 19–20
lacewings 190
  Bt maize trials 188–9
lactase deficiency 97
lactic acid bacteria 110
  bacteriophage 115
  genetically modified 110
  probiotic 129
Lactobacillus 115
Lactobacillus bulgaricus 128–9
Lactococcus 116
Lactococcus lactis, genetically modified
  124, 125–6
  colonization of gnotobiotic mice 127
  plasmid complement loss 127–8

lactulose 48
laurate canola 71–2
lauric acid, increased in canola oil 71, 72
LEAR oil (low-erucic acid rapeseed oil)
  70
  average/upper limit intakes 70
  toxicology 70
Leiurus quinquestriatus hebraeus 204
leukocytosis, eosinophilic 100
linoleic acid 148
linolenic acid 156
  GT73 canola 148
  low-erucic acid rapeseed (canola) oil 71
looper, alfalfa 204
low-erucic acid rapeseed oil *see* LEAR oil
  (low-erucic acid rapeseed oil)
LqhIT toxin gene 204
LqhIT toxin specificity 205
α-lycopene 85
Lymantria dispar 180
Lymantria dispar nucleopolyhedrovirus
  (LdMNPV) 201, 207
lysogeny 114

maize
  fungal toxins 193
  identity-preserved sources 56
  insect-resistant and glufosinate
    ammonium tolerant 48
    labelling of food 55
maize, Bt 1, 6
  aphid – lacewing studies 188–9
  Bt protein levels 190
  comparison with chemical insecticides
    189
  Cry protein quantity 188
  fungal toxins 193
  pollen dispersal 189
  pollen effects on monarch butterfly
    185–6, 189–90
malabsorption
  carbohydrates 99
  gluten-sensitive enteropathy 101
malabsorption syndrome, food-induced 97
marker genes viii, 110–30
  antibiotic resistance 12, 16
  mobile genetic elements 120
  non-expression 120
  rDNA 120
mast cells 95, 100
metabolic degradative pathways 119
metabolome 77, 89
  profiling 81–2, 83, 84–5
metabolomics 78
milk, fermented 128–9
milk allergy 104

mites
  plant pests 23
  straw-itch 203, 204
    toxin 205
molecular genetics 89
molluscs, plant pests 23
monarch butterfly 185–6, 189–90
monosodium glutamate 98
mRNA fingerprinting 78–9
mutagenesis
  high oleic acid canola preparation 71
  insertion 75
mycoproteins 48
mycotoxins 31–2
myristic acid, increased in canola oil 71, 72
myrosinase 140–1
  inactivation 141, 155

N-linked oligosaccharides 81
National Academy of Sciences (US) 4–5
National Environmental Policy Act (US
    Congress, 1969) 2
National Environmental Research Council
    (NERC) Institute of Virology (UK)
    203
National Institutes of Health (NIH) 2
National Research Council (US) 4–5, 41
natural diversity 76
NeoCheck-S™ 201
Neodipirion sertifer nucleopolyhedrovirus
    (NsMNPV) 201
neonates, overfeeding 97–8
neurotoxins 98, 204
Nitzia pungens 98
non-regulated status
  determination by USDA 24–5, 26, 27–8
  extensions 28–9
non-resistance-based marker genes 16
non-target organisms
  Bt crops 165, 187–91, 194, 195
  Cry proteins 188–9, 190–1
  plant-incorporated protectants impact
    36–7
  safety of Bt to invertebrates 179–83,
    187–91
Novel Foods Regulation (EC) 45, 47–51,
    52, 53–61
  application submission 49–50, 60
  approval procedures 49–51, 52
  authorization decisions 50–1
  dietary supplements 48, 49
  enforcement 61
  exemptions 48
  guidelines 49, 57–60
  labelling 48, 54–7
  marketing consent 51

post-market monitoring 61
pre-market approval of novel foods 75
scope 47–9
seeds directives 51, 53
simplified procedures 53
transparency 60
nptII kanamycin-resistance gene 112
nuclear magnetic resonance (NMR)
    spectroscopy, high-resolution
    proton (^{1}H) 81, 82
nucleopolyhedroviruses, feral 208–11
nucleopolyhedroviruses, recombinant 201,
    204–5
  bioinsecticide formulation 206
  environmental stability 208–11
  invertebrate selectivity 207–8
  killing speed 210
  vertebrate safety 205–6
nut allergy 96, 98, 104
nutrient changes with genetic modification
    77, 78
nutrients, key, application of substantial
    equivalence 67, 68
nutritional composition of plant
  FDA requirements 13
  metabolic perturbations 75
nutritional evaluation of GM food crops
    9–16
  analytical data requirements of FDA 12
  decision analysis 10, 11
  documentation requirements of FDA
    10

oestrogenic compounds 78
Office of Science and Technology Policy
    (US) 2–4
oils, modified 11
  federal regulation of content in food
    crop 6
oils, vegetable, LEAR oil comparison 70
oilseed rape
  anti-nutrients 140–1
  commercial processing 141
  cultivation history 140–2
  double-zero varieties 141
  erucic acid 74, 140, 141
    low level varieties 141
  glucosinolates 74
  herbicide tolerant 53
  see also canola; GT73 canola; Westar
    canola
oleic acid 148
  high levels in canola 71
onchocerciasis 182
oral allergy syndrome 96
oral vaccines 110

Organisation for Economic Co-operation and Development (OECD) 5, 76
  safety assessment of biotechnology-derived products 63
  substantial equivalence 63, 76–7, 139
  Task Force for the Safety of Novel Foods and Feeds 66
    science-based consensus documents 68, 69
  Working Group see Working Group on Food Safety and Biotechnology (OECD)
Orygia pseudotsugata nucleopolyhedrovirus (OpMNPV) 201, 207
overfeeding of neonates 97–8

palmitic acid 148
papaya ringspot virus watermelon strain (PRSV-W) 219, 220, 227, 228–9
  viral coat protein 231
paralytic shellfish poisoning 98
peanut allergy 95, 96, 98, 104
peptic ulcer disease 98
pesticides
  aggregate exposure risks 232–3
  biological 29–30
  chemical 29–30
  expression by modified plants 30
  FIFRA registration 29, 211–12
  microbial
    field testing 212
    genetically modified 212–13
  residue tolerances 38
  see also plant-incorporated protectants (PIPS); plant-pesticides; tolerance levels for pesticides
phospholipidcholine 174
photoallergic reactions 102
plant breeding methods
  alternative technologies 16
  conventional vii
  genetic transformation 74
plant-incorporated protectants (PIPS) 30–1
  commercial approval 35
  data requirements
    of EPA 36–7
    of USDA 35
  definition 31–2
  ecological effects 36–7
  environmental risk assessment 35
  exemptions from regulatory scrutiny 33
  heat stability 38
  human health effects 38
  product characterization 36
  registered 31

registration requirements 34–8
  development 39–40
  reporting 35
  residue tolerances 38–9
  resistance management plan 37, 40
plant-pesticides
  definition 31–2
  GM crops expressing 5
  renaming 39
  see also plant-incorporated protectants (PIPS)
plant pests 23
  definition 22, 23, 24
  USDA list 22
plant protection, speed of kill 202
plant sterols, margarine-based product 48
  consumption increase 59
plant varieties, new
  FDA regulation 7, 8
  routine chemical analysis 7
  safety evaluation 7
plasmids 113, 115–19
  carriage 124–5
  conjugative 115–16
  encoding of environmentally important traits 119
  genetically modified 118
  loss from genetically modified L. lactis 127–8
  maintenance by bacteria 119
  mobilization 118, 128
  replication 119
  retrotransfer 118
  transfer
    antibiotic effects 122, 124
    from bacteria in vivo 122, 123, 124–9
    by conjugation 118–19
    dietary factors 128–9
    donor cell concentration 128–9
    frequency 119
    Lactococcus lactis modified strains 125–6
    between microflora of human gut 121, 122
    system 120
pollen allergy 81
potatoes, alkaloids 74
potatoes, Bt 193
Pre-market Biotechnology Notification (PBN) 9, 10, 16
  guidelines on consultation 16–17
  presubmission consultation 20–1
  public disclosure 21
  submission 17
pre-market consultation process 9

pre-market review 12
probiotics 110, 111, 129
product characterization 12
promoters, plant pests 24
proof of safety process 203
protein hypersensitivity 99
proteins from genetic modification
    allergenic potential 15, 38, 55
    canola 145–6
    FDA safety requirements 11
    insecticidal expression in food crop
        federal regulation 6
    pre-market review 12
    presence in food 55, 56, 57
proteome filing 80–1
proteomics 78
protozoan plant pests 23
Pseudomonas ice-minus strain 2
public relations activity for recombinant
    baculoviruses 203
Pyemotes tritici 203, 204

quail, bobwhite 191
Quorn mycoprotein 48

random block effects 82, 85
rapeseed oil *see* canola; oilseed rape
rDNA
    marker genes/systems 120
    monitoring fate in ingested food 120
    stability in GMOs 124, 127–8
    transfer from GMOs to human gut
        microflora 111
        monitoring 119–22, *123*, 124–9
rDNA technology 1
    agricultural products 5
    bioengineered foods 9
    development 2
    federal guidelines/policies 2
    food products 40–1
        safety assessment 63
    precision 41
    product regulation 3
    regulatory requirements 1
    research into potential applications 2
refuge area planting 28
registration
    baculovirus with EPA 201, 211–15
    Bt crop with EPA 195
    GM crops vii
    microbial pesticides with EPA 212–15
    pesticides with FIFRA 29, 211–12
    requirements for PIPS with EPA 34–8
        development 39–40
        international criticism 40
        reporting 35

regulatory decisions, access to information
    5
regulatory requirements vii
    rDNA technology 1
residue tolerances, plant-incorporated
        protectants 38–9
resistance
    herbicide vii
    insect vii, 1
    management plan for plant-incorporated
        protectants 37, 40
    virus vii
    see also antibiotic resistance; disease
        resistance
respiratory disease 101, *102*
retrotransfer 118
reverse transcription polymerase chain
        reaction (RT-PCR) 78–9
rhinitis, allergic *95*, 96, 100
Rhodopalosiphum padi 188
rice, food protein gastroenteropathy 99
risk assessment
    baculoviruses 203–11
    communication 203
    environment *35*
    genetically modified plant crops 76
    transgenic plant comparison with
        traditional plant 223
RNA levels in transgenic plants 227, *228–9*
Roundup Ready™ Canola (RCC) 138
    substantial equivalence premise 139, 156
    see also GT73 canola
Roundup® herbicide
    canola tolerance 138–59
    safety 139

safety assessment vii–viii, 57–8
    animal studies 57–8, 76
    GM food crop evaluation 1
        decision analysis 10, *11*
        documentation requirements of FDA
            10
        expressing plant pesticides 5
        FDA requirements 9–16
    novel foods 58
    proof of safety process 203
Scientific Committee for Food (EC) 50
scombroid poisoning 98
scorpion toxin gene 203, 204
seeds
    allergies 96
    marketing in EC 51, 53
serum testing, allergic patients 14, 16
sesame seed 104
sex pheromones 116
shellfish allergies 98, 104

shellfish poisons 98
sinapines 151, *152*
skin contact, harmful effects 74
skin reactions, food allergy *95, 96*
sodium channel inappropriate modulation
    204
soil microorganisms 190–1
a-solanine 80
somaclonal variations 76
soy protein 99
  allergy 104
soya
  glyphosate-tolerant 48
    labelling of food 55
  herbicide-tolerant 6, 48, 55
  identity-preserved sources 56
  modified with *Streptomyces* gene
    sequences 1
  transgenic 94
sphingomyelin 174
Spodoptera, Bt insecticides 179–80
squash 74
  biology 220–1
  cucurbitacin production 235–6
  CZW3 line 221, 222, 224–5
    cucurbitacin absence 235–6
    new virus appearance likelihood 227
    tolerance exemption 234
  disease resistance breeding 219–20
  free-living 221, 224–5
  gene exchange between free-living/
    cultivated 221
  genetic engineering 221–2
    allergenicity 232
    allergenicity of donor organism 236
    FDA food safety analysis 234–7
    field trial results 222
    introduced DNA toxicity 233
    new proteins produced by introduced
      DNA 232–3
    recombination 225–7, *228–9*
    RNA presence 227, *228–9*
    safe use history 230
    transcapsidation 225–7, *228–9*
    USDA-APHIS exemption 222–7,
      228–9, 230, *231*, 232–4
    viral coat protein production 225–7,
      228–9, 230, *231, 232*
    viral sequence introduction 225–6
    weed potential 224–5
  tasting for bitterness 235–6
  virus coat proteins 237
  virus disease 219
  ZW20 line 221–2, 224–5
    cucurbitacin absence 235–6
    new virus appearance likelihood 227

tolerance exemption 230, 234
  viral coat protein *231, 232*
Standing Committee for Foodstuffs (EC)
  50–1, 61
Staphylococcus 117
steatorrhea 100, 101
Streptococcus 117
Streptococcus thermophilus 128–9
Streptomyces modified soybeans 1
stress gene induction 87
substantial equivalence viii, 63–73, 76–8,
  89
  application 66–8
    canola 67
    to compositional data 67, 68
    FAO/WHO Expert Consultation
      recommendations 67
    guidance documents 68, 69
    to key nutrients/toxicants 67–8
    level of single fraction of material 67
    level of unprocessed food product
      (multiple fractions) 67
    to unintended effects of novel foods
      68
    WHO workshop (1995) 66
  Canadian guidelines 139
  canola (rapeseed oil) case study 69–72
    breeding programmes and development
      69–70
    canola meal 71
    comparators 71
    fatty acid changes 71
    low-erucic acid 70–2
  Codex *Ad Hoc* Intergovernmental Task
    Force view 66, 72
  comparators
    identification 67
    nature of 72
  concept definition 58, 64
    clarity lacking 72
  effectiveness in food safety assessment
    65–6
  flexibility and levels of application 66–7
  GT73 canola 139–40, 156, 158
    seed compositional analysis 156, 158
  historical background 63–4
  interpretations 65, 66, 72
  limitations 67, 68
  misinterpretation and criticism 65, 72
  not as measure of absolute safety 65
  OECD Task Force view 66
  Roundup Ready™ Canola (RCC) 139,
    156
  safety assessment structure 65
  Working Group on Food Safety and
    Biotechnology (OECD) 64, 65

'substantially equivalent' term 53
sulfapyridine 101
sulfiting agents 98

T cell epitopes 103
tartrazine yellow 98
tobacco budworm 204
tolerance levels for pesticides 1, 30, 212
    canola 6, 70–1
        for Roundup® 138–59
    EPA assessment 33–4
    residues 38–9
    soya for glyphosate 148
tolerance requirement exemption
    Bt formulations 179
    transgenic squash 230, 231, 232–4
    WMV2 and ZYNV coat proteins 232
α-tomatidine 86
α-tomatine 80, 86, 87
    CAT-Tox(L) assay 88
tomatoes
    antisense RNA exogalactanase 75
    chemical fingerprinting 82, 83, 84–5
    compositional changes due to genetic
        modification 84–5
    cytotoxicity assays 86
    Flavr Savr™ 6
    gene expression 79–80
    genotoxicity endpoints 86
    N-glycans 80–1
    insect-resistant 75
    mRNA fingerprinting 78–9
    post-translation modification of
        transgenic 80
    transcription factor assessment 87
tomatoes, Bt 75
    safety in mammals 192–3
tox34 gene 204, 214–15
toxicants
    analytical data requirements of FDA 12
    application of substantial equivalence 67,
        68
    changes with genetic modification 77,
        78
    genetically modified crops 9
    identification 11
    natural 7
        novel food 58
    plant-incorporated protectants 38
    safety evaluation 10, 11
toxicological profiling of genetically
        modified crop plants 85–8
toxicological testing
    LEAR oil (low-erucic acid rapeseed oil)
        70
    new foods/crops 64

toxin genes 202, 203, 204
toxins 98
transcapsidation 225–7, 228–9
transcription factors 87
transgenic crops 175–7
    coat protein production exemption by
        EPA 230
transposons, conjugative 117, 118
tree nut allergy 95, 98, 104
Txp-I paralysis dosage 205

United Kingdom voluntary approval
        process for novel foods 45
United States Department of Agriculture
        (USDA) 3
    Advisory Committee for Agricultural
        Biotechnology 29
    authority 21–2
    commercialization requirements 24–9
    environmental impact assessment 25
    experimental use permit 34
    federal regulation 6
    field trials performance standards
        26
    mandate 22
    non-regulated status determination
        24–5, 26
    notifications 25–6
    permits 25
    PIPS registration requirements 34–8
    role 22
United States of America (USA)
    federal guidelines/policies 2
    federal regulations for agricultural
        biotechnology 5–6
    registration of GM crops vii
    regulatory approach 41
    safety evaluation of GM food crops 1
urticaria 100
    contact 102

vectors, plant pests 24
vegetable oils, LEAR oil (low-erucic acid
        rapeseed oil) comparison 70
viral coat proteins 230
    infant/child exposure 234
    production in transgenic squash 225–7,
        228–9, 230, 231, 232
viral proteins encoding genes 33
virulence factors 111, 119
virus-resistant crops 219–37
viruses
    plant pests 23
    resistance vii
Volta River Basin, blackfly control
        182

watermelon mosaic virus 2 (WMV2) 219,
    220
  coat protein genes 222, 230
  resistance introduction into squash
      221–2, 224–5, 226–7, *228–9*
  viral coat protein 230, *231*, 232, 233,
      234
    similarity to squash coat proteins
        236–7
weeds
  potential for development 26
    transgenic squash 224–5
  volunteer 26, 36
Westar canola 156, 158
  compositional analysis 145, 146, *147*,
      148–51
  glucosinolates 149–50
  processed meal 151, *152*, 153, 155, 158
  sinapines 150
  study method 142–5
wheat allergy 104
wheezing 97
wildlife
  adverse effects 27
  plant-incorporated protectants impact
      36–7

Working Group on Food Safety and
    Biotechnology (OECD) 63–4
  case study presentations of novel foods
      64
  objectives 64
  scientific approach to food evaluation 64
  substantial equivalence application 64,
      65
  see also substantial equivalence
World Health Organization (WHO) 76
  substantial equivalence premise 139

xenobiotic response element (XRE) 87

yield vii

zucchini 74
zucchini yellow mosaic virus (ZYMV) 219,
    220
  coat protein genes 221, 222, 230
  resistance introduction into squash
      221–2, 224–5, 226–7, *228–9*
  viral coat protein 230, *231*, 232, 233,
      234
    similarity to squash coat proteins
        236–7